The Roland x0x Series

The Roland x0x Series: A History of Early Roland Synthesizers and Drum Machines from SH-101 to TR-909 offers an authoritative but accessible investigation of how the design of these most influential musical instruments is shaped by technological factors, institutional contexts, market forces and imagined users and uses. Drawing on the perspectives of designers and engineers along with musicians and producers from around the world, the book explores how these iconic instruments both enable and constrain the creative work made with them.

Providing detailed analysis of these seminal instruments, this book combines archival research that draws on contemporary reviews, instruction and service documentation, advertising and marketing materials, and technical analysis of the underlying analogue and digital circuits, with discussion of creative composition and performance practice, hardware and software modifications and modern remakes, emulations and clones. *The Roland x0x Series* is dedicated to demystifying these mainstays of popular music, and revealing the 'hidden histories' of the creation, lives and afterlives of these most important, recognisable, yet often misunderstood, synthesizers and drum machines.

This is an essential reference for music producers and music technology enthusiasts, as well as those studying music technology, instrument design and human interaction in the context of electronic music.

James Newman is Research Professor and Senior Teaching Fellow at Bath Spa University, UK. He is also an External Affiliate of the School of Public and Applied Humanities at the University of Arizona and a Fellow of the Royal Historical Society. Over the past 20 years, he has written widely on histories of technology, media preservation, digital games, sound and music and has worked with museums across the world on exhibition and interpretation. He is an avid collector and player of synthesizers, drum machines and music technology.

Histories of Electronic Musical Instruments

Series Editor: James Newman

Histories of Electronic Musical Instruments cut through the hype, hyperbole, mythologies and misinformation associated with electronic musical instruments, to offer authoritative and accessible histories of the design, operation and creative uses of seminal instruments.

From the TR-808 to the Prophet-5, Fairlight CMI and Yamaha DX7, the books in the *Histories* series offer insights into how these instruments have been used, adapted and reimagined by subsequent generations of musicians, producers and DJs. Series authors consider how the authenticity of the instruments interacts with the myriad clones and emulations in hardware and software and, ultimately, what the musical and cultural meanings of these instruments are today.

Titles in the *Histories of Electronic Musical Instruments* series will be of interest to a range of audiences, including advanced undergraduates, postgraduates and researchers of Music Technology and Music Production, as well as those studying the history of these fields. Further to this, some titles will be of significant interest to professional music producers, musicians and hobbyists working with the instruments in question, as well as the large communities of fans that remain fascinated by particular instrument manufacturers and their products.

The Roland x0x Series
A History of Early Roland Synthesizers and Drum Machines from SH-101 to TR-909
James Newman

For more information about this series, please visit: www.routledge.com/Histories-of-Electronic-Musical-Instruments/book-series/HEMI

The Roland x0x Series

A History of Early Roland Synthesizers and Drum Machines from SH-101 to TR-909

James Newman

Routledge
Taylor & Francis Group
LONDON AND NEW YORK

Designed cover image: James Newman/implant4

First published 2026
by Routledge
4 Park Square, Milton Park, Abingdon, Oxon OX14 4RN

and by Routledge
605 Third Avenue, New York, NY 10158

Routledge is an imprint of the Taylor & Francis Group, an informa business

British Library Cataloguing-in-Publication Data
A catalogue record for this book is available from the British Library

ISBN: 978-1-032-44921-0 (hbk)
ISBN: 978-1-032-44919-7 (pbk)
ISBN: 978-1-003-37451-0 (ebk)

DOI: 10.4324/9781003374510

Typeset in Optima
by SPi Technologies India Pvt Ltd (Straive)

For Martha, Susan and Terry

Contents

Acknowledgements

My sincerest thanks go to Susan and Terry for firing my initial enthusiasm and continuing to support my growing interest in electronic music. Thanks also to them and to Lucy for putting up with hours, months, years of drum programming, filter sweeps and all available flat surfaces being covered in cables, connectors and pieces of equipment in varying states of disrepair that I keep promising to fix.

And a special thank you goes to Martha who has provided continual inspiration and insight and has introduced me to new music and new ways of thinking about music I thought I already knew. Perhaps the greatest compliment I can pay is that as her drumming improves, she's sounding almost as good as a drum machine. Maybe one day she'll replace the TR-606 in the band.

Turning the original germ of an idea into this book would not have been possible without the encouragement, expertise and support of Hannah, Zoe, Emily and Jen at Taylor and Francis, my copyeditor Patricia, and the reviewers whose comments and insights were invaluable at every part of the process from the initial proposal through to the final manuscript. My special thanks go to the wonderful folk at implant4 in Osaka for providing the images of the TB-303, TR-808 and TR-909 and for letting me play with the instruments! I cannot thank you enough.

There can be little doubt that the SH-101, MC-202, TB-303, TR-606, TR-707, TR-808 and TR-909 are among the most important and influential electronic musical instruments, indeed the most important musical instruments, ever made. And so my thanks go to the legion of developers, sound designers, software engineers, players, performers, producers, modders and hackers who have individually and collectively helped shape the meanings of these instruments and whose creative work I have the great pleasure of attempting to document.

Preface

Throughout the early 1980s, among my most treasured possessions was a collection of printed brochures that I acquired from various musical instrument stores in and around London. In the days before the Internet, these free catalogues were some of the best, and certainly the cheapest, ways of finding out which synthesizers and drum machines were currently available and, most excitingly, what was going to be coming next. I spent hours poring over the glossy colour pages, dreaming of one day owning, or even just getting my hands on and playing, the growing array of instruments that filled the pages. I studied every minute detail of the images and practically memorised the specifications of each keyboard and rhythm machine. As well as reading the brochures from cover to cover to absorb every grain of information, I treated them as a birdwatcher might use their identification guide. I would scour the liner notes of cassettes and vinyl records for information about the equipment that had been used so I could read up on it or identify new catalogues I needed to hunt down. I would surround myself with my brochures as I sat watching music TV programmes like 'Top of the Pops' and 'The Old Grey Whistle Test' hoping to get a sighting of a rare synth in the wild so I could hear it, size it up and tick it off my list.

Of all the brochures I collected, the Roland ones were unquestionably my favourites. Roland Product Catalogues were almost impossibly thick because there was just so much in the company's range, and while individual instrument brochures were thinner, they were more densely packed with detailed information about specific synthesizers and drum machines. There was breadth and depth, and I was hooked.

I still have most of the brochures today and, in a very real sense, they provided a large part of the inspiration for this book. Most obviously, they were a rich source of information and, along with Owner's Manuals, Service Notes, advertisements and magazine features and reviews, they proved invaluable in considering how individual instruments were introduced, described and framed. Indeed, as much of this book arises out of a desire to investigate the interplay between the design of musical instruments and the meanings, and music, that are made with them, being able to consult these archival documents was hugely helpful. But these brochures proved useful in other ways, too.

My interest in electronic musical instruments began at the time when digital technology was first making an impression. Initially, digital control of analogue oscillators unlocked rock solid tuning with no drifting, no wavering, and, crucially, no embarrassing mid-performance out-of-tune notes and hasty recalibrations. Before long, however, analogue oscillators, precise or otherwise, gave way to wholly digital sound sources. Drum machines became sample playback devices with recordings of 'real drums' replacing synthesized facsimiles, and synthesizers like Yamaha's DX7 and Roland's own D-50 offered more realistic emulations of acoustic instruments and a new palette of glassy, crystalline sounds that simply could not be created with analogue circuits.

Just as the soundscapes changed, so too did the look and feel of instruments that created them. Gone were the inviting rows of sliders, switches and knobs that gave direct access to the analogue sound sculpting circuits and which encouraged sonic experimentation and manipulation. In their place came altogether sleeker, more economical, and very often inscrutable, interfaces built around alphanumeric displays and a handful of data entry and increment/decrement keys. If their pared down interfaces rendered these new instruments less intuitive to program oneself, the technologies of digital storage offered a solution by giving players Bank upon Bank of readymade presets emulating pianos, strings, guitars, brass and more. Each Bank sold separately, of course.

Digital technologies were clearly the future and analogue synthesizers and drum machines looked almost unimaginably outdated. And so, with each successive 'Roland Keyboards' or 'Rhythm Machines' catalogue, new digital instruments stole the limelight of the cover and opening spreads, while the previous generations of premium instruments moved further towards the back pages. I clearly recall tracing the journey of the Roland System 100m modular synthesizer through successive brochures until it finally ended up in the bottom half of the last page of a brochure otherwise dominated by modern, forward-facing instruments. With its panels full of faders and knobs all of which were partially obscured by a viper's nest of patch cables, the System 100m looked unwieldy, impractical and, above all, hopelessly out of date. It was obviously yesterday's instrument, and it sat uneasily in a brochure that mapped out the future of electronic sound. Even at the time I remember wondering whether it was really included as a serious sales proposition or if its presence served as a reminder of just how far we, or rather Roland, had come in just a few years.

Today, the System 100m is a highly prized, much sought after instrument just like the SH-101, MC-202, TB-303, TR-606, TR-808 and TR-909 that form the basis of this book. Collectively known as the 'x0x' or 'Mid-O' series by virtue of their product names and numbers, this collection of early-1980s synthesizers and drum machines had similarly journeyed through and eventually disappeared from the pages of Roland's catalogues as each was comprehensively superseded by the next generation all-digital instruments like the D-50 and TR-707 along with the host of anonymous-looking sound modules designed to be rackmounted in the corner of a studio and remotely controlled by the personal computer that was becoming increasingly central.

In part, this book, charts that turbulent period in electronic musical instrument design in which transformative digital technologies were seemingly rendering obsolete the analogue sounds as well as the interfaces used to control them. The book aims to understand the creative and technical decisions that underpinned the design of these transitional instruments that bridge the analogue and digital worlds. It aims to explore their contemporary reception and, crucially, the meanings and music that have been made since.

Crucially, this book resists the temptation to uncomplicatedly present the recollections of famous users of these synthesizers and drum machines. Of course, the perspectives of musicians, producers and DJs are vital in helping unravel the histories of these instruments. As such, the book draws extensively on interviews and reflections on composition and performance practices as well as my own deep investigation of the instruments, their documentation and the myriad hardware and software recreations, homages and clones. Central to the methodology underpinning this research is the assertion that the ultimate meanings of these instruments are revealed by exploring the intersections between design intentions and musical applications, between commercial and corporate aspirations and the responses and reactions of the marketplace, and between the foundational technological systems and the materiality of the interfaces that shape and frame the creative, experimental and improvisational practices of musical expression. As such, the insights of users and designers sit alongside the discourses of advertising and marketing materials, the instructions and imagined users and uses inscribed in owner's manuals, and the documentation of the technological substrate of components and calibration procedures captured in service notes and circuit diagrams. In drawing together this variety of materials and providing what I hope is a comprehensive account of the social, cultural, technological, economic and political contexts in which the production and consumption take place, the book seeks to unpick the mythologies and taken-for-granted narratives that have grown around the x0x instruments.

Emphasising the centrality of the instruments themselves, a few organisational qualities of the book are noteworthy. First, each instrument is given its own chapter. Most obviously, this allows space to encounter and explore each instrument on its own terms while also making connections between other members of the x0x family. Moreover, this structure also allows scrutiny of the full range of x0x instruments rather than just the most widely known or popular. Plainly, the TR-808 and TB-303 have attained a level of musical and cultural visibility unmatched by the comparatively obscure and commercially unsuccessful MC-202. However, not only is the MC-202 an important transitional instrument that speaks to emergence of computer and sequencer-based electronic making practices, it is also the prototype 'groovebox' which is a category of instrument that dominates the electronic musical instrument marketplace today.

It is also important that, while all the x0x instruments explored in this book were initially released in the first half of the 1980s, they were each designed and came to market separately and sequentially. Clearly, this meant that they eventually co-existed, but it is important to remember that the instruments demonstrably built upon one another and that their designers, and users, learned from their experiences, successes and failures as they created them and created music with them. Of course, while they were released sequentially, the x0x instruments were not released in numerical order. As such, readers glancing at this book's contents page might feel that starting with the '808' and ending with the '909' (with a coda that explores the '707', '626', '505' and '1000'), is a little odd. However, in order to facilitate the discussion of the accumulation of features and knowledge, the book follows the chronology of the instruments' design and release rather than the order of names and numbers.

With such a variety of instruments, functions, features, histories and meanings, finding a format and structure is something of a challenge. Nonetheless, each chapter follows a loose pattern in opening with an overview of the history, impact and influence of each instrument along with an image of its front panel and an illustration of its key circuits. The remainder of the discussion digs into the detail of the design, operation, features and functions of the sound synthesis, sequencing and connectivity, before exploring the musical applications, marketing discourses and the ongoing lives and afterlives of the instruments as they age and are forgotten and remembered, remade, modified and re-imagined in new configurations of hardware and software.

Ultimately, by documenting the complexity of their long histories and considering them both as old and contemporary instruments that continue to exert influence and inspire, this book is, in part, an attempt to update and extend my collection of Roland 'Product Catalogues' and brochures by exploring what happened to all those instruments after they dropped off the last page.

James Newman, May 2025

1 We Design the Future

A History of Innovation

It is hard to overstate the impact and influence of Roland on electronic musical instrument design. Indeed, as Gordon Reid noted in a 2004 company retrospective, '[t]he Roland name is almost synonymous with music technology.' (Reid 2004a). Since the company was founded in 1972 in Osaka, Japan by Ikutaro Kakehashi, Roland's product range has gone on to cover almost every category of electronic musical instrument from synthesizers, drum machines and samplers through digital pianos, organs, accordions and drum kits along with mixers, amplifiers, effects processors and a host of other live and studio-based tools. Over more than five decades, the company has operated at the forefront of innovation and design with 1973's SH-1000 being, not only Roland's first synthesizer keyboard, but also one of the first compact analogue synthesizers manufactured in Japan. Fewer than 15 years later, Roland's D-50 'Linear Arithmetic' synthesizer pioneered the combination of digital synthesis and sampling technologies and set the blueprint for synthesizer design for the next decade and more.

In between these two influential instruments, Roland collaborated with Dave Smith of Sequential Circuits in the US to develop the MIDI (Musical Instrument Digital Interface) protocol that allowed compatible instruments to communicate with one another. Publicly demonstrated for the first time at the 1983 Winter NAMM (North American Music Manufacturers) show, where a Roland Jupiter-6 and Sequential Prophet 600 were connected, it is no exaggeration to say that MIDI revolutionised musical instrument design and use. Not only could keyboard-based instruments like the Jupiters and Prophets be used to control one another as in the NAMM demo, MIDI made possible the development of new categories of keyboard (and other) controllers along with rack-mounted and desktop sound modules whose parameters could be remotely manipulated. Perhaps the most transformative impact of MIDI, however, was its incorporation into interfaces that allowed the computer and sequencing software to become the centrepiece of the contemporary studio. Where there had been no standardised means of connecting and synchronising electronic musical instruments, with each manufacturer developing its own proprietary (and incompatible) system, MIDI provided an instrument and manufacturer-agnostic lingua franca. In recognition of the magnitude of their achievement, Smith and Kakehashi were jointly awarded a Technical Grammy for the development of MIDI (Stewart 2014). Given the speed and enthusiasm with which protocols and interfaces are superseded and rendered obsolete within the computer and technology industries, that the MIDI specification remains the backbone of electronic musical instrument communication today, and a Jupiter-6 can still control a modern device manufactured decades later, verges on unbelievable and stands as a testament to its visionary design.

DOI: 10.4324/9781003374510-1

Throughout the 1980s, Roland routinely used the phrase 'We Design the Future' across its English language marketing and promotional materials. Indeed, as Doershuk and Milano (1987) noted, by 1987, nearly half of Roland's total workforce was involved in engineering research and development. In the face of such long-lasting innovation, it is very tempting to take this at face value rather than simply dismiss it as over-exuberant advertising hyperbole.

We might be forgiven for thinking that co-developing one of the most enduring communication protocols and transforming musical instrument design would be the crowning achievements for any company. However, as impactful as they are, these inventions barely scratch the surface of Roland's innovation and influence. Indeed, while the shift from analogue to digital sound generation, the emergence of computer control, and the coming of MIDI as the dominant communications and synchronisation method provide the backdrop, this book focuses on a series of even more influential instruments.

Putting the Mid-O in x0x

This book focuses on some of the most influential electronic musical instruments ever created. It is primarily about three synthesizers and three drum machines, each of which was released within the first few years of the 1980s and each of which was discontinued and out of production long before the decade was over. It is about six instruments that went on to shape the sound of popular music across almost every genre and that, in numerous cases, defined the sound and form of those genres. It is about the history and design of those instruments, the assumptions and intentions that are inscribed into their designs and the complex responses and reactions that have seen them used, ignored, recovered, reinterpreted and reimagined through complex, intertwining processes of shifting trends in technology, production and taste, creative improvisation, corporate necessity and commercial opportunity.

This book primarily focuses on six instruments that have shifted in and out of fashion, have been available for just a fraction of their original retail prices, yet whose subsequent use and scarcity means they have now become fetish objects commanding ever-growing price tags in the collectors' market. As such, this is a book about six instruments that, despite the enormity of their influence, might never be encountered in their original incarnations, yet which through remakes, reissues, clones, replicas and the embedding of their sounds in sample libraries and the factory presets of a host of modern hardware and software instruments made by Roland and others, are now effectively in the hands of more musicians, DJs and producers than ever before. It is about the rich histories of these instruments and the ways in which their meanings, workflows and sounds came to be and how they have been transformed over the course of more than four decades.

Those six key instruments are the SH-101, MC-202, TB-303 synthesizers and the TR-606, TR-808 and TR-909 drum machines with the TR-707 marking the next phase of Roland's development.

Taken individually, each instrument sets out new ways of creating music with palettes of previously unheard sounds and a suite of compositional and performance opportunities and limitations. Each has a distinctive physical form with a materiality and tactility arising from its unique combination of controllers, buttons, switches and faders. Analysing the interfaces and workflows of each instrument tells us much about the designers' aims and the uses and users they imaged. As we shall see, however, many of those users ended up creating ways of understanding and working with these instruments in ways that significantly deviate from, and even actively work against, those initially inscribed intentions.

Of course, as the rare and increasingly unobtainable original instruments are reimagined in modern hardware and software clones and remakes, and as their sounds are extracted and transplanted into new instruments, their musical, cultural and technological meanings are further transformed. Remade interfaces are subtly or drastically altered, circuits (whether real or digitally emulated) are augmented and modified, and aspects of the original instruments are broken out into separate processes and products or entirely forgotten. Even the original instruments are sometimes modified to add new features and functionality. This, then, is a book about the complex practices of reverence and reimagination, reduction and reintegration into new production and performance contexts.

Crucially, this is also a book about a family of instruments that are in dialogue with one another, that build on one another and that are, in some but not all cases, designed to interface and work with one another. In recognition of their familial and technological connections, collectively, these instruments are usually referred to as the 'x0x' or 'Mid-O' series (the latter being more prevalent in Japan). The x0x/Mid-O monicker derives from Roland's product naming, or more precisely, its numbering convention. Where previous Roland products had born names such as SH-09, TR-77, CR-78 and MC-8, a new generation of instruments needed a new designation. Interestingly, while the origin of the iconic x0x/Mid-O numbering has been the subject of much speculation, engineer Tadao Kikumoto who led the development of many of the instruments, has recently revealed the altogether more prosaic and pragmatic reason. 'Kakehashi-san put zero in the middle to differentiate it from previous models, and thus the new Mid-O Series like the TR-808, 606, 909, TB-303, MC-202, SH-101 was launched.' (Kikumoto 2020).

As well as being differentiated from previous products, the x0x/Mid-O instruments speak eloquently to the challenges faced by Roland's engineers and designers in a time of considerable turbulence. In the design of these instruments, we can trace the shift from analogue to digital sound creation and control. Importantly, this is not a smooth transition and by analysing these instruments, we reveal palpable tensions between technological opportunity, the need to respond to competitors and market trends, and a continuing commitment to personal and corporate creative and engineering visions. We see instruments that predict, and even directly inform, the development of future approaches and products yet which are, simultaneously, rooted in the past. Indeed, as we shall see, the technology underpinning the fully digital D-50 synthesizer that earned Roland a coveted TEC award for technical innovation in 1988, can be directly traced back to the foundational research, analysis and experimentation undertaken during the development of unabashedly analogue instruments like the TR-808 (which was released in 1980 and was the first of the x0x instruments despite the numerical implications of its name).

Of course, this is far from the only impact the TR-808 and the other x0x instruments have had. And just as it is hard to overstate Roland's influence on electronic musical instrument design, so too is it near impossible to overplay the effect of Roland's x0x instruments on contemporary popular music. Over four decades, the sounds of the x0x instruments have been heard across almost all genres of popular music including Hip Hop, Soul, Industrial Postpunk and Pop. Furthermore, genres such as House, Techno and Acid, among others, are in no small part defined by the sonic fingerprint, modalities, groove and feel of instruments such as the TR-808, TR-909 and TB-303. Indeed, in the world of electronic music, the sounds and workflows of the x0x instruments were – and still are – absolutely central. As Graham Massey of the band '808 State' notes, 'The Roland gear began to be a kind of Esperanto in music.' (Beaumont-Thomas 2014).

Quite simply, these are some of the most widely used electronic instruments. By way of example, it is often reported that the TR-808 has appeared on more hit records than any other drum machine. While this is almost impossible to substantiate, the sound of the instrument's distinctively synthetic Snare, Cowbell and, in particular, its booming Bass Drum are instantly recognisable and have become staples of music production. Even for those unfamiliar with the specific instrument, its use in recordings as diverse as Marvin Gaye's 'Sexual Healing', Whitney Houston's 'I Wanna Dance with Somebody', The Beastie Boys' 'Paul Revere', Afrika Bambaataa & the Soulsonic Force's 'Planet Rock' all ensure that the sounds will be familiar to myriad listeners. Beyond this, genres such as Trap and Miami Bass are positively defined by the sound of the TR-808 Bass Drum. Indeed, the deep sub bass tones that provide the foundation of these genres and much contemporary Hip Hop are simply referred to as '808s' within production circles and are even referenced in song lyrics. However, while they might resemble the tonality of the original instrument, these 808s almost always derive from heavily processed samples or even dedicated '808' plugins and are usually created within the contemporary Digital Audio Workstation (DAW). As we shall see, such are the transformations of the sounds and the reintegration into new production and music making contexts, the '808' has now comprehensively transcended the TR-808.

And we should not think it is just the TR-808 that has achieved such genre-defining status as it has been used and reused. While Massey's comment on the ubiquity of the x0x instruments in 1980s and 1990s electronic dance music is telling, it is worth reflecting for a moment on the name of his band and its catalogue of tracks namechecking '101s' and '202s'. Similarly, Fatboy Slim's 'Everybody Needs a 303' and Daft Punk's 'Revolution 303' all speak to the degree to which Roland's x0x instruments are embedded in the culture of popular music. Of course, there are plenty of musical instruments that have inspired artists, but there are few drum machines, or portable Bass Line synthesizers that have been so extensively referenced in album titles and song lyrics or after which artists have named themselves. The x0x synthesizers and drum machines can claim all of this and more. Over four decades, despite their uses, meanings and material forms being reimagined, they have become legendary instruments.

Hype, Hyperbole and Ghosthunting

And herein lies a problem. The complexity of their journeys has meant that the x0x instruments have been so comprehensively mythologised that it is difficult to disentangle the actualities of their histories from the rumour and hearsay that have built up around them.

Moreover, although the release of modern analogue and digital hardware recreations as well as the incorporation of x0x sounds in preset banks, sample collections and virtual instrument plugins (some created by Roland themselves) might ensure that the sounds of these instruments find a home in the palettes of contemporary musicians and not just wealthy collectors, we could also argue that the myriad emulations, clones and homages serve to further venerate and perhaps even add further mystique to the originals.

More intriguingly, the continued valorisation occurs as the result of a discourse that has grown around the actual impossibility of virtually recreating Roland's instruments. According to this way of thinking, these are not simply designs that can be reanimated by consulting schematics and documentation. The vagaries of analogue components have given rise to a discourse in which a kind of indescribable magic flows through the circuits of these instruments. In similar manner, commenting on the operation of the filter circuit that is so characteristic of the Moog 'sound', engineer Jim Scott suggests that,

> The best analog components in sound nearly always have this quality of not being quite understood and nearly always involve some not quite specifiable resonances and distortions that occur at high frequencies beyond the audible range but that produce audible effects.
>
> (Scott cited in Pinch and Trocco 2002: 261)

To compound matters, with so many of the components used in the circuits of instruments such as the Minimoog and Roland x0x series no longer in production, even the most dedicated creators of hardware facsimiles have their work cut out in attempting to replicate the original circuits, which only adds to the mystique around those designs and implementations.

> Although the fabric of lore surrounding the design, inception, and use of the 808 lends a richness to its overall mythology, they also give credence to a blithe sort of analog fetishism. The device's ingenious and satisfying properties are often attributed to mere circuit element nonlinearities. In addition to being inaccurate, this mindset directs attention away from a more interesting story.
>
> (Werner et al. 2014c: 1)

That said, calibration ranges and the operational and manufacturing tolerances of the components in a device mean that any individual SH-101, TB-303, TR-808 etc. will have its own unique character and personality. Furthermore, as components age over decades and their behaviour alters, these differences can become even more pronounced so that, even if all TR-808 circuits were created equal, no two instruments would ever have sounded alike. Different oscillator tunings or envelope times make Cowbells higher pitched, cymbals more metallic, and some Bass Drums boom even deeper and longer.

It is notable, too, that while the x0x instruments are now thoroughly celebrated as important and impactful, this has not always been the case. Indeed, as Reid (2004b) notes, the TR-808 did not even warrant an inclusion in Roland's 25th Anniversary commemorative catalogue, published in 1997. The reasons for such ambivalence towards the instruments are manifold but are surely informed by the strength of Roland founder Ikutaro Kakehashi's corporate vision for Roland as a future-facing company dedicated to innovation. Speaking two years after Kakehashi's passing, Roland's current CEO, Jun-ichi Miki, observed that,

> Our founder Mr. Kakehashi always said, 'Never chase a ghost,' and I really understand his meaning. 'Chasing the ghost' of the original JUPITER-8 or TR-808 does not make sense as we will never catch them, and this effort would not align with our vision for the future. Roland has developed many legendary products throughout our history that came from our drive to always design the future by applying the latest technology and our unique know-how from each era. This is our DNA which will never change.
>
> (Jun-ichi Miki 2019)

While it remains the case that Roland has not fully recreated any of its vintage instruments with analogue circuitry, the last few years have seen the company revisit its attitude towards the ghosts of its past. Its 'Roland Cloud' service offers software emulations of instruments and effects for use within Digital Audio Workstations, while the 'Boutique' series digitally remake synthesizers and drum machines including the Jupiter-8 and TR-808, along with others from the Juno, JD and TR lines, in hardware.

Based around Roland's 'Analogue Circuit Behaviour' (ACB) modelling technologies, Boutique instruments such as the JP-08, TR-08, SH-01a and TB-03 clearly reference their points of inspiration in their names and numbers, albeit in truncated form. However, each also meticulously replicates not only the sound of its predecessor but also its workflows and physical interface, awkward anachronisms and all. That said, most of the Boutiques also offer additional modes of operation that allow those counterintuitive procedures and operations to be sidestepped if desired, thereby creating ways of working that will be more familiar to musicians and producers four decades on. Of course, as thoroughly modern machines, the Boutique instruments also add a full complement of digital connections such as MIDI (which was absent from most of the original x0x instruments) and audio over USB allowing them to fit seamlessly into the kinds of computer-based music-making contexts that were unimaginable in the 1980s.

One other notable feature of the Boutiques is their 'flattening' effect. While each recreation bears the distinctive colour scheme and a facsimile of the switches, buttons and knobs of the original user interface, they are now miniaturised. Although the diminutive size of their controls can render them hard(er) to use than the originals, the decision to package all the digital versions in identically sized cases could be argued to lead to an erosion of their individual identities. As such, the Boutiques represent what is perhaps Roland's clearest nod to the expression of its legacy in the form of specific instruments, yet they simultaneously bestow each with a curious uniformity. This could be read as a desire to emphasise familial resemblance and continuity, though by placing recreations of the analogue SH-101 and digital D-50 in the same case, it perhaps says more about profit margins in the contemporary musical instrument marketplace and the economies of scale of mass production.

Even the flagship TR-1000 Rhythm Creator, released late in 2025, is not an exact recreation of either the TR-808 or TR-909 despite including interface elements and, crucially, sound generation circuits derived from both. Nonetheless, it is difficult to understate the importance of the TR-1000 as it marks a significant shift in Roland's attitude and commercial approach to its legacy. While recent instruments like the TR-8, TR-8S and even TR-6s had explicitly referenced the lineage of the 1980s x0x instruments, their sonic emulations were, like the Boutiques, wholly digital. The TR-1000, by contrast, sees Roland returning to analogue sound generation circuitry for the first time since the release of the TR-909 more than four decades earlier. However, despite considerable fanfare and an overwhelmingly positive initial response from reviewers and early adopters alike, the TR-1000 is a curious beast as it focuses only on the analogue circuits of the TR-808 and TR-909 (eschewing the TR-606, for instance) and, even then, offers only a subset of the TR-808 and TR-909's sonic palette. Alongside these analogue circuits, the TR-1000 also draws on Roland's ACB modelling technology to provide considerably extended emulations of all of the sounds from the TR-808 and TR-909 along with the TR-606, TR-707, TR-727 and CR-78. Without doubt, the TR-1000 is a fascinating instrument partly in terms of its sound, sequencer and functionality but as much in terms of what it says about Roland in 2025 and beyond and the company's relationship with its history and 'ghosts'.

Recent years have also seen Roland keen to memorialise its x0x instruments through celebrations of '303', '808' and' 909' Days among others on 3 March, 8 August and 9 September, respectively. Events featuring notable artists reflecting on the importance of the instruments in shaping their careers, branded merchandising including trainers, hats and T-shirts, and collections of sounds and samples, all seek to reinforce the legacy of individual instruments within the wider x0x brand, and embed them in contemporary music making and popular culture.

For all the corporate fanfare directing attention to what these instruments have become, it remains the case that the most oft-repeated story of the x0x series is a collection of instruments that were each released to a resounding chorus of disapproval. In the face of such indifference, each of the successive x0x instruments is uncomplicatedly coded as a commercial failure. With retailers unable to sell them in the face of more desirable competition, the taken-for-granted narrative places the unwanted x0x instruments in pawn shops and classified ads where they change hands for comparatively small amounts of money. Certainly, there is some merit in the broad arc of this narrative and we will encounter examples of instruments changing hands for a small fraction of their retail prices just a short while after their initial retail release. Indeed, it is a central argument in this book that the availability of these instruments is crucial to understanding their histories. It is their comparative cheapness that puts them in the hands of many of the musicians, producers and DJs who are influential in shaping and reshaping their uses and meanings and who would, in turn, use those instruments to help define a cultural and political identity for themselves as creative practitioners. Whether we could go as far to suggest that there is a democratising effect at work here is a matter of debate but we should certainly bear in mind that these economic conditions were the unanticipated consequences of the marketplace and we can only speculate as to whether other commercially unsuccessful instruments would have taken their place in an alternative universe. What we can be certain of, however, is that the growing cultural and monetary value more recently placed on the original x0x instruments, along with the emergence of the collectors' market fuelled in part by Roland's continued veneration, has created a circularity that ensures the original instruments now exist far beyond the reach of all but those with the greatest means.

While we must be acutely aware of the realities of political economy and the intersection between designerly intent and the logic of the marketplace, there is a tendency in x0x histories to erase nuance in favour of a simplified story. There is often more complexity in the trajectories of the different x0x instruments with the SH-101 performing rather better than credited on launch. Similarly, there are varying reasons for putatively underwhelming responses. In the case of the TR-909, it was Roland's US division that slashed the price at retail after just a few months in an attempt to clear inventory as they felt the instrument's design and largely analogue sound meant that it was unlikely to perform well against competing products from other manufacturers and, most importantly, their own forthcoming fully-digital drum machine, the TR-707.

Moreover, although the overarching narrative of discarded instruments later discovered by receptive creators makes for a compelling origin story, it leaves questions unanswered. What can we learn about those users who did make use of the instruments upon their initial release, for instance? Similarly, what should we make of Roland's designers, engineers and marketers? Did these teams simply misjudge their designs and messaging? Or perhaps, armed with the benefit of the hindsight that reassures us that each of the x0x instruments would go on to find an enthusiastic user base who positively embraced their sounds and features, we might re-conceive the question and ask not what was wrong with the designs but what was it that led to their initial audience failing to appreciate them? Implicitly adopting a version of this perspective, some extant histories present the x0x instruments not so much as failures but rather as designs that were ahead of their time and that simply needed to fall into the right hands so that their genius could be recognised and harnessed.

Continuing this line of thinking, it has become almost a cliche when writing about the x0x instruments to suggest that the 'right hands' used the instruments in the 'wrong way'. So much of the discourse surrounding the TB-303, for instance, centres on how 'misusing' the instrument leads to the revelation of the distinctive sound of Acid House. While there can be no doubt that the sound of the TB-303 and the manipulation of its filter circuit is utterly central to Acid, how might we distinguish between 'use' and 'misuse'? After all, the creation of the characteristic 'Acid' sound is achieved via the front panel controls and requires no internal modification of the instrument's circuits.

As Zappi and McPherson (2014) have noted, improvisations like these and the adoption of specific sounds or techniques should hardly surprise us, as musicians have long used instruments in ways that might have been anticipated, surprising or even challenging to their designers. Giving examples of jazz saxophone, the harnessing of electric guitar distortion as a creative effect rather than an undesirable consequence of recording equipment and technique, and the transformation of the vinyl turntable from a domestic playback device into the instrument of scratch DJs, Zappi and McPherson (2014) describe how musical practice often reveals 'hidden' affordances of instruments or situates them in new contexts that break with the traditions and accepted norms of composition and performance. Indeed, as Trevor Pinch has suggested of synthesizer technology, unanticipated affordances and opportunities might come from a variety of sources. 'Broken modules were sometimes sources of frustration, but with experimentation I found they could produce even more interesting sounds.' (Pinch 2007: 167).

Similarly, considering the way in which the use of a musical instrument might be understood as the revelation of hidden affordances, Tahıroğlu et al note that:

> When instruments are put in the hands of performers, a further probing occurs. Musicians often use musical instruments in ways that the original designers never intended, probing for hidden affordances.
>
> (Tahıroğlu et al. 2020: 65–66)

Noting that the sometimes idiosyncratic techniques developed by performers can often come to define an instrument's identity, we might go further by questioning not only the notion of 'intended' and 'actual' use, but also what the function of instrument design and the purpose of an instrument, electronic or otherwise, might be. 'If a performer uses a new musical interface in an unexpected way, should this be considered a design success or failure?' (Zappi and McPherson 2014: 455).

It almost goes without saying that Roland's engineers could not anticipate every use of their instruments or the genres that emerged around the specific uses and combinations of sounds and performances. This would be no more possible than expecting George Beauchamp to have predicted Heavy Metal when designing the first commercially available electrically amplified string instrument in 1931.

> After 1992, I was happy to see artists in the studio tweaking the knobs of our instruments. I remember DJ Pierre with the TB-303, Jeff Mills with the TR-909, and others. They were all surprises, but so fun for me.
>
> (Kikumoto in McCabe 2020)

The slippage between the intended and emergent uses and cultural meanings of the x0x instruments is often codified in terms of 'failure' and 'misuse'. For example, in discussing the TB-303, Tahıroğlu et al. (2020: 66) note that, 'It turned out to be awkward to programme and essentially 'failed' at what the designers wanted it to do.' Similarly, Cameron (2022) celebrates the creative misuse that brought a new sound from the instrument, 'when specific knobs are turned farther than the synthesizer's manufacturers likely intended.' (Cameron 2022).

If we continued along this line, we might find ourselves editing Roland's famous marketing slogan to read 'We Design the Future ... by accident'. Yet, rather than simply conflating any perceived gap between intended and emergent uses with the apparent commercial failure of the instruments in their initial releases, perhaps we might adopt a different line of analysis. Following Zappi and McPherson and Tahıroğlu et al, we might instead ask whether the true success of the x0x instruments is actually found in the richness of their latent potential and the hidden affordances which are revealed as they are explored and probed by users. With just six controls available on the machine's front panel, each of which is carefully calibrated in the design and manufacture process to deliver a specific filter characteristic and behaviour, is it reasonable to assume that these sounds were not heard in Roland's research and development labs and were only discovered years after the instrument had been publicly released and initially dismissed? Perhaps what we see are instruments suffused with possibilities even if these possibilities and the creative contexts of their use are not immediately evident to performers or designers.

To be clear, none of this takes away from the imagination of the musicians who gave life to the unmistakable sound of Acid and who cemented the cultural and musical meaning of the TB-303 through their practice. Rather, these questions invite us to further investigate the interplay between the practices of design and use, to ask how we can know the intended uses of an instrument or what are the 'right' or 'wrong' ways to play? And, perhaps even more pertinently, how these putatively intended uses come to be communicated to potential users and from what sources and by what means might alternatives emerge?

Instruments as Platforms

As someone who has spent more than 20 years analysing and writing about digital games, I am perhaps primed to see electronic musical instruments through the lens of game studies. However, I feel there is much merit in exploring the translation of some of the methodologies, approaches and conceptual frames that have developed in relation to one playable, playful media form to another. Most obviously, this book is founded on the principles of co-creativity and the (re)configurative nature of play (Newman 2008). Rather than conceiving of a game as a sealed unit, replete with meanings, this approach recognises the role of the player is not merely to enact the game and give it life, but rather is to construct and configure, reconstruct and reconfigure it through exploratory performances of creative play. In this formulation, game designers and developers do not so much create singular 'games' as they offer frameworks, opportunities and contexts in which games can be created. Seen in this way, their creations are plastic, malleable sets of resources to be played and played *with*.

It is hardly surprising to find that the configurative and exploratory nature of play frequently gives rise to unexpected and 'emergent' performance possibilities. These new configurations of the material often travel in directions unanticipated by designers and sometimes even actively explore, or 'probe', glitches and exploit bugs and errors in the underlying systems. These new affordances are not simply revealed through acts of play but are better understood as being created by them. Indeed, we might argue that it is only through the operation and performance of particular styles of exploratory and investigative play that the materials of 'the game' are fashioned into these new configurations. The multifaceted nature of the game is seen in the stark differences between the 'imagined' playing as set out in an instruction manual or introductory tutorial and the reconfiguration seen in the practices of 'speedrunning' or 'superplay' that adopt techniques such as moving 'out of bounds' into the netherworlds beyond the spaces intended to be visualised on screen, or that skip portions of the game's narrative by breaking its putatively prescribed linear sequences. Importantly, these different configurations are not in competition with one another, nor does any have a greater claim on being 'right' or command more legitimacy. Rather, each of these possible playings (and many more besides) help to constitute the full potential of the game.

In one sense, we might apprehend the game both as a 'design object' infused of intentions and inscribed with imagined uses, and as a 'played object' constituted through exploratory acts of performance and given meanings by individuals and groups acting upon and reconfiguring the resources sometimes in new and unpredictable ways. It follows that, the richest understanding of the game comes from the synthesis of these two objects of study and an interrogation of the spaces between. Through the architecture of the game along with other prefigurative materials such as tutorials and instructions, the designer suggests ways of playing and points towards specific outcomes and possible pleasures, yet in reality, they can do little more than this.

Ultimately, it is players who take up, act upon, interpret, ignore, reimagine, misunderstand, discard and even reject some of what is, or appears to be, inscribed. Here, we might draw parallels with Madeleine Akrich's (1992) work on technology design. Akrich points to the ways in which designers imprint their visions of a device's purpose, the scope of its operation, the manner in which it should be used, the degree of a priori knowledge required and the expertise that the imagined user might be expected to acquire in order to successfully use the device. For Akrich, these designerly assumptions are inevitably 'scripted' into the device. A digital game, just like a TB-303, may therefore be thought of as a rich suite of latent potentials and hidden affordances that are produced and revealed through the creative practices of playful design and designerly play. The true meaning of the game or instrument is found in the interplay between these practices. It is formed from the coincidence and collisions between intentions, interpretations, uses and (re)imaginings all of which are performed and mediated through the material of the interface and circuitry.

Drawing on these approaches, a central theme of this book is the conceptualisation of the various xOx drum machines and synthesizers as instruments that are implicitly and explicitly inscribed and scripted with the intentions, aims and assumptions of their designers and which are brought to life, given meaning and explored through the creative performance practices of musicians, DJs and producers. Crucially, these configurative, improvisational and interpretative practices that construct particular visions and versions of the instrument are not fixed but rather are historically, socially and culturally situated. The meanings and uses of an xOx drum machine or synthesizer, like any other instrument (or game) accrue and change over time. Sometimes this is a gradual accumulation of meaning and sometimes it is the product of self-consciously diverging from previous uses and conceptualisations.

In addition to the exploration of these designerly processes and the configurative performance practices of use and play, this book further draws on game studies in its conceptualisation of the relationship between technological systems and their interactions with creativity, expression and cultural production which it inherits from Platform Studies. Again, while this approach has become widely deployed in game studies following its introduction in the work of Montfort and Bogost (2009), its application outside that domain remains comparatively novel. As such, it is worthy of some commentary.

While the term 'platform' is one that is already in widespread usage in the parlance of gaming and broader computing, it is relatively unfamiliar within discussions of musical instruments. However, it has considerable utility in this context provided we are clear in our use of the term. One concise definition favoured by Bogost and Montfort comes from Marc Andreessen, who states that,

> Definitionally, a 'platform' is a system that can be reprogrammed and therefore customized by outside developers – users – and in that way, adapted to countless needs and niches that the platform's original developers could not have possibly contemplated, much less had time to accommodate.
>
> (Andreessen in Bogost and Montfort 2009: 3)

As such, while we might be used to thinking about platforms as stable collections of hardware and software layers upon which to build, a key quality is actually their facility to be altered. In computational terms, because the platform may be programmed and reprogrammed, it is both configurable and reconfigurable. To think of drum machines and synthesizers in this way is to think of them not merely as instruments, objects or pieces of abstract technology, but rather as sites for different kinds of expressive work. In this sense, the term 'platform' also bestows a potentially useful openness that speaks to the range of uses and reuses, interpretations and reinterpretations, that are not fixed in time, space or musical application and genre.

To some extent, Platform Studies is a critical and methodological response to a number of competing approaches to the study of technology, particularly in the context of computational media. On one hand, Platform Studies seeks to challenge the centrality of technology as the determinant of social and behavioural change. Such 'technologically deterministic' thinking very often suffuses popular and academic investigations of digital media and computing. At its most reductive, 'hard' variants of technological determinism propose a cause-effect model in which technologies are seen to develop beyond or outside social factors and influences. Similarly, the effects of technology on culture seem to occur autonomously and without social intervention. As their name implies, 'soft' versions still place a high value on the impact of technology as a force for change but are more accommodating of processes of reception and discrimination and perhaps tend towards positioning technologies as major rather than the sole determinants of change.

In this regard, the desire in Platform Studies to reframe the relationship between technology, society and culture will remind us of social constructivist approaches which similarly assert that technologies do not simply arise but rather are the products of social and cultural factors, funding regimes, and the orientation around specific visions and objectives. However, where Platform Studies shifts in focus from the social constructivism of proponents of Social Construction of Technology (SCOT) theories, is in the detail of its analyses of the technologies themselves. For some critics of SCOT, there is an insufficient interrogation of the material construction and design of the technologies under consideration. For commentators such as Langdon Winner (1993), this leaves technology not merely as an inscrutable 'black box' but one that, for Winner at least, is found to be analytically empty on lifting the lid.

By placing the reciprocal connection between the specificities of design and expressive cultural practices at its heart, Platform Studies investigates the relationships between the hardware and software design of computing systems (platforms) and the creative works produced on those systems. As Bogost and Montfort (2009) note, the advantage of this approach is that it recognises the role of designers and users in constructing the meanings and uses of the platform. Taking inspiration from this, key to the approach adopted in this book is the synthesis of an investigation into how the design of the instrument-as-platform both enables and constrains the creative work undertaken on it, with an analysis of the influence of social, economic, cultural, technological and other factors on the designers' decisions to create their technological systems in particular ways. As such, Platform Studies recognises that it is not only the user experience of the system that is culturally embedded but also the form and function of its interface design, underlying code and circuitry. In this way, we can begin to conceive of the platform as an in the amalgamation, or perhaps even collision, of the imprinted expectations of designers and the unexpected creative practices and revelatory aesthetics of users. By continuing to explore the ongoing lives and afterlives of these instruments as they are celebrated, remade, modified, cloned and emulated, we see these expectations and probings folding back on one other. Over the decades, the accumulating practices of making meaning and making music have created complex feedback loops that provide new frames for interpreting and approaching instruments like the TR-808 and TB-303. It follows that a central aim of this book is to identify, critique and disentangle these new, and sometimes contradictory, contexts and visions of the x0x series.

Creativity, Capacitors and Culture

As Bogost and Montfort note, the desire to uncover relationships between technology and culture makes the detailed discussion of the technical details essential and not merely a matter of 'geek value' as they put it. As such, a key tenet of Platform Studies is the foregrounding of the 'material substrate' of media forms. In relation to computing, and particularly in relation to the study of games, this might refer to the central processing units, sound chips, graphics cards, storage media and hardware interfaces such as joysticks and controllers. The specifications of these individual components arranged in particular combinations are, in one sense, what define the capability and affordances of any given gaming platform. The graphics card, for instance, operates at a particular resolution, allows the display of a certain number of colours, and might be designed to efficiently process 3D geometry or manipulate 2D sprites and backgrounds. Similarly, any CPU will be optimised for certain operations rather than others and will, of course, have a finite computational capability. The choices that inform the selection of individual components and their arrangement into specific configurations and systems are important in helping set the boundaries of any given platform but also because they are key to differentiating one platform from another. This is the reason why the 'same game' might look, sound and feel different when released on a Nintendo Switch, a Sony PlayStation or an Apple iPhone. As Montfort and Bogost note,

> Sometimes the influence is obvious: A monochrome platform can't display color, a video game console without a keyboard can't accept typed input.
>
> (Bogost and Montfort 2009: 5)

Of course, in addition to these immediately obvious features, there is far more complexity and low-level detail in the design of the platform that both constrains and supports the expressive and creative work made with it. The analogue drum machine with its particular set of fixed (and perhaps overtly synthetic rather than 'realistic' sounds), the digital drum machine with its samples that freeze one specific cymbal recorded in a specific room, the monophonic synthesizer with its ability to play just a single note at a time, the frequency range of an oscillator, the response of a filter, and even the values, manufacturing tolerances and behaviours of specific resistors, capacitors and transistors in a circuit, all impart a distinctive sonic quality and define one instrument or platform from another. Moreover, the design of an interface whether it be the implementation of a sequencer or the physical shape, tactility or movement of buttons, switches or sliders, all make possible certain kinds of expressive creations while seeming to dissuade or preclude others. And let us not forget that these instruments are not simply the coming together of the available technologies of their era but are, like all design objects, the products of a series of compromises that balance potential capabilities with the costs of research, development and manufacturing.

The x0x drum machines and synthesizers analysed in this book, then, are understood as creative design responses that are situated within specific social, cultural, technological and economic contexts. These instruments are, in turn, responded to, navigated and negotiated by musicians, DJs and producers operating in their own distinctive economic and socio-cultural contexts, who create aesthetics and methodologies through the exploratory practices of their expressive work. Crucially, the instrument designs, as well as the advertising, marketing and technical documentation that frame them, reveal certain ways of thinking about their utility and purpose, yet once they are played, played with, explored and probed, they are reconstituted as malleable platforms for creative expression.

In fact, Heermans' (2016) work encourages us to go even further in thinking about the nature of the TB-303 not merely as 'a transformative object that enables new forms of creativity' but rather as an 'active agent' in a creative process. This position draws on Barnett and Boyle (2017) whose co-edited collection 'Rhetoric, Through Everyday Things', argues for moving beyond the taken-for-granted view of objects as passive, inert tools or the background for human activity, and conceiving of them as vibrant agents. Considering the active role the x0x instruments play in the process of music making might seem initially unusual, yet perhaps makes considerably more sense in the context of hearing users' discussions of their encounters. Indeed, as Heermans (2016) notes, it is telling, yet should surprise us little, that DJ Pierre describes the distinctive use of the TB-303's controls that gave rise to the sound of 'Acid' not as an act of 'invention' but rather as a process of 'discovery'. Viewed in this way, the synthesizer or drum machine shifts from being a canvas or backdrop to being a vital actor. To fully understand the x0x instrument is to apprehend it as a design object, as a playable and played object, and as a distinctive concoction of intended, unintended, emergent, hidden and revealed affordances that support particular styles while simultaneously stymying others.

Given the importance of the 'material hearts of the machine' as Altice (2016) puts it, this book unashamedly dives deep into the circuit level of design as well as considering the more immediate user interfaces and surfaces. Combined with a detailed discussion of the technical documentation, advertising, marketing, as well as the commentaries of reviewers and critics, it is in the flow of voltage through the transistors as well as in the physical controls offered to the player that the specific affordances and essence of the x0x instruments are to be found.

In considering the enormity of their influence, LeRoy (2023: 285) poses an interesting and seemingly unanswerable question of Roland's xOx drum machines. 'Did those machines make the music or did the music make the machines?' My hope is that by exploring the intertwined nature of the design and uses, the revelation of affordances and the economic, cultural and musical contexts in which these instruments were created, understood and probed, this book offers one possible answer. People made the machines with which people made the music that made the machines.

2 Before the 808

Figure 2.1 Roland CR-68 CompuRhythm (image credit: James Newman).

DOI: 10.4324/9781003374510-2

Tape Loops, Spinning Discs and a Boatload of Programmable Percussion

Of course, important and influential though the instruments we will explore in this book went on to be, they were far from Roland's first forays into such territory and the company had already produced a number of synthesizers and rhythm units before the TR-808 inaugurated the xOx series in 1980. Moreover, even Roland's earliest instruments are predated by the creations of other companies and individuals by more years than we might imagine. As such, instruments like the TR-808 both inherit and respond to some important conventions and contexts in terms of design, audience and musical usage.

It may not surprise us too greatly to learn that trying to pin down exactly what constitutes the first drum machine is a somewhat difficult task that has exercised commentators for many years. Even Leonardo Da Vinci has a potential claim for a 15th century automated drum that, to the untrained eye, looks as much like a grass roller for tending to a lawn or sports pitch than a musical instrument. Heading further back to 1206, however, we find the publication of Ismail al-Jazari's 'The Book of Knowledge of Ingenious Mechanical Devices' which documents a host of inventions including a number of musical and rhythmic automata. One particularly notable creation describes a model boat containing a robotic harpist, flautist and two percussionists that was designed to float across a lake performing music as an entertainment at royal drinking parties. The copper figures were operated by a system of pegs and levers that were powered by the movement of water. Intriguingly, because reconfiguring the arrangement of pegs led to the creation of new rhythmic patterns, this could even be described as a fully programable percussion instrument.

While purely mechanical musical devices continued to be developed with instruments such as street organs often being embellished with automated percussion elements such as drums and bells throughout the 18th and 19th centuries, it was innovation in electro-mechanical engineering that would give rise to a breed of instruments that more obviously exert influence on contemporary drum machines. Indeed, it is the 1931 Rhythmicon that is most often cited as the first drum machine. The instrument came about through a collaboration between US composer Henry Cowell and designer Léon Theremin. Theremin was already established in the music technology field as, a couple of years earlier, he had patented his eponymous instrument which made use of two antennae that sensed the position of the player's hands to control pitch and volume without actually touching the device. Cowell's proposition was to create an instrument that would allow a player to perform complex polyrhythms with comparative ease and metronomic accuracy albeit with the potential for variation and expressivity that arises from a human performer triggering the contrasting sequences against one another. Based around the mathematical relationships between rhythmic patterns and the harmonic series of pitches, the Rhythmicon used a series of rotating discs to interrupt a beam of light that was detected by a photoelectric sensor and translated into sound through a built-in tube amplifier.

Intriguingly, although perhaps betraying the centrality of pitch in the underlying sound generation method, the Rhythmicon was controlled with a truncated piano-style keyboard. Each note triggered a sequence of pitched pulses and subdivisions related to the fundamental rhythmic base and, as the instrument was polyphonic, pressing multiple keys triggered patterns simultaneously with potentially complex and unpredictable results. The output of the Rhythmicon is firmly located in the world of experimental sound creation and is perhaps best described as an acquired taste. The instrument was used by Cowell and others in a handful of performances and only three units were produced.

Nonetheless, the influence of the Rhythmicon is considerable. Most obviously, it informs the design of Moog's Subharmonicon released in 2020, although here the rhythmic patterns of the instrument's built-in sequencer are derived from the subharmonic series rather than overtones. More than this, however, the Rhythmicon speaks to a desire to automate the process of rhythm creation yet, with its interface, it offers an extremely high level of interactivity that places expression and live performance at its heart. Indeed, it can only be played in real time and it is an inherently tactile instrument. As we shall see throughout this book, given how often drum machines are embroiled in discussions about inhuman, machine-like sterility and claims of undermining musicianship, it is important to bear in mind these crucial foundations.

If the Rhythmicon was an unashamedly experimental instrument intended to help explore complex polyrhythms and challenge traditional notions of musical metre, Harry Chamberlin's Rhythmate was an altogether more mainstream proposition. Unlike Cowell and Theremin's creation which created complex rhythmic patterns with comparatively abstract sounds, the Rhythmate was intended to replicate the sound of a drummer playing an acoustic drum kit. It achieved this by using the sound of a drummer playing an acoustic drum kit. The Rhythmate is based around a series of tape loops each with a recording of a live performance. These drum loops could be replayed at different tempos by adjusting the speed of the tape moving across the playhead, albeit with a resultant change in pitch. If the basic principle of the Rhythmate sounds familiar, Chamberlin achieved greater success with a version of this technology based around a musical keyboard and with recordings of acoustic instruments such as strings, flutes, French horns, harpsichord and accordion. In fact, Chamberlin had originally developed a prototype of the Rhythmate in the late 1940s but returned to the idea in the 1960s following the success of the keyboard instruments.

Interestingly, the tape-based keyboard would eventually transform into the Mellotron after a Chamberlin sales representative presented the technology to UK company Bradmatic (later renamed Mellotronics) who enthusiastically began developing, refining and selling their instrument. As this arrangement was undertaken without Chamberlin's knowledge or consent, an eventual settlement was reached that saw the Mellotron sold in the UK and the Chamberlin in the US (Epand 1976). Even the refined Rhythmate did not make a significant commercial impact with estimates suggesting that just a few tens of units were manufactured. And, while they had the distinct advantage of being able to play back recordings of drum and percussion tracks, tape loops were ultimately cumbersome, unreliable, liable to break and wear out and required considerable effort to calibrate let alone change if the player wished to hear different rhythms. But even if the technology did not blaze a trail to follow, the purpose of the instrument certainly did. Where the Rhythmicon and even al-Jazari's musical machines were essentially solo instruments, with the former perhaps supported by an orchestra, the Rhythmate was designed as an automated accompanist for a soloist performing at the organ.

It was this application of the rhythm machine that really propelled its development although it would not be Chamberlin that would take things forward as he was otherwise engaged with the Mellotron dispute and an ongoing threat of litigation from musician's unions concerned that tape recordings of acoustic instruments would put musicians out of work (Epand 1976: 28; also Crossley [2023] and Jones [2024] on the UK Musicians' Union response to synthesizers and drum machines). Instead, it was the organ manufacturer Wurlitzer whose Side Man instrument, released in 1959, that would become the first commercially available drum machine. As the name of the instrument already suggested, the proposition was a clear one. 'Now you can be a combo all by yourself! Flip a switch … and there's a full rhythm section at your side!' boasted the headlines of advertisements. It is in the detail, however, that the distinctiveness of the instrument is to be found. The copy continues by proudly proclaiming that 'The Side Man uses no tapes or recordings.' Of course, this nod to the Rhythmate could be read as a step backwards. After all, the benefit of tape was found in the authenticity of recorded sound. Wurlitzer's advert, however, focuses on the limitations of the medium and, in particular, the pitch change that unavoidably accompanies a change in tape speed. 'Changing speed does not change the sound or alter the pattern – and it plays for any length of time *you* desire.' The advert is notably silent, though, as to how the Side Man generates its tones and what they sound like.

> Unlike the Chamberlin Rhythmate's tape loops, the Sideman's rhythms were created by vacuum tubes and a valve amplifier. Its deep, rounded bass drums, brushed noise of its cymbals and shrill toms anticipated the sound of Roland's TR series, yet the preset rhythms were much the same as the Rhythmate – popular dance standards like the samba, bolero and tango.
>
> (Wilson 2016)

The Side Man does not play back recordings of drums but, rather, generates its sounds in real time through circuits of discrete components. Similarly, rather than pre-recorded patterns on tape, the Side Man's sequences are generated by an ingenious mechanical system involving a spinning disc covered in contacts that trigger the instrument's individual sounds. The mechanism is somewhat similar in principle to a musical box and bears more than a passing resemblance to al-Jazari's mechanical automata with carefully positioned contacts literally mapping out the patterns to be performed. Varying the speed of the disc's rotation changes the tempo without affecting the tone of the drum and percussion sounds thereby overcoming one of the inherent limitations of tape.

The sounds of the Side Man certainly have an undeniable charm but their imitations have little of the authenticity of the Rhythmate's recordings of live drum and percussion performances. We might even suggest that Side Man's sounds and the built-in patterns with names like Rock 1, Bossa Nova, Rhumba, Beguine, Waltz and Swing are the first steps along the path to what now might be charitably, and perhaps even positively, described as 'cheesy'.

For all the differences, however, the Side Man shares some key aesthetic similarities with the Rhythmate. Both instruments are housed in large, heavy, floor-standing wooden cabinets with a built-in speaker and a control panel on top for selecting different rhythms and tempo. If we only analysed the ergonomics and industrial design, we would be left in little doubt that both instruments were designed to sit next to – or along*side* – an organ or piano. Listening to them, as Wilson suggests, while they take wildly different technological approaches, they are both clearly dedicated to delivering the same dance rhythms from a selection of popular music genres. The Side Man's technology offers an intriguing possibility, however. In addition to the predefined patterns, the instrument's top panel includes ten small buttons that allow the individual sounds to be played in real time. Although the interface resembles a button accordion rather than any percussion instrument (nor even the Rhythmicon keyboard let alone the modern drum pad), its inclusion arises from the dynamic nature of the sound generation and, crucially, introduces the possibility of live performance.

A few companies initially continued to refine and develop the rotating disc technology, most notably the Korg (then known as Keio Gijutsu Kenkyujo) with their snappily named 1963 'DA20 Donca Matic Disk Rotary Electric Auto Rhythm Machine' which added more preset patterns and a small musical keyboard. This bore more than a passing resemblance to the keys on the Rhythmicon but, like the Side Man, it was to trigger the individual sounds rather than sequences or patterns. Other manufacturers explored the potential of the comparatively new transistor technology to miniaturise and make more efficient and practical instruments. Removing the large speaker and relying on plugging the instrument into external amplification (much like an electric guitar or synthesizer) was another way of shrinking the devices and making them more portable. Additionally, companies like Seeburg in the US also did away with the cumbersome mechanical wheels for pattern storage by making use of an innovative diode matrix to store preset beats in their Rhythm King and Rhythm Prince instruments. Korg would also implement transistor-based circuits in their later Donca Matic and Mini Pops instruments released from the mid-1960s. Among the first to create entirely transistorised, solid-state instruments was another Japanese company called Ace Tone. Designed in 1963 by company founder Ikutaro Kakehashi, the Ace Tone 'R-1 Rhythm Ace' was showcased at the 1964 NAMM Convention (National Association of Music Merchants) in Chicago, Illinois. In 1972, Kakehashi would resign from Ace Tone to found Roland.

Making, Remaking and Bending Beats

As poetic as it would be to write that the first drum machine designed by the founder of the company that would go on to make some of the most influential instruments and change the face of contemporary popular music was a roaring success, the truth is that it was a comprehensive failure. Ironically, it was the R-1's implementation of a live performance interface in the style of the Side Man that proved to be the downfall. Six electronically-generated sounds could be triggered in real time from the dedicated buttons on the front panel of an instrument that was designed to be attached by its included brackets to the underside of an organ. On paper, it sounds promising, and the idea was, perhaps, partly inspired by the stops and controls underneath an organ keyboard that allow the player to switch between different settings and registers with a flick of the thumb without moving the hand too far from the keys themselves. For momentary switches that are infrequently activated to change parameters, this is well suited, but for the performance of live drumming while also playing the keys of an organ, this was all but unworkable. To compound matters, these awkward buttons were the only way to play the R-1. Where the live performance controls of the Side Man, Donca Matic et al could be used to augment the built-in pre-programmed patterns, the R-1 had no built-in sequencer and therefore no ability to act as an automatic accompanist. The R-1, then, was a rhythmic performance instrument in a growing world of drum machines.

The lack of sequencer in the R-1 was clearly a serious omission, but Kakehashi would not make this mistake with the follow-up. Returning to the drawing board, the next instrument implemented a newly-patented diode matrix for pattern storage that also brought the ability to combine multiple patterns for hybrid play back options, along with buttons to mute some of the sounds, which afforded personalisation beyond the standard beats. The resulting FR-1 did not quite propel Ace Tone to become a household name (not least because the instruments were sold under a bevy of different brand names including Multivox in the US and Bentley Rhythm Ace in the UK). The true impact of the instrument, however, is ultimately found in the way it set off an unlikely chain of events that led directly to the development of the Roland TR-808. Most immediately, the FR-1 came to the attention of the Hammond company who endorsed and marketed it alongside their organs. The success of this venture led the two companies to enter a partnership that, among other things, saw Ace Tone distribute Hammond products in Japan and Kakehashi take an active role in embedding rhythm units in future organs (Reid 2004a). The Hammond connection also brought Kakehashi into contact with a figure who would become central not only to the future of Roland but also to the development of drum machines, synthesizers and electronic music more broadly.

Don Lewis had been hired by the Hammond company to demonstrate its organ products at the 1969 NAMM show. Among the instruments that Lewis showcased was the Ace Tone FR-2L. This was essentially the same as the earlier FR-1 but housed in a casing to allow it to sit on top of an organ rather than being mounted underneath. However, this was far from the only change that had been made to this particular unit. In fact, having witnessed the performance, Kakehashi was sufficiently intrigued that he introduced himself to Lewis saying, 'This looks like my rhythm unit, but it doesn't sound like my rhythm unit.' (Hayes 2019). The reason was quite simple as Lewis explains:

> I had modified my Ace Tone to death, changed all the rhythms because none of them fit my style of playing. I also wired it through the expression pedal of the Hammond, so I could get [percussion] accents, which no one was doing then.
>
> (Lewis cited in Wolbe 2013a)

As well as being a virtuoso musician, Lewis had trained as an electronics engineer and so was equipped with a perfect combination of technical skills and a keen sense of what would be musically useful (Wolbe 2013b). As such, through a process of creative analysis and reverse engineering (with no schematics or documentation), Lewis had essentially circuit bent the drum machine not only to reprogram it by adding new patterns but also to add entirely new features and functionality unavailable in the original instrument. So impressed was Kakehashi with this feat of creative engineering that he invited Lewis to act as a consultant on the development of future products. Initially, this would involve working on the incremental development of further Ace Tone instruments including the FR-7L that added new preset patterns and features such as volume controls for individual sounds. When Kakehashi resigned from the company in order to form Roland in 1972, Lewis went with him and began advising on new products.

In truth, Roland's very first products were not actually that new. The TR-77 (the initials standing for Transistor Rhythm and reflecting the continued focus on current solid-state technologies) was essentially a rebadged version of the Ace Tone FR-8L while the TR-55 was an updated version of the FR-6. However, after these initial releases, Roland's CR-68 and CR-78 CompuRhythm instruments broke important new ground. Released in 1978, at first glance, the instruments look somewhat similar to previous generations of rhythm units with a wooden enclosure and two rows of colourful buttons that are used to select from 20 preset patterns. Offering Rock 1-4, Disco 1-2, Waltz, Shuffle, Slow Rock, Swing, Fox Trot, Tango, Boogie, Enka, Bossanova, Samba, Mambo, Chacha, Beguine and Rhumba, these offer few surprises and speak to the organ/combo accompanist roots of the instrument.

Sonically, the CompuRhythms are entirely analogue with a selection of tones including Bass Drum, Snare Drum, Rim Shot, Hi Hat, Cymbal, Maracas, Claves, Cow Bell, High Bongo, Low Bongo and Low Conga. These are reminiscent of those found in earlier Roland and Ace Tone instruments although they are now augmented with intriguingly-named Metal Beat, Tambourine and Guiro preset patterns and sounds that can be added to or removed from the mix with their own dedicated faders. While there are no specific tone shaping controls for any of the individual sounds, a Balance fader affects the overall mix of timbres by accentuating either the high or low frequencies. Preset Variations, Fills and Rolls can be triggered manually or automatically after a specific number of bars. So far, at least, the CompuRhythms appear to be well-specified but ultimately unremarkable instruments.

However, the 'CR' designation and the word 'CompuRhythm' emblazoned on the front panel attest to these being the first drum machines with microprocessors at their heart. The microprocessor in question was an NEC-designed μPD8048 which is a clone of the widely used Intel 8048 microprocessor. This chip was also used by Roland in the 1978 Jupiter-4 'Compuphonic' synthesizer whose name already suggests a familial connection with the CompuRhythms but whose front panel leaves no doubt with its row of identical multicolour selector switches. For the CR-68, the NEC microprocessor controlled the playback of its preset patterns but in the case of the CR-78, it also opened up the potential for programmability. This meant that, while the individual sounds were fixed, the user was no longer limited to the preset patterns that shipped with the instrument. Programmability was a feature that Don Lewis had been requesting since he first met Kakehashi, 'I begged him for years to come up with a programmable rhythm unit because I got tired of having to change out the patterns the hardware way – it was hard!' (Wolbe 2013a).

It is difficult to overstate the significance of programmability in the realm of drum machines. Most obviously, programmability frees the musician from the 'tyranny of the preset' (Douglas 2023). Even with the ability to combine two patterns simultaneously, as was the case with some of the earlier instruments from Roland/Ace Tone and other manufacturers, there remains a finite number of combinations. These patterns can soon become recognisable and even indelibly associated with certain artists or tracks. While they will be rediscovered and reused by subsequent generations of musicians and producers, the preset patterns of the Korg Mini Pops 7, for example, will forever be connected by some listeners with Jean Michel Jarre's 'Oxygène' (1976) album. Programmability allows artists to adapt instruments to their specific needs and the feel and genre of their music and, while the CR-78 manual might be stretching the point in suggesting that 'An infinite number of rhythm patterns are yours because of the adoption of a micro-computer in this instrument.' (Roland 1980a: 1), it is certainly the case that this creates an immeasurably more flexible instrument.

The CR-78 was not the first programable drum machine, however. That distinction had been earned by PAiA Electronics' 'Programmable Drum Set' although, as Vail notes, that instrument was based around TTL (Transistor-Transistor Logic) rather than the integrated circuits of Roland's design (Vail 2014: 84). Nor was the CR-78 programmable straight out of the box. While preset patterns could be played using the buttons on the instrument's front panel, programming was undertaken using optional pedals. In fact, there were two methods of programming the CR-78. The first used a single switch-style pedal much like the sort that would be used as a sustain pedal for a synthesizer or electric piano. This was connected to the somewhat confusingly labelled 'Write TS-1' input on the rear of the unit. By moving the instrument from 'Play' to 'Memory' mode, it was possible to use the pedal to tap in beats in real time while the CR-78 was playing. Full patterns could be built up by successively recording a single drum sound into each of the four available Tracks. One potential snag here was that there was limited quantisation or time correction which meant that playing had to be almost perfect which was made harder still as the instrument did not include an audible metronome or click to play along with. In practice, this actually made it rather difficult to accurately record new patterns in real time because, as Roger Linn put it, 'what it effectively did is always turn your beats into a mess.' (Linn 2012).

The alternative was to use the WS-1 Write Switch to program the instrument in step time. This optional, battery-powered device was sold separately and comprised a rotary dial to select the number of steps from 8–48 and, crucially, two large buttons that could be used to enter either notes or rests. Once connected to the CR-78, these two buttons made it possible to move through a Track one step at a time and build up sequences by entering either a note or a rest. To enter a simple 'four-on-the-floor' Bass Drum pattern with a note on every beat of a 4/4 bar, for instance, the programmer would select the appropriate Instrument using the selector on the CR-78, and then use the WS-1 to select a 16-step pattern and enter 'note-rest-rest-rest-note-rest-rest-rest ...' and so on. Selecting the Snare Instrument and inputting 'rest-rest-rest-rest-note ...' would place a hit on the second beat of the bar. Roland's instructions for the WS-1 hint at the way this step time programming helps to overcome the limitation of the CR-78 and the potential sloppiness and imprecision of real time recording.

> The WS-1 divides the rhythm pattern into steps of either 8-12-24 or 48 and since it is programmed by one step at a time, writing rhythm can be done accurately and easily.
> (Roland 1980a: 5)

The CR-78 enjoyed a period of mainstream chart-topping popularity and can be heard in the opening of Blondie's 1979 'Heart of Glass' (with a mix of the Mambo and Beguine presets), along with Soft Cell's 1981 'Tainted Love' (Foxtrot), Phil Collins' 1981 'In The Air Tonight' (Disco 2), Daryl Hall & John Oates' 1981 'I Can't Go For That (No Can Do)' (Rock 1), Ultravox's 1980 'Vienna' (Rock 4) and 1986's 'Lady In Red' by Chris De Burgh (Enka). The instrument also found favour with more avant-garde and experimental artists such as Throbbing Gristle and John Foxx who even named a track 'Metal Beat' (1981) in reference to the instrument. Although it might seem as though the lure of programmability would render the WS-1 a near-essential accessory, Roland made only 2000 of the units to accompany the 8000 CR-78 instruments it manufactured. As such, as Douglas (2023) notes, they were a comparative rarity even at the time which might help explain the extensive use of presets in many of the songs listed above, or perhaps the operation was just too awkward.

Even though it introduced programmability into Roland's lexicon and achieved some success in its own right, the CR-78 is perhaps best understood as a transitional instrument. This sense is somewhat detectable in contemporary advertisements for the instrument and its preset-only CR-68 relation pictured at the beginning of this chapter. The marketing copy somewhat uneasily, and perhaps even cautiously, places it in the familiar setting of the rehearsal room and at the centre of a new computerized future of electronic music making.

> The Compu-Rhythms are two astounding devices that produce accurate and dynamic rhythm tracks that can be used in recording, performing, or just practicing. The Compu-Rhythms will find a lot of places in your music-from writing music to laying down

> click tracks for multi-track recording. They can even control of synthesizers of digital sequencers. The Compu-Rhythms may not replace the drummers of the world, but they're going to make it a lot easier to live with their little inconsistencies.
>
> (RolandCorp USA 1979)

Ultimately, the real impact and influence of the CR-78 is felt in the way it mapped out a path for future innovation. In truth, the instrument was really only partially programmable as only a small selection of sounds fell under user control and there were just four memory slots to save original patterns. Given how crucial programmability would be in the development of the drum machine, Goldstein (1985: 92) might be a little unfair in casting the CR-78 as 'a preset device that had pattern-programmability tacked on as something of an afterthought', although it is harder to argue with it 'employing an incredibly unwieldy programming system'. Nonetheless, by 1985, for all the forward-looking aspirations, it was clear that critics felt that the CR series was yesterday's instrument. 'Their wooden casings and brightly-coloured buttons make them seem old-fashioned today.' (Mills 1985: 8). Unintuitive and anachronistic though it may have seemed, it is unarguable that by introducing programmability, the choice of manual and automatic fills, and a workflow based around incrementally building new patterns one sound at a time, the CR-78 is a pivotal instrument. It is not, however, the uncomplicated 'missing link' between the organist's auto-accompaniment box of preset rhythms and fully-programmable drum machines like the TR-808.

If any instrument warrants such a status it is the DR-55 'Dr. Rhythm'. Usually absent from drum machine histories and even Roland's own corporate retrospectives, the DR-55 deserves far greater recognition than it enjoys. Detroit Techno pioneer Juan Atkins is clear about the importance of the instrument, however, describing it as, 'the most quantum innovation that ever happened', and asserting that, 'If I had to pick one machine, that would be it.' (Barker 2012).

Released shortly after the CR-78, under the 'Boss' brand usually reserved by the Roland Corporation for guitar effects pedals, the DR-55 is the breeding ground for many of the ideas introduced in the CR-78. One certainly could be forgiven for dismissing the DR-55. After all, at first glance it looks more like the optional programming accessory for the CR-78 than a fully-fledged drum machine in its own right. In fact, the similarity to the WS-1 is not a coincidence as the DR-55 essentially melds the functionality, interface and even the looks of the programmer peripheral with some (very) simple analogue sound synthesis to create a surprisingly capable, and comparatively inexpensive, instrument.

The original inspiration for the DR-55 had come from Kakehashi's desire to create a miniaturised version of the CR-78 that could be used by guitarists in practice sessions (LeRoy 2023: 162). This might begin to explain the Boss branding though, while it addressed Kakehashi's aims admirably, the functionality and cost of the instrument led to the DR-55 being used by a number of artists on record and in live performances with Depeche Mode and Juan Atkins among the particularly notable users.

The DR-55 is built around four sounds. The Bass Drum, Snare Drum and Rim Shot are fully programmable while a Hi Hat can be set to play either a constant 8th note or 16th note pattern or silenced altogether. Sonically, the DR-55 continues the tradition of the earlier Roland instruments and includes a tone control similar to that found on the CR-78 that rolls off or accentuates the instrument's high frequencies. Functionally, however, the DR-55 is really rather advanced, especially given its diminutive size. It has 16 memory locations for patterns (compared with four on the CR-78) and implements the same step programming method with two large yellow buttons familiar from the WS-1 used for inputting notes and rests at each 16th note step of the bar-long patterns. Patterns are built in exactly the same way as on the CR-78+WS-1 combination with four programmable tracks and an instrument selector. Here, the larger rotary control of the CR-78 is replaced with a four-position slider for the three programmable sounds and an Accent.

One notable feature of the DR-55 is its implementation of Variations. Where the CR-78 offered two different preset patterns per rhythm style (two versions of Disco, Waltz and Boogie beats, for instance, as well as half a dozen preset and one programmable Fill-Ins that could be manually or automatically triggered after a certain number of bars), the DR-55 simplifies things by having a single three-position switch that selects between either Pattern A, Pattern B or Pattern A and B alternately, creating a 2-bar pattern. The simplified, but extremely flexible and musically useful, AB Variation feature introduced in the DR-55 is just one aspect that is carried over wholesale to Roland's next major rhythm machine. That this next instrument would go on to be the most important and influential drum machine in the world reminds us just how important the unassuming Dr Rhythm is.

Development of the DR-55 was led by Tadao Kikumoto, a relatively recent recruit to Roland. Building on this foundation, Kikumoto would go on to have the most transformative impact on the company, its drum machine and synthesizers designs, including the TR-808 and the family xOx instruments that followed.

3 TR-808 Rhythm Composer

Figure 3.1 Roland TR-808 Rhythm Composer (image credit: implant4).

DOI: 10.4324/9781003374510-3

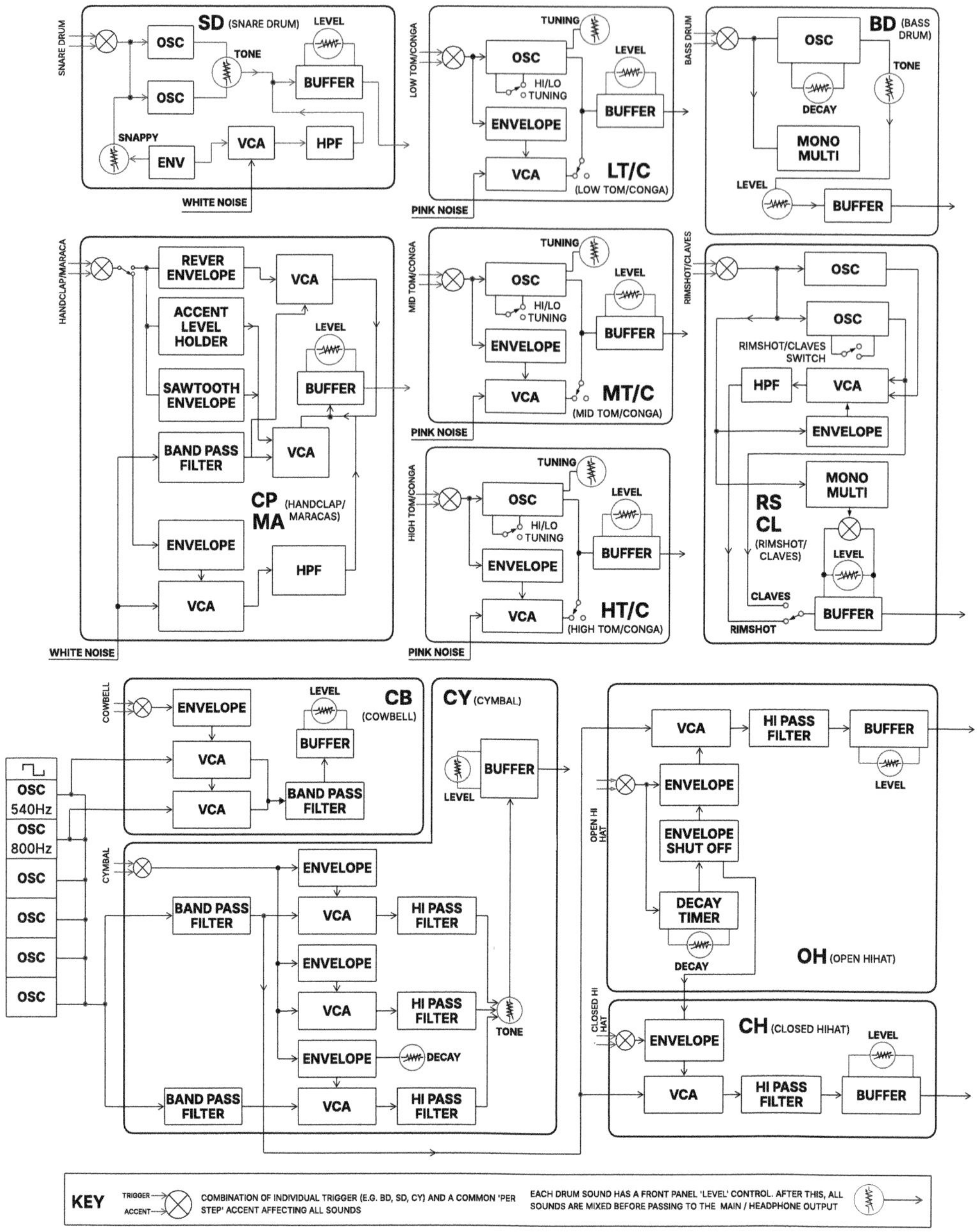

Figure 3.2 Roland TR-808 Simplified Block Diagram (adapted from Roland 1981a TR-808 Service Notes).

TR-808 Rhythm Composer

Released
1980 (discontinued 1983)

Sound source
All sounds generated by analogue circuitry
Bass Drum (BD); Snare Drum (SN); Low, Mid and High Toms (LT, MT, HT); Low, Mid and High Congas (LC, MC, HC); Rim Shot (RS); Claves (CL); Hand Clap (CP); Maracas (MA); Cow Bell (CB); Cymbal (CY); Open and Closed Hi Hat (OH, CH)
12 note polyphonic (simultaneous sounds)

Real time controls
Individual Level Controls for all sounds
BD: Tone and Decay
SD: Tone and Snappy
LT (LC), MD MT (MC), HT (HC): independent Tune
CY: Tone and Decay
OH: Decay
Overall Accent Level

Tempo
Tempo control is labelled '0-10' with 40 click-stop positions plus a separate Fine control for more precise adjustment

Memory
Measure: 1-32 steps
Basic Rhythm 12 x A/AB/B; 4 x Intro/Fill In A/B
Rhythm Track: 64 Measures 12 Tracks (768 Measures)

Connectivity
Main Audio output (monoaural)
Multiple Outputs: BD, SD, LC (LT), MC (MT), HC (HT), CL (RS), MA (CP), CB, CY, OH, CH
Trigger Out: CB, CP (MA), AC
Start/Stop Pedal Input
Intro/Fill In Pedal Input
DIN Sync with Input/Output Selector

Dimensions
508 (W) x 305 (D) x 105 (H) mm

Weight
5 Kg

Notes
According to Roland (2018) ('A legend in the making'), 12,000 TR-808s were manufactured.

Suggested listening
Afrika Bambaataa and the Soulsonic Force 'Planet Rock' (Tommy Boy, 1982)
Marvin Gaye 'Sexual Healing' (Columbia/CBS, 1982)
Charanjit Singh 'Raga Bhairav' (The Gramophone Company of India, 1982)
The S.O.S. Band 'Just Be Good to Me' (Tabu, 1983)
Whitney Houston 'I Wanna Dance with Somebody (Who Loves Me)' (Arista, 1987)
Kanye West 'Love Lockdown' (Roc-A-Fella and Def Jam, 2008)

Shake the Room

Even if successfully matching them to the instrument from which they originate might prove too great a challenge for the average listener, it is almost certain that the sounds of the TR-808 will be immediately recognisable. The booming Bass Drum, piercingly snappy Snare, bitingly metallic Cymbal and Hi Hats and extraordinarily unrealistic but utterly iconic Hand Clap have provided the rhythmic foundation for a staggeringly diverse range of popular music that spans almost every conceivable genre and has even played a major role in originating and defining the sonic landscape of many. Released in 1980, the TR-808 was not an immediate commercial or critical success. However, through its use in the eclectic synthetic pop of Yellow Magic Orchestra, the power ballads of Phil Collins' 'One More Night' and Paul Young's cover of 'Wherever I Lay My Hat' (That's My Home), the upbeat R&B of Whitney Houston's 'I Wanna Dance with Somebody (Who Loves Me)', the Argentinian Rock of Charly Garcia's 'Clics Modernos', the futuristic electronic soundscape of Afrika Bambaata and the Soul Sonic Force's 'Planet Rock', and the provocative soul of Marvin Gaye's 'Sexual Healing', the instrument became a mainstay of popular music production and helped to shape the sound of the decade. Beyond the music charts, samples of the TR-808 even made their way into the soundtracks of videogames such as 'Streets of Rage'. While its use was undoubtedly wide, the TR-808 sounded the loudest in the emerging genres of Electro and Hip Hop, where pioneering artists such as LL Cool J, Public Enemy and the Beastie Boys along with producers such as Rick Rubin put the instrument centerstage and harnessed the full range of its rather unusual sound design capabilities.

What is so fascinating about this situation and the breadth of musical styles to which the TR-808 has lent legendarily metronomic stability of rhythmic timing, is that the instrument produces some of the least convincing emulations of traditional drum sounds it is possible to imagine. The sounds of the TR-808 are unequivocally, almost wilfully, electronic. As we shall see, while each of the 12 separate drum and percussion voices is undeniably unrealistic, the Hand Clap, Toms, Congas and Cowbell are notable for bearing only the slightest sonic resemblance to their real-world namesakes. Distinctive even iconic, without doubt, but judged purely as imitations, they are strikingly weak. Perhaps this should not come as too much of a surprise as, despite having a CPU at its beating heart and proudly bearing the legend 'Computer Controlled' on its front panel to establish its achingly futuristic credentials, the sound creation methods were deeply rooted in the past. In terms of tone generation, the TR-808 is an entirely analogue device and, while analogue synthesis is supremely well-suited for certain musical tasks (as we shall see with the SH-101, MC-202 and TB-303), the authentic replication of acoustic drum and percussion sounds is not its greatest strength.

And so, notwithstanding some absolutely ingenious and efficient circuit designs that actively harness the imprecision of component manufacturing and are built around exploiting the high tolerances of individual electronic parts, the net result is a sound palette that offers approximations rather than simulations. In fact, as we shall see in more recent interviews with some of Roland's engineers, the creation of sounds that captured the impression or essence of specific drum and percussion sounds rather than emulate them in minute detail was partly born of necessity but was also a conscious aesthetic and artistic choice. Indeed, in retrospect, analysing the sounds in this manner might also help us explain their enduring appeal.

Nonetheless, in 1980, the TR-808's sound swam against the tide of fashion. Approximations were out and samples were in. Roger Linn's LM-1 Drum Computer released in the first year of the new decade saw to it that an instrument based around analogue synthesis looked, and sounded, like a blast from the past. The LM-1 boasted digital recordings of 'real drums' and marked a step change in drum machine design and sonic aesthetics. Perhaps we might not go quite as far as Sean Montgomery, a product manager at Roland, who suggested of the TR-808's designers that, 'They did it the best they could with the analogue technology, and it sounded shit' (quoted in Beaumont-Thomas 2014), but it is certainly the case that, upon its initial release at least, the TR-808's sounds had more in common with what was viewed as an old, outdated generation of automated rhythm machines than the new breed of sample-based instruments.

In the wake of the LM-1 and its even more successful LinnDrum successor, the market was soon awash with drum machines with sampled sounds. Premium instruments like the Oberheim DMX (1980), E-Mu Drumulator (1983) and SP-12 (1984) and Sequential Drumtraks (1984) sat alongside comparatively inexpensive models including the Yamaha RX series (1984 onwards). By 1986, the Casio RZ-1 offered onboard user sampling for £399. Demonstrating the penetration of samples and sampling, Cheetah's 'SpecDrum' peripheral could turn a ZX Spectrum home computer into an 8-Bit sampling drum machine for less than £30. Given the prevailing market conditions, it was almost inevitable that the TR-808 would be superseded and so after just three years in production and with around 12,000 units sold according to Roland ('A legend in the making'), Roland discontinued the instrument and eventually replaced it with the TR-909 (1983) and then the TR-707 (1985) which were partly and wholly sample-based respectively.

And yet, more than four decades after it was released, and in the face of the onslaught of sample playback and even sampling instruments, it is barely contentious to claim that the TR-808 is the most influential and enduring drum machine of all time. In fact, many contemporary uses of the TR-808 rely not on the original machine which has become almost unobtainable due to its scarcity and high price on the second-hand market, but on the transformation of its sounds through creative sampling techniques. New genres and sub-genres of dance music and Hip-Hop have built their signature sound around the manipulation of elements of the TR-808 kit. The syncopated snares and triplet Hi Hat rolls of Trap and Southern Hip Hop are just two of many examples that continue to breathe new life into the instrument and its sounds.

As such, the technology that seemed destined to put paid to the TR-808 and its analogue drum synthesis in the early 1980s transpires to be precisely the technology that has continued to breathe new life into it as new generations of musicians, DJs and producers record, bend and stretch its sounds, transforming them beyond just drums and percussion and into a new release of bass and sub bass. It is a reflection of the enduring nature of its sonic palette and the cycles of musical fashion that the very same digital samplers that cast the shadow of obsolescence and irrelevance over the TR-808 are now harnessed in the service of attempting to recreate and reanimate the sounds of its analogue circuits and components.

Today, in addition to the core sounds forming the basis of countless libraries of sample packs and drum machine and synthesizer preset banks, the term '808' is just likely to refer to an ultra-low, often extremely distorted, pitched bassline created by sampling and sequencing the TR-808 Bass Drum than to the original instrument itself. These new, creative uses of the instrument both as a palette of sounds with their own power, punch and history and as raw material for new compositional and sound design practice, speak to the versatility of the instrument but also help us appreciate it as more than just a drum machine. Indeed, the term '808' is deeply imprinted with cultural connotations that arise from and inform its use in band names '808 State', album titles such as Kanye West's '808s and Heartbreaks' and song lyrics such as Public Enemy's 'Power to the People' which name checks the instrument, and the Beastie Boys' 'Everybody getting down make no mistake nothing sounds quite like the 808.' In fact, so common is it to reference the boom of the instrument's Bass Drum that lyrics such as 'heart beating like an 808' have now become almost hackneyed. Beyond this, and further reflecting the cultural penetration of the instrument, the TR-808's iconic colour scheme has even found its way onto clothing and trainers for satisfy those for whom merely listening to or performing with the instrument is not sufficiently intimate.

The brave new world of sampling drum machines and the fashion for 'real drums', as Linn's LM-1 advert famously put it, might have led the TR-808 to sound dated on first listen in 1980 but, as Beaumont-Thomas puts it, over four decades later, it sounds like the future. However, as Werner (2015) cautions, with an instrument of this stature and status, mythologising and the (sometimes inadvertent) retelling of apocryphal or incomplete histories is commonplace. As we shall see, the origin of the TR-808 is a complex one that builds on the work of a number of comparatively unsung engineers and musicians such as Don Lewis and which has its roots in experimentation and what we might today call 'circuit bending'.

This is Not a Drum Machine

If the CR-78 was a hearty starter and the DR-55 an amuse-bouche whose small size belied its tastiness, the TR-808 was undoubtedly the main course. Released in 1980, this was the apotheosis of Roland's accumulated knowledge and experience of drum machines. With development led by a newly promoted Kikumoto riding high on the somewhat unexpected success of the DR-55 and now acting as chief engineer, the TR-808 would draw together everything the company had learned thus far while adding some important new twists and features. In this way, the TR-808 is both a logical continuation of Roland's TR and CR series, along with the fledgling Boss DR unit, and a shocking departure that seems almost to be a self-conscious reaction to that heritage.

The first thing to note is that the TR-808 is that it is an imposing instrument. Compared with the compact CRs and positively diminutive DR-55, the TR-808 is physically large and there is not a single piece of wood, real or otherwise, anywhere in sight. This is a substantial piece of metal with chunky plastic end cheeks and visible screw heads that speak in an unabashedly industrial design language. This is not a piece of furniture designed to blend in with the decor like a wooden-clad organ add-on. This is a solid and imposing piece of music making equipment.

For all the black metal, however, the TR-808 is an undeniably colourful object, with plenty of buttons in gloriously bright hues along with clusters of rotary controls capped in white and orange. All of this riotous colour certainly harks back to the CR-78 and its predecessors but there is an altogether different organisation at work here. A single row of 16 'Step' buttons dominates the lower third of the instrument with the controls elegantly arranged into groups of four by virtue of the colourful sweep through a spectrum that fades from red through orange and yellow to white. In fact, as Kikumoto (2020) notes, the plastic caps for these Step Buttons were borrowed from Roland's flagship polyphonic synthesizer, the Jupiter-8. Before the TR-808 is even powered up, it is clear that its design connects it to the company's highest-end instrument.

These Step buttons, complete with integrated red LEDs to indicate their status, are simply and clearly labelled with two sets of numbers reflecting their dual function in the instrument's writing and playback modes. There is plenty of space around the relatively small rotary controls that occupy the top two thirds of the instrument. This appears to provide ample space and access for two-handed operation and tweaking, but it is what is absent that is just as revealing. There are no references to preset rhythms anywhere on the front panel and, while the TR-808 is perfectly capable of generating Waltzes, Foxtrots, Rhumbas and ChaChas if so desired, such words are nowhere to be seen. In fact, the TR-808 did ship with a selection of pre-programmed patterns which are also notated in the Operating Manual. However, this collection of 'Rock', 'Samba' and 'Mambo' patterns could be erased and replaced with original material with the flick of a switch and the tap of a couple of buttons.

If presets and the associated musical genre names that might serve to unduly pigeonhole the instrument were relegated to the back pages of the manual, some other phrases are rather more conspicuous. In the middle of the front panel, the instrument's model name 'TR-808' is extravagantly underscored by the words 'Computer Controlled'. Both of these, however, are dwarfed by the legend 'Rhythm Composer'. The potency and meaning of these phrases might be somewhat lost over four decades later but, in 1980, they unambiguously signalled a break between this and previous Roland instruments and, crucially, made it clear that the TR-808 was fully programmable and made use of the latest technologies. Indeed, while they were always colloquially referred to as 'drum machines', 'Rhythm Composer' is the term that Roland would continue to use from this point to describe all but one of its x0x percussion instruments. Other manufacturers might make drum machines, but this phrasing indicated Roland's keenness to communicate this as the first of a new category of instruments capable of generating more than just a few simple patterns.

Indeed, Roland's marketing materials centred on just this point. Simultaneously underlining the futuristic nature of the new computer-controlled instrument while reassuring the reader and potential purchaser. Under a headline reading, 'We Want You To Understand the Future', the copy foregrounded the flexibility of the TR-808 while essentially explaining its name.

> A feature that sets the TR-808 apart from any other rhythm device is its ability to record an entire compositions percussion score which we call Composing the Rhythm Track.
>
> (RolandCorp USA ad [1981])

In fact, the TR-808's change in functionality, and design, was also reflective of an important shift in the perceived audience for the new instrument and the opening of new possibilities. Speaking to journalist and author Mark Vail, Kakehashi revealed that,

> It used to be that our customer was the home-organ player. Then people in the music industry started to pay attention to our rhythm machines. Such a musician was agreeable to programming by himself. That's why we developed the step-writing system, so you could slow the tempo down, enter your rhythm events, and then speed it up and hear the realistic rhythm pattern you had just created.
>
> (Vail 2000: 282–283)

Nonetheless, programmability, especially of rhythm units, was still unfamiliar territory and the TR-808's Operational Manual treads carefully in describing the novelty of the process and its associated terminology.

> As we mentioned earlier, your TR-808 will not only play back Rhythms but also allow you to write them as well. Because we are not just writing the Rhythms but are storing them in Computer Memory as well, we call this process Programming.
>
> (Roland 1980b: 13)

In truth, it is arguable that the description in the manual actually makes the process of programming seem more complicated than it really is. Part of this arises from the need to explain the new emerging terminology of computerisation, memory and storage. However, as we shall also see when we explore other instruments in this book, part of the complication is a result of the number of assumptions made by the writers of the manual about the approach a user might take to the instrument as well as their familiarity with the conventions of Western musical notation and contemporary US terminology (such as 'sixteenth note' rather than semiquaver, for instance).

> The Program Score is a transcription of the music score using a system that indicates the minimum number of notes and rests. However, it is not completely musically accurate in that a quarter note, for example, when transcribed to the Program Score from a music score, is written as a sixteenth note, followed by three rests. This also follows for whole notes, half notes and eighth notes as well.
>
> (Roland 1980b: 13)

There is a palpable sense in the manual of a workflow based around the transcription of rhythmic ideas into the language of the computer. Indeed, its pages are filled with illustrations of stave-based classical drum notation which is translated into a visualisation of the TR-808's grid-based programming score and explanations of how this maps onto the instrument's row of multicoloured buttons. Having become familiar with this conceptual model and interface especially as it has influenced so many subsequent designs, we might now think of the TR-808's buttons and even the instrument as a whole, as an improvisational space. However, Roland's manual frames the hardware as the means by which pre-existing musical ideas can be transferred into the machine. Here, programming seems much more akin to a process of data entry rather than the more tactile experience that might define the approach of today's musician. Again, this is a salient reminder of the transitional period into which the TR-808 is born and is, at least in part, responsible for creating. Indeed, the positioning of Western musical notation as the lingua franca to be translated into the new language of computer-controlled, programmable music technology, is something we will continue to see not only with Roland's successive drum machines but also with synthesizers and sequencers such as the MC-202.

For all the apparent anxiety over the ease of translation and transcription, the TR-808's underlying rhythmic programming methodology is little short of a work of inspired genius. With each measure of music split into sub-divisions or 'steps' that are each represented in the hardware interface with their own dedicated Step Button, rhythmic patterns are immediately visualised on the front panel of the instrument itself. Where the CR-78's somewhat clunky programming method gave no visual feedback whatsoever, the TR-808 laid bare the unique combination of hits and rests for each sound. There is a direct connection between the 'step' and the 'Step Button' which creates an interface of unrivalled clarity. Moreover, while it was programmable, the CR-78 offered no practical means of editing or tweaking patterns once they had been entered. As the CR-78 manual notes, if the player made a programming error or simply wanted to alter or experiment with an alternative pattern, the only option was to entirely erase a track and re-enter all of the steps again. With the TR-808, however, all steps in a measure are immediately visible with the red LEDs integrated into the multicoloured Step Buttons making it instantly clear which will trigger a sound and which will remain silent. As such, edits are made freely and the results communicated through the same hardware interface in real time. Even the colour coding of the 16 Step Buttons helps to split the measure and its subdivisions into clear beats or quarter notes which makes navigation even easier.

Of course, being based around a hardware interface of buttons, some compromises have to be made. As we shall see below, the TR-808 includes 16 sounds (12 of which can be played simultaneously on any given step). Although the TR-808 is a physically large instrument, it would be utterly impractical to include a row of buttons for each sound. Indeed, it would be another five years before Roland would release the TR-707 which allowed the simultaneous visualisation of multiple sounds and which required the use of a completely different display technology. For the TR-808, however, the solution came directly from the CR-78 and DR-55. A large rotary 'Instrument Select' control switches between the different sounds and resets the display of the Step Buttons' LEDs accordingly. As such, while each sound has its own dedicated measure of steps in the sequencer, only one is ever visible in the hardware interface. By selecting the appropriate sound, patterns can be built up incrementally, reviewed and edited with the Step Buttons and LEDs dynamically reflecting the selections.

It should be noted that, in addition to being able to program patterns in step time, as Kakehashi intimated above, the TR-808 also offers real time recording. By using the 'Tap' button located to the right of the row of Step Buttons, the selected sound can be played live with the relevant step LEDs illuminating accordingly. A crucial development from the CR-78's real time input method is that beats tapped into the TR-808 are quantised. That is, they are moved to the nearest step. Unlike the CR-78 where the resolution was comparatively high, leading to sloppy patterns, the TR-808's quantisation typically sees playing corrected to the nearest sixteenth note/semiquaver. Importantly, notes entered either directly via the colourful Step Buttons or played live with the Tap button are immediately available for editing via the same interface.

Perhaps inevitably, the TR-808 inherits the concept of 'modes' from its programmable CR and DR predecessors and the instrument must be set to either Write or Play in order to create or replay patterns respectively. Play mode reveals yet more flexibility and efficiency in the interface as it transforms the row of Step Buttons into Pattern selectors. One notable feature of the Play and Write modes, however, is that switching between them is not seamless. Rather, playback must be stopped before the mode can be switched from Pattern Play to Pattern Edit or vice versa. This is a feature of most (though, intriguingly, not all) of the x0x instruments explored in this book and is one that renders programming and performance as two discrete, separate processes. Most importantly, the design decision to create a modal interface inscribes an implicit workflow based around the transcription and playback of existing ideas rather than enabling one that presents the instrument as a site for live improvisation and on-the-fly writing and rewriting. This separation of editing and playback processes is a key differentiator between the original x0x instruments and more modern devices including, as we shall see below, the host of recent recreations and clones of the TR-808 which typically elect to sidestep this behaviour.

Parts, Variations, Intros, Fills, Tracks and other Mysteries

Although we have concentrated thus far on the creation of single Patterns that each comprise a bar or measure of rhythmic material, the TR-808 offers considerably more flexibility and depth in its programmable sequencer. Each Pattern memory slot has the potential to hold two 'Parts' as well as A and B 'Variations'. Parts and Variations operate slightly differently and can be used to unlock a range of different musical behaviours.

In reality, what we think of as a single Pattern in the TR-808 is always comprised of two Parts. However, the default setting sees the '1st Part' allocated 16 steps and the '2nd Part' none, meaning that only the data stored in Part 1 are ever heard. Extending the step count of the 2nd Part to 16 allows for the creation of a two-bar Pattern comprising 32 sixteenth notes. Of course, as the 1st Part's step count can be reduced below 16 and the 2nd Part can set to any number between 0–16, it is perfectly possible to create an aggregate Pattern in a variety of time signatures such as 3/4, 7/8 and so on. Once the 2nd Part's step count is set to anything other than 0, pressing Start on the TR-808 will automatically play both the 1st and 2nd Parts in a continuous loop. A small red LED to the left of the row of Step Buttons indicates which Part is currently heard with the LEDs in each of the individual Step Buttons changing to reflect the status of the selected sound. In truth, many TR-808 users are unaware of the existence of Parts and focus exclusively on creating multiple one-bar Patterns. This is perhaps unsurprising given that the feature is deactivated by default and is rather hidden behind an interface that does not foreground its functionality ('1st and 2nd Part' are hardly the most intelligible terms for what is essentially a doubling of the step count). More than this, the Step Button interface maps so elegantly onto a single measure that it is hard to imagine that this is not the atomic unit of the instrument.

Variations are given rather more prominence on the front panel. A three-position selector lever allows the playback of either 'Variation A' or 'Variation B'. The lever may be used in real time to arbitrarily choose which Variation will play at the beginning of the next measure while a middle position, labelled 'AB', plays 'A' then 'B' in a continuous loop. Although naming this feature 'Variation' implies that the two Patterns might be musically related, there is nothing that actually requires 'A' and 'B' to bear any resemblance to one another. Essentially, the two Variations are independent Patterns that can contain their own step data. Moreover, as each Variation can also comprise two Parts of up to 32 steps in total, setting the Variation selector to 'AB' effectively allows the creation of a Pattern of up to two-bars. With careful programming and by making use of both the Parts and Variation features together, a single 'Pattern' can be set to play four unique bars worth of material.

To round out the performance-oriented features, the TR-808 also includes Fill and Intro features. These perhaps hark back to the altogether simpler Fade in and Fade out feature found on the CR-78 and earlier Roland preset TR instruments. Here, however, rather than simply increasing or reducing the volume of a Pattern as it starts or stops, Fills and Intros are separately programmable sequences that can be manually triggered or set to play after a specified number of bars. These elements are created in exactly the same manner as the basic Patterns while in Write mode and offer all of the same programming possibilities and access to sounds. Switching to Pattern Play mode, different Fills or Intros are selected by the four white Step Buttons labelled 13–16 and are triggered with the Tap button (or a pedal connected to the instrument). Again, the hardware interface is put to highly efficient use with most buttons serving a dual function depending on the mode to which the instrument is set.

The very presence of Fills and Intros speaks to the ways in which song structure is demonstrably at the heart of the design of the TR-808. The CR-78's design with its limited number of genre-specific preset beats suggests that a single pattern might be expected to accompany any given performance of a song throughout its entirety. Perhaps this might be set to fade in and out at the appropriate moment but, aside from changing pattern (and potentially genre or even time signature) there is no facility for specific variations for start and end or any other section of a song. The TR-808's provision of distinct and fully programmable Variations, Fills and Intros goes some way to explaining the instrument's soubriquet and the shift from accompanist or even drum machine to Rhythm Composer.

More than this, it is notable that these elements can all be selected dynamically. With the instrument set to Pattern Play mode, Patterns of multiple bar lengths can be triggered on the fly with the Step Buttons. Variations and Fills are selected and activated automatically or manually by the performer. As such, entire songs can be improvised by piecing together pre-programmed constituents just as a well-rehearsed piece with a defined structure can be performed on the fly from the front panel interface. This is clearly a level of flexibility unknown in competing instruments. As such, although the ability to program individual Patterns is among the most celebrated innovations of the TR-808, that these can be arbitrarily chained together in real time to create altogether more complex and varied song structures is just as significant.

That said, it is important to remember that while the real time chaining of material can bestow elasticity and improvisational potential on the structure of a performance, the modality of the TR-808 interface means that the fundamental building blocks of Patterns, Parts, Fills and Variations each have to be pre-programmed. Any editing of the elements requires the performance is temporarily halted while the sequencer is stopped and the instrument's mode is changed. Again, the design of the TR-808 offers clear insights into the series of presuppositions and conceptualisations about how the instrument might be used and compositionally complexity clearly trumps live performability.

While the ability to dynamically play with and create song structure is perfectly possible in Pattern Play mode, Roland's notion of Rhythm Composition is actually found in the instrument's 'Track' modes. With all the elements of Patterns, Variations, Intros and Fills having been created, auditioned and edited in Pattern Write and Play modes, Track Compose allows specific combinations of these to be memorised. Essentially, a 'Track' in Roland's TR-808 terminology is what is often called a 'Song' in more modern electronic musical instruments and sequencers. As we saw, such longer-form Tracks or Songs can be effectively performed on the fly by selecting different Patterns, Fills, Variations and so on in Pattern Play mode. The TR-808's Track Compose and Track Play modes mean that these can be stored, edited and automatically played back without the need for further real time interaction. The TR-808 offers 12 Track locations which each have a capacity of 64 measures. These can be used separately or chained together to make longer compositions up to a maximum of 768 measures if all are used. It is very clear that Roland's ambition for the TR-808 was considerably greater than the original brief of an inexpensive drum machine (McCabe 2020) and its design with its focus on structure and variation locates it firmly as an integrated part of a complex performance rather than simply as an expanded metronome or unwavering accompanist.

Of course, how widely these features are actually used is another thing and it is interesting to note how many of the notable uses of the TR-808 are based around looping one-bar Patterns or, as with Marvin Gaye's iconic 'Sexual Healing', a two-bar sequence that makes use of the comparatively intuitive 'AB' Variation option.

TR-REC

The preponderance of one-bar Patterns in the oeuvre of popular TR-808 songs, speaks to impact of the Step Button interface. Indeed, the one-to-one mapping of hardware controls to musical steps is found in all subsequent xOx instrument as well in numerous interfaces adopted by other manufacturers. Roland, themselves, have gone on to dub this the 'TR-REC' mode and the phrase has not become synonymous with accessible and intuitive programming. Roland's 2014 flagship TR-8S, its smaller sibling the TR-6S and the 2025 TR-1000 each sport dedicated TR-REC buttons glowing brightly on their front panels. Even the SP-404MK2, Roland's sampler and effects unit received a TR-REC mode in a 2023 firmware update. This is particularly notable as the instrument is primarily based around the entirely different interface of a 4x4 grid of real-time playable drum pads popularised by the Akai MPC instruments from 1988 onwards.

Moreover, while TR-REC is a Roland descriptor, the underpinning interface principles of the TR-808 and beyond have been translated into the instruments of countless other manufacturers including Elektron and Korg to the point where it is almost *de rigeur* in drum machine design. Speaking about the creation of the interface model, Kikumoto has noted that it began its life as part of Roland's research and development process.

> The idea of TR-REC originated as a tool for the early rhythm box development. It was a matrix panel with rows and columns for 16 instruments over 24 steps and other was used to program the rhythm patterns by plugging points into the panel.
>
> (Kikumoto in Bjørn 2022: 20)

Given this, it might come as something of a surprise to learn, then, that Roland did not actually invent it. There is no doubt that the TR-808 is the instrument that popularised and refined it, but the 16-step interface actually appeared almost a decade earlier. Released in 1972 by Italian company Eko, the Computerhythm is a large instrument. With a raised front panel that looks more like a 1980s computer terminal, it is perhaps unsurprising to find that it featured as a prop in the 1978 science fiction film, 'La guerra dei robot' ('The War of the Robots') directed by Alfonso Brescia. But while its six rows of 16 flashing lights made it a perfect starship control surface, they also allow for the intuitive programming of drum patterns. Despite the physical differences in the hardware and the style of the buttons, the interface model is strikingly similar to the TR-808. A row of switches breaks down a measure into 16 subdivisions with steps being activated and illuminated to trigger the built-in sounds. Unlike the TR-808, however, the Computerhythm has only six sounds, but it does include a full row of 16 step buttons for each. As such, while the instrument is considerably less complex, it does present a visual overview of an entire pattern. It is estimated that fewer than 20 instruments were made (Réveillac 2019) though, despite its rarity, it found use by electronic music pioneers Jean-Michel Jarre and Manuel Gottsching and its (very CR-78-like sounds) can be heard on the 'Oxygène' (1976) and 'New Age of Earth' (1977) albums respectively.

Although the 16-step interface is the most immediately arresting feature of the Computerhythm, the instrument is also notable for its storage method which makes use of punched cards. This physical medium was used to store information through the presence or absence of physical holes punched into thick card stock and created a machine-readable format for storing digital data. Although relatively commonplace in more general computing throughout the 1950s and 1960s, this form of storage media had been largely replaced by magnetic tape by the 1970s and was extremely unusual in to find in musical instrument design at any point. That said, even though punched cards might have seemed comprehensively obsolete even at the time, the ability to permanently store and retrieve data at all put the Computerhythm ahead of the TR-808. The carefully programmed and more complex contents of Roland's machine were backed up by battery so that information could be retained while the instrument was powered off, but there was no means of exporting Pattern or Track data. Lamenting the situation in a mid-1980s review, Tim Goodyer reflects on the popularity of the TR-808.

> Nor is it to do with quick and easy program storage and retrieval; there's no dump-to-tape facility, let alone disk storage. It seems a laughable omission now, but the only way to preserve your diligent hours of TR808 rhythm programming beyond the power-down memory stage is by writing it all out on a piece of paper.
>
> (Goodyer 1986)

Of course, the situation we arrive at here is a curiously circular one that is again based around transcription, only this time the contents of the instrument must be transferred back to paper. Indeed, just as Roland's manual documents the translation of written musical scores into the instrument, it also provides photocopiable templates of the 'Program Score' so that musicians can create a permanent record of their Patterns.

Ultimately, it is testimony to the influence and near-ubiquity of the 'TR-REC' step recording method that Roland perfected in the TR-808 that the manual's discussion of entering rests or silences between rhythm notes by leaving buttons un-selected seems so self-evident as to be redundant to the contemporary drum machine programmer. However, the presence of this text reminds us just how unusual the concept of programmability was in the early 1980s. Interestingly, it also draws attention to something that Kikumoto would later consider a limitation not only of the TR-808 but of drum machines and automated rhythm programming in general. Speaking about the development of the MIDI specification that came after the TR-808's release and would set out the protocols by which electronic musical instruments would communicate up to the time of writing, Kikumoto noted that, 'When I was writing the MIDI protocol in 1980, I wondered why the notation of rhythm did not indicate Gate Time and Key on the score, except for the Open Hi-Hat.' (McCabe 2020).

Kikumoto's reference to Gate Time is particularly notable here as it sets out a key principle of the TR-808's sequencer. Each step simply defines whether any given drum sound will be triggered but does not define a duration or pitch. The triggers that comprise each individual step are simply pulses that activate the analogue sound circuitry rather than signals that have variable length or duration (the 'Gate' that to which Kikumoto refers). In terms of duration, this is much like only registering the initial press of a key on a keyboard rather than recognising the length of time it is held down to sustain a note, for instance. This might make intuitive sense for a drum machine but it is worth considering the different tonalities that arise from acoustic drums, percussion and cymbals being struck in different ways. A cymbal, for instance, may be momentarily tapped with the tip of a drumstick creating a clear attack transient while also ringing out, or it might be struck with the full length of the stick, thereby altering some the pattern of vibrations and resonances to create a different tone. The same stick and the same cymbal but a multitude of different sounds. There are countless other examples in which short, staccato strikes elicit markedly different sounds than longer ones.

What Kikumoto highlights as a comparatively limited MIDI specification for rhythm programming is something we also see in the TR-808 although, while it might be presented as a restriction, it is a key part of what keeps the instrument so accessible and easy-to-use (despite what the manual might have some users think). Each TR-808 step simply defines a 'strike' or 'hit' of a fixed duration with the musician/programmer having to think no more or contend with any more complex an interface. Any given drum or percussion element either sounds or does not without any complications as to how it might be virtually struck, by what material or for how long.

An Accent feature allows some degree of control over the heaviness of a hit. The Accent has its own 16-step sequence and is programmed just as any other sound. However, rather than making a sound itself, all of the other sounds are affected by an Accented step. The provision of an Accent was considered important enough by Kakehashi in allowing for programming flexibility that it also made the journey across from the CR-78. The comparative simplicity of the step triggers and absence of any form of Gate time does not, of course, mean that the sounds of the TR-808 do not have different, and controllable, durations, but all of these variations and nuances are handled and sculpted within the sound synthesis circuits rather than the sequencer.

Similarly, the TR-808 sequencer does not permit the programming of pitch information. Instead, variations in tuning are achieved through manipulation of the synthesis circuits via the range of parameters brought to the surface of the front panel. While separating out the patterns of rhythmic beating from the tuning of individual drums might appear to mimic the design of the drum sets the TR-808 seeks to replicate, it disallows per-step pitch variations that later come to define characteristic '808' uses and aesthetics beyond the original hardware.

Sounds and Circuits

The TR-808 offers a total of 16 sounds. The overall palette continues in the tradition of the CR-78 and its organ accompaniment predecessors in comprising a mixture of rock drums and percussion sounds. The full complement of sounds is:

> Bass Drum; Snare Drum; Low, Mid and High Toms; Low, Mid and High Congas; Rim Shot; Hand Clap; Claves; Maracas; Cowbell; Cymbal; Open and Closed Hi Hats.

The TR-808 does not have full polyphony meaning that some sounds are mutually exclusive and cannot be played together. Five switches on the front panel of the instrument allow switching between Rim Shot and Claves, Hand Clap and Maracas, and Toms and Congas (with the Low, Mid and High sounds individually switchable to create a mixture if desired). Just like Roland's earlier instruments, all the TR-808's sounds are produced by analogue synthesis circuits comprised of discrete components with the microprocessor so proudly advertised on the instrument's front panel playing no role in generating audio. Rather, the 'Computer' is responsible for controlling the timing of the sequencer along with Pattern and Track storage.

The analogue voice circuits were designed by Hiro Nakamura who had also designed the sound generation hardware for the CR-78 (Roland 2020). Figure 3.2 shows a simplified Block Diagram of the TR-808 sound circuits. Each of the sounds has a dedicated Level control on the front panel which enables their relative volumes to be adjusted to create a mix of the overall kit within the instrument and without need for further outboard processing. Some sounds also offer further sculpting capabilities via front panel controls allowing them to be tuned, for instance (as with the Toms and Congas), have their decay lengthened or shortened (Bass Drum and Cymbal). Some sounds offer more control over their sonic properties such as the Snare Drum whose 'Snappy' setting adds even more high frequency punch to an already tight and crisp sound, for instance. The array of sound shaping controls occupy most of the top section of the instrument and offer an unparalleled degree of control compared with Roland's earlier instruments.

In addition to a 'Master Out' that presents the mix of sounds set by their front panel Level controls, 11 separate 'Multi Outs' allow the signals of individual drum sounds to be separately treated, mixed and processed with external effects.

While the roster of sounds might, in name at least, have sounded familiar to those present in instruments from the previous decade, Nakamura employed some extremely inventive and unusual circuits in the TR-808. As Werner et al. (2014a) note, it is often suggested that the TR-808's cymbals make used of filtered Pink or White Noise (see also Reid 1999 who infers their use in the TR-808). This is understandable given the frequency with which forms of White Noise were (and still are) used in percussion synthesis. The sound of cymbals in countless 1970s auto-accompaniment drum machines is the sound of various flavours of processed, but still undeniably hissy, Noise. Similarly, Kraftwerk make extensive use of filtered and processed Noise circuits to create the insistent, locomotive rhythms that power their 1977 'Trans-Europe Express' album.

Of course, while it is eminently suitable for the creation of novel machine-like tones, in truth, Noise does not make an especially good basis for the emulation of metallic sounds like cymbals. Consequently, these rank among the most challenging acoustic percussion sounds to replicate with analogue synthesis techniques (Reid 2002a). Indeed, consulting Roland's TR-808 Service Notes (1981a, First Edition) we learn that Nakamura's TR-808's Cymbal takes a different approach and does not use Noise. Rather, as the 'CY' section of Figure 3.2 shows, the Cymbal circuit is actually based on a foundation of six square waves. Each oscillator is independently tuned to create a discordant and somewhat harsh spectrum that is rich in mid and high range tones. This does the job often achieved with White Noise but adds a metallic tonality rather than a hiss.

Two of the six oscillators can be calibrated with internal trimpots not accessible to the user and, according to Roland's 'TR-808 Service Notes' (1981a) these should be tuned to 540Hz and 800Hz to create a slightly out of tune Perfect Fifth interval. The remaining oscillators are hardwired in the circuit to generate specific frequencies. In their detailed technical analysis and emulation, Werner et al. (2014b) give these nominal frequencies as 205.3, 369.6, 304.4, 522.7Hz. As many involved in analysing, repairing and reanimating the TR-808 have noted, the frequencies of these oscillators varies considerably from instrument to instrument due to component values and tolerances, with further changes occurring over time. This means that no two TR-808 Cymbals might ever have sounded exactly the same and, many decades on, might have changed considerably.

Even if perfectly factory-calibrated, when heard on its own, the mess of atonal square waves does not sound much like a Cymbal and might seem to hold little promise. Indeed, in isolation, the six square waves are enough to set one's teeth on edge so jarringly brash are they. However, the ingenuity of the circuit comes in the way this cluster of tones is subsequently processed. The clashing mass of square waves is first split, the two bands with the higher and lower frequencies being contoured differently. While the lower frequency band can be controlled by the player using the 'Decay' control on the instrument's front panel to increase or decrease its duration, the high frequency component is further separated out with different ranges running through voltage-controlled amplifiers (VCAs) so that they each decay at varying rates. Finally, a front panel Tone control alters the mix of the high and low frequency bands so that the sound may be tailored yet further. As such, from a comparatively simple and somewhat unorthodox starting point, the interplay of Nakamura's parallel and cascading processing circuits transforms six square waves into a remarkably complex, multifaceted sound full of life and movement. Above all, the TR-808 Cymbal has the unmistakably metallic ting and jangle absent in filtered Noise circuits.

As if this circuit were not already efficient and ingenious enough, it is further exploited in the production of the similarly metallic (and immediately recognisable if not even remotely authentic) Cowbell. Wonderfully described by Beaumont-Thomas (2014) as a 'corroded chime', the Cowbell makes use of two of the six square wave oscillators from the Cymbal cluster. As the 'CB' circuit in Figure 3.2 shows, the 540Hz and 800Hz oscillators are tapped separately and are separately processed through VCAs to contour their level and a Band Pass Filter to remove and aceturate different frequency ranges. As the frequencies of these two oscillators can be adjusted (by those willing to open the case of their TR-808 at least), the Cowbell may be tuned for creative effect or retuned if the circuit needs recalibration and returning to its factory state.

For all the creativity and efficiency in design that is demonstrated in the TR-808 circuit, sometimes the magic ingredient relies on being alert to unexpected opportunities. As Don Lewis recalls when working with Kikumoto on the prototype of the instrument in Tokyo,

> That day he had a bread board of an 808 and was showing me what was going on inside – he sort of bumped up against the breadboard and spilled some tea in there and all of a sudden he turned it on and got this pssh sound – it took them months to figure out how to reproduce it, but that ended up being the crash cymbal in the 808. There was nothing else like it. Nobody could touch it.
>
> (Lewis in Wolbe 2013b)

Although it was not a feature of the Cymbal or Cowbell, Noise is key to a number of the TR-808's sounds including the Snare Drum with its characteristic bite. Indeed, 'sizzle' is a word often used to describe the TR-808's overall sound and it is fascinating to note that this distinctive quality of sound actually comes from an unexpected source. Kakehashi recalls that he came across a batch of out-of-specification transistors that imparted a particularly desirable sound quality. Manufactured by companies such as Toshiba and Panasonic, the components were not faulty, per se. Rather, their performance fell outside the parameters and tolerances defined in their specification. Realising the advantage these brought, Kakehashi actively sought to acquire these 'rejected' components for inclusion in the TR-808. Interviewed in 2015 for Alexander Dunn's '808' documentary, Kakehashi estimated that the yields from a production run of 10,000 transistors might leave approximately 2–3% out of specification in the way he sought. As a result, the 2SC828-R transistor has attained an almost mythical status in TR-808 aficionado and emulation circles. This is partly because as the efficiency and reliability of semiconductor production improved, the number of out-of-specification parts diminished to the point where they became unavailable (see Sue and Jones 2018a and 2018b). As Kakehashi notes, this made it impossible to continue the production of the TR-808 and uncomplicatedly gives this as the reason why, after manufacturing (Roland n.d.) units, it had to be cancelled. While the exact impact of these components is hard to assess, they are often taken to impact on the sizzling high end with Roland now cryptically describing them as a 'secret weapon' in the circuits in their 40th anniversary promotional materials. Certainly, Kakehashi's version of events is rather at odds with the more widely recounted history that positions the TR-808 as a commercial failure that was abandoned.

Perhaps the most interesting sound in the TR-808 arsenal, however, is the Bass Drum. This is partly because it has gone on to become one of the most influential sounds in all of electronic music, but also because it is based around another distinctive and efficient circuit design. Consulting Roland's TR-808 Service Notes (1981, First Edition) documentation, we can see that the Bass Drum, for instance, makes use of what is known as a 'Bridged T network'. This is quite unlike the method often used to synthesise a bass drum.

Typically, a Bass Drum circuit created with analogue electronics would be modelled after a traditional subtractive synthesizer architecture such as might be found in the Moog Minimoog or Roland SH series. In such cases, a fundamental low frequency tone such as a Sine wave forms the basis of the sound. For analogue synthesizers like the Minimoog, that do not offer a Sine wave, this tone is often generated by a self-oscillating Filter. The volume of this sound is contoured by an Envelope that gives it a sharp attack to simulate the strike of the beater on the drum head with a decay to represent the tailing of the sound to silence. Very often this might be accompanied by a very short burst of Noise to accentuate the initial transient of the drum's membrane being struck while the Sine wave might swoop in pitch from high to low as it decays to further emphasise the attack and body of the sound. All of this could be readily synthesised using a keyboard such as Roland's own SH-101 and is, in fact, the method used by Roland in their later TR-909 albeit with some slight tweaks to the generation of the basic waveform. However, in the case of the TR-808, things are rather simpler as the 'BD' circuit shown in Figure 3.2 attests. Rather than build a mini analogue synthesizer of oscillators, filters and envelopes, Roland's designers built this iconic sound around a single 'Bridged T' network. In discussing the circuit, Gordon Reid describes the 'Bridged T' as follows.

> Now cast your minds back to junior school. Did you annoy your teacher by holding your ruler hard against the desktop before flicking it to go 'booooiiiiiinngggggg'? If you did, you were using an oscillator that, once excited, produced a sound that decayed to silence, yet did so without any contour generators or amplifiers.
>
> (Reid 2002b)

The key difference between the Bridged T networks used across many of the TR-808's voice circuits and more traditional oscillators is that, while the latter generate sound continuously and need to be contoured by amplifiers and envelopes to create a decay to silence, the former naturally decay as part of the their operation. The length of the decay can be varied by creating a positive feedback loop (to the point where it will sustain indefinitely just like a standard oscillator). As Werner et al's technical analysis of the circuit remind us, the degree of sonic flexibility belies the simplicity of the design.

> It could be tweaked via stock user controls to sound like a fairly realistic kick, or extended beyond recognition to a multi-second-long decaying pseudo-sinusoid with a characteristic sighing pitch. Its clicky attack could cut through a mix, but could be dialled back with a passive tone control.
>
> (Werner et al. 2014c: 1)

While the Tone knob is certainly an important sound shaping control that softens or sharpens the attack of the Bass Drum, it is the Decay that is most transformative. At its maximum setting, the Bass Drum sustains for a quite unrealistic amount of time. And, as Werner et al note, a deliberate leakage in the circuit leads to a slight pitch drop over time that sees the sound go very slightly flat as it 'sighs'. As Reid (2002b) puts it, 'Earlier Roland rhythm products (such as the CR68 and CR78) did not do so, and this is one reason why analogue drum machines sounded so much better from the TR808 onwards.'

In fact, there is much more to this circuit than the Service Notes and Block Diagram suggest. The effect of triggering the Bass Drum on an Accented step does more than just increase the volume of the sound. Rather, some additional circuitry that Roland themselves refer to as a 'trick' alters the frequency and resonance of a filter which creates a punchier, snappier sound. If we were in any doubt, these are living, breathing circuits.

When critics, commentators (and lyricists) talk about the 'Boom' of the TR-808, it is the Bass Drum to which they refer, particularly with the Decay extended towards or at its maximum. Here, the sound has as much in common with orchestral timpani as the bass drum of a drum set. As we shall see, the decision to include such an 'unrealistic' decay that exceeds what might be necessary to emulate an acoustic bass drum was hard won by Kikumoto and, although it was not one that met with unanimous approval among the design team, Kakehashi gave his approval to include the feature in the final instrument. Perhaps more than any other aspect of the TR-808, this one control reveals the underlying spirit of experimentation and speaks to an instrument whose design intention exceeded the straightforward emulation of acoustic drums and percussion.

But Not Necessarily the Right Sounds and Circuits

Ultimately, as Werner et al. (2014c: 1) note, the design of the TR-808 'blended ingenuity and efficiency, creating circuits with great detail and complexity, but a part count low enough to be amenable to mass manufacture.' However, as inventive and economical as they may have been, it is enlightening to read some of the criticality in contemporary assessments of the TR-808's sound. Overall, response was far from overwhelmingly positive. Speaking about the Cowbell, Goodyer's assessment was damning. 'Clumsy, clonky and hopelessly underpitched, it sounded more like a sick monosynth than a cowbell' (Goodyer 1986: 71) while Goldstein and Twigg were more succinct though no less critical in calling the sound an 'ill-tuned monstrosity' (Goldstein and Twigg 1984: 52). Perhaps even more surprising are the criticisms of some of the other sounds – 'the bass drum is weak and the toms far from powerful' (Mills 1985: 11). Indeed, more recently, Nick Batt of Sonic State has talked of the 'Bibble Bobble' Toms in reference to their somewhat inauthentic, highly pitched analogue tone (Batt 2019). Back at the low end, Goodyer notes that,

> The bass drum is also sloppy and lacks attack, especially when set up against the tightly gated, punchy kick drum samples that are as fashionable now as they were when the 808 was first released.
>
> (Goodyer 1986: 71)

It is clear that at least part of this critical evaluation is informed by prevailing sonic fashions and it is unarguable that the TR-808 Bass Drum is neither tight nor punchy and clicky. It is deep, powerful and almost overwhelming in its low end boominess. Or rather it is in the right context, and it is also worth noting that some of the TR-808 Bass Drum's true character is only revealed when heard on particular sound systems such as those in clubs or high-end studios. Indeed, Kikumoto has an amusing and telling admission that sheds some light on the situation,

> Truthfully, the audio system in the Roland lab was so poor that we could not listen to the very deep bass of the TR-808 … So we actually did not even notice the signature deep bass … and we did not expect the deep bass to become the essence of the 808 and cornerstone of new music creation to come.
>
> (LeRoy 2023: 165)

As those early reviews note, it was not so much what the TR-808 was that sealed its fate but rather what it was not, as very shortly after Roland released their flagship instrument, Linn Electronics unveiled the LM-1. What distinguished Linn's instrument was that it was based around cutting edge digital sampling technology rather than analogue synthesis. The initial advertising left no doubt about the innovation of the instrument. Under a huge splash headline reading 'Real Drums', the copy foregrounded the various features of the instrument including its ease of use, the ability to play 'entire songs (intro, verse, chorus, fills, ending etc)' and the storage and tape backup facilities that meant patterns could be archived. However, for all the flexibility and functionality, the central message was clear. 'Digital recordings stored in computer memory' (Linn Electronics, Inc. 1980) brought unrivalled authenticity.

It is difficult to overstate the impact of the LM-1. Although the high cost of the instrument meant it was available only to a select few, the introduction of sampling brought an entirely new palette of sounds to popular music. It was soon heard on tracks by The Human League, Stevie Wonder, Michael Jackson and Peter Gabriel, but it is perhaps Prince who is still most closely associated with the instrument. The LM-1 provides the backbeat to countless tracks including '1999' (1982), 'When Doves Cry' (1984) and 'Purple Rain' (1984) to name just three. As the LM-1 included separate outputs for each sound, just like the TR-808's Multi-Outs, it was common to further process Linn sounds with external effects. Ironically, as Carlozo (2019) notes, long-term collaborator and studio engineer Susan Rogers has revealed that Prince mostly used Boss guitar effects pedals for this purpose which were, of course, made by Roland. In addition to external Flanging and Distortion, Prince's recordings make considerable use of the LM-1's ability to tune its samples. In particular, pitching down the relatively low fidelity samples adds a gritty, grainy texture to them that, paradoxically, had the effect of making them sound more synthetic than 'real'.

Even though contemporary assessments were ebullient in their praise, 'The Linn is so far ahead of anything else in its field at the moment that it's just not worth drawing comparisons.' (Cann 1981: 58), the LM-1 had its share of limitations. Computer memory remained extraordinarily expensive and even with samples stored in 8-Bit 28KHz format, there was no room for lengthy Ride or Crash cymbals. In fact, all of the LM-1 sounds have a remarkably clipped feeling to them caused, at least in part, by the comparatively low sampling rates but also by the need to keep the recordings as short as possible to conserve every last byte of expensive memory. Exploring contemporary reviews immediately reveals the sense of revolution.

> The quality of its sounds? They're going to be a shock for people used to the artificialness and hitherto general wetness of rhythm machines. True, the obviously synthetic nature of their sounds can be very appealing but it is eventually very limiting.
>
> (Cann 1981: 56)

Aside from its sound, the LM-1's timing features would prove especially influential. Where the TR-808 could be programmed in any combination of step or real time, the LM-1 required the player to play their patterns by hand using the hard plastic pads on the instrument's front panel. Inputs could be quantised to tighten up timing and bring performances onto the musical grid as with TR-808. However, in addition to this 'Timing Correction', the LM-1 also introduced a 'Shuffle' feature. By delaying certain notes and pushing them off the grid, Shuffle introduced a musical swing to playback. Cryptically described as 'Human Rhythm Feel made possible by special timing circuitry' in the instrument's advertising and promotional brochure, Shuffle marked another important point of distinction between the LM-1 from the TR-808. Interestingly, while the TR-808 is often talked about as having or imparting a 'groove' or 'feel' (see Iverson 2021), it was designed to play with metronomic, quantised precision. As Mike Banks of Detroit Techno collective Underground Resistance notes, 'The 808 is stiff, it has no swing to loosen the timing.' (Rubin 2017). It would be three years before Roland would first implement such a feature in the TR-909. The implementation of Shuffle can be seen as another way in which the LM-1's design sought to 'humanise' the drum machine.

The success of the LM-1 led to a host of other companies releasing sample-based instruments with the Oberheim DMX (which also included Swing/Shuffle features), Sequential Circuits Drum-Traks and Tom, E-mu Drumulator and the Yamaha RX11 among many others. Indeed, by the mid-1980s, instruments like Casio's £300 RZ-1 not only brought the price of sampled drums down to less than one tenth of the $5000 of the LM-1, but also allowed users to easily record their own samples to replace the built-in sounds. By 1985, and for just a tenth of the price of the RZ-1, UK company Cheetah even had a £30 package combining hardware and software called SpecDrum that transformed a ZX Spectrum home computer into a sampling drum machine.

There can be little doubt that sampling and differing approaches to 'human feel' defined the aesthetics and design of drum machines throughout the 1980s. Indeed, by 1989, Roland had consolidated its product line around the R-8 Human Rhythm Composer. One consequence of this dual focus on authentic sound capture and the accurate reproduction of the imperfections of timing, was that the dominant mode of the drum machine further consolidated around imitation as both digital recording and programming methods centred on recreating 'real' drums and the nuances of the 'real' performances of 'real' drummers. In short, the dominant design imperative was to sound more 'drum' and less 'machine'.

Even in the analogue realm, a 1983 review of the now largely forgotten Korg KPR-77 drum machine in *One Two Testing* magazine reminds us, judgements of sounds are both highly subjective and subject to change over time as fashions, musical tastes and applications shift. 'But the burning question is how does the KPR-77 compare with its obvious rival, the Roland TR808? Comparisons are odious, but soundwise I think the Korg has the edge.' ('Korg KPR-77 review' 1983: 44). Regardless of the instruments' future trajectories, it is clear that in the context of early 1980s mainstream music reviewing, the TR-808 was swimming against the tide.

In truth, the aim of creating 'realistic' drums had informed instrument design for some time and was not the product of Linn or the LM-1. After all, the Rhythmate was based around taped recordings of real drummers really playing real patterns on real drums. However, with the availability of digital samples, the LM-1 brought a new way to combine sonic verisimilitude with the flexibility of programmability.

Of course, it is clear that there were important underlying technological conditions that made the LM-1 possible, albeit at an initially high price. The increasing availability of computer memory and integrated circuits as well as Roger Linn's proximity to the knowledge and expertise of Silicon Valley provide a crucial enabling context. However, the instrument and its clarity of focus and new approach to delivering realism and authenticity was also a clear critical response to the plethora of previous instruments and their attempts to synthetise drum sounds in analogue circuitry. As Linn has famously noted of the inspiration for the LM-1, 'I wanted a drum machine that did more than play preset samba patterns and didn't sound like crickets.' (Vail 2000: 289).

In fact, this is perhaps a veiled reference to one of the most infamous, yet apocryphal, criticisms of the TR-808. In a review of the LinnDrum, the cheaper (though still $3000) successor to the LM-1 that benefitted from the falling costs of computer memory and included, among other things, cymbal samples, Dominic Milano notes that,

> The unit is a big improvement over electronic-sounding drum machines that only do 4/4 time and sound like marching anteaters (unless of course you prefer marching anteaters and being locked into 4/4 time).
>
> (Milano 1982)

Over the years, it has been so widely understood that this was a barbed reference to the TR-808 that the phrase 'marching anteaters' has become almost indelibly enmeshed in the narrative of the instrument (see Hoffman 2014; Marsden 2008, for instance). However, as Werner (2015) notes, subsequent interviews have revealed that Milano was actually referring to the generation of auto-accompaniment instruments that preceded the TR-808. Ultimately, while it is amusing that one of the most often repeated quotes about the TR-808 is actually unconnected with it, the willingness to apply its sentiment to the instrument speaks eloquently to acceptance of the broader narrative of a lacklustre initial response and the unfashionable nature of its sound.

More Real than Real Drums

There is another taken-for-granted assumption about the early years of TR-808 that is worthy of closer analysis. Reviews and commentaries often imply that Roland's engineering and design teams were blindsided by Roger Linn and the LM-1. There is an implication that Kikumoto and his colleagues simply ploughed on with their analogue synthesis blissfully unaware of the existence of sampling technology let alone its embodiment in an imminent commercial product. However, as more recent interviews have revealed, the Roland team were well aware of sampling.

> PCM sampling was expected to become the mainstream of next-generation musical instrument technology, but for us synthesizer engineers, simply recording and playing back the sound of acoustic instruments is a very easy method, and it goes against craftsmanship. There was a spirit and pride.
>
> Therefore, since Roland is a synthesizer manufacturer, we decided to develop a 'drum synthesizer' that allows users to create their own sounds by making each percussion instrument into a synthesizer rather than creating realistic sounds, and Kakehashi agreed.
>
> (Kikumoto 2020)

In fact, Kikumoto (2020) notes that both he and Kakehashi were not only aware of sampling but also of the existence and sound of the LM-1 which the Roland founder had heard during a US trade show visit. As LeRoy (2023: 163) recounts, following a later trip Kakehashi returned with a cassette recording of the pre-production unit which he gave to Kikumoto who conducted analyses on the sounds to understand their quality and how they might be recreated. As such, far from being surprised by the LM-1, the instrument provided a key point of reference for Roland's continuing development. What is notable, however, is that the Roland team took a self-consciously different approach to sound generation. As Kikumoto explains,

> The TR-808 was designed with the idea of emulating an acoustic drum kit, utilizing the latest Roland analog synthesis technology. Roland was a synthesizer manufacturer. It was solely their brand identity at the time. Therefore, when we developed the 808, we provided the ability for expanded sound design. For example, people criticized the long delay of the TR-808's kick sound as unrealistic. But it provided the user with the ability to create something new and different. Plus, if they didn't like it, they could just give it a short decay. I never understood what the commotion was about.
>
> (Kikumoto in McCabe 2020)

Indeed, even Roland's next drum machines including the TR-909 would not shift entirely to the use of samples and continued to incorporate analogue sound synthesis circuitry. It might be tempting to consider the ongoing use of synthesis as the result of production schedule that saw the TR-909 already too far in development to be retooled yet work on its successor only began once the TR-808 had been completed. In fact, the team resisted internal calls to shift to sampling technology even though work on their analogue systems were delaying the product coming to market. Rather, Roland's commitment to analogue sound synthesis in the TR-808 and TR-909 is just that, a commitment. It is reflective of a deeper, philosophical approach to the creation of sound that is based around the limitations of replication versus sonic exploration.

Kikumoto has often spoken of the connection between the TR-808 and Japanese woodblock prints, known as ukiyo-e. In particular, he describes both technical and creative parallels between the approach to sound synthesis in the TR-808 and that of artist Katsushika Hokusai whose 'Thirty-Six Views of Mount Fuji' series produced in the early 1830s includes 'The Great Wave off Kanagawa'.

> In that piece of work, he threw off the photographic details into abstraction that boldly enhanced the great waves by outlines and made contrast with the peak of Mt. Fuji. This manner, derived from its own limits, is something impossible with realistic paintings that put priorities on the details. In this same way, I feel the 808 provide the impact over the reality.
>
> (Kikumoto in Bjørn 2022: 19)

Conceptually, then, what we arrive at with the analogue circuits in the TR-808 are representations of drum sounds which, by harnessing their limitations, seek to capture essences rather than details. If PCM sampling was like photography in seeking to perfectly reproduce the sound of a drum, Roland's analogue synthesis circuits were the ink and line drawings that sought to encapsulate, and even accentuate, the fundamental qualities of sound. For Kikumoto, synthesis was not only a matter of pride, but like the ukiyo-e of Hokusai and others, it was the foundation of an artistry that was more real than real, more real even than 'real drums'.

The Dream of a Drum Synthesizer

Forty years after the TR-808 was released, chief engineer Tadao Kikumoto has begun to reveal a little more about the design of this influential instrument and the creation of the circuits that gave rise to those now iconic sounds. Reflecting on the initial brief from Kakehashi and Don Lewis to create an inexpensive, realistic sounding drum machine that would, in Kakehashi's estimation, lower the cost of music production because musicians would not have to hire drummers, Kikumoto has spoken of his vision for an altogether more complex instrument (Kikumoto 2020). In fact, rather than simply making another drum machine, Kikumoto's ambition had been no less than a drum synthesizer. Recognising the range of nuances that went into the production of drum, percussion and cymbal sounds, Kikumoto had imagined a system with a huge number of sound shaping parameters. 'My ideal snare drum for the TR-808 had controllable diameter, depth of the shell, and tension of the membrane.' (LeRoy 2023: 165).

The practicalities of 1980s analogue sound synthesis, however, meant that this was simply impossible. As Kikumoto notes,

> In order to set the price under a thousand US dollars, we had to limit the electronic components both in kinds and quantity to be used. We also had specification limits with dynamics and stabilities of semiconductors. The biggest problem was that deploying numerous potentiometers to allows users to synthesize sounds will consume so much space on circuit boards that we will quickly run out of rooms to place electronic components that comprise the electronic circuitries. This greatly forced us to lose the freedom on designing the product.
>
> (Kikumoto in McCabe 2020)

Even setting aside the prohibitive cost of the circuits and front panel control hardware, the simple truth is that analogue synthesis is not especially well-suited to the emulation of analogue percussion sounds. Indeed, while Kikumoto's criticism of the CR-78 might seem inevitable given that the TR-808 was intended to improve on it, his assessment of the Bass Drum in the earlier instrument speaks to the focus in Roland's new approach. 'There was no impact, partly due to the very weak attack.' (LeRoy 2023: 165). This frustration is surely born of Kikumoto's sound analysis which had ascertained that much of the sonic fingerprint of a percussion sound came from its initial transient when the interaction of stick and head create a brief period of distinctive harmonic complexity. And so, although the TR-808 Bass Drum is most widely celebrated for its long, booming sustain, for Kikumoto at least, the first few milliseconds were just as important.

Kikumoto has been rather more scathing in his assessment of some of the TR-808's most celebrated and iconic sounds including the Hand Clap, however. The sound had been included at the personal request of Don Lewis who had originally modified his CR-78 Clave sound to create his own approximation. The TR-808 Hand Clap comprises a comparatively complex circuit which shares many of its components with the Maracas rather than the Clave which morphs into a Rim Shot as Figure 3.2 shows. The underpinning sound source for the TR-808 Hand Clap is provided by the shared White Noise generator which fed into a 1000Hz Band Pass Filter to shave off the extreme high and low frequencies. The key to the sound, however, is found in the dual VCAs and envelopes that work in parallel to create the two identifiable elements of the sound. A complex multi-stage sawtooth envelope creates the initial 'snap' and three more distinct attacks that simulate the natural flaming effect of multiple people attempting to clap simultaneously. This is combined with a more traditional envelope contour that more gradually fades from high to low volume to create a more sustained 'reverb' effect. Apparently unsatisfied with the result on the TR-808, and presumably referencing the sound of the juddering sawtooth envelope, Kikumoto has described it as sounding like 'a bunch of bamboo sticks'. (LeRoy 2023: 165).

What is interesting, of course, is that it would actually be the alien otherworldliness and 'unrealistic' nature of the TR-808 sound palette that would become central in defining the appeal and longevity of the instrument. Looking back several decades after its initial release, Joe Mansfield, author of 'Beat Box: A Drum Machine Obsession' (2014), has noted, "It sounded futuristic, what you thought a computer would sound like if it could play the drums." (Beaumont-Thomas 2014). More than 15 years earlier, Chris Carter of the band Throbbing Gristle, offered similar sentiments noting that the appeal of the TR-808 was precisely that it sounded neither like a real drums nor auto-accompaniment anteaters.

> Whereas drum machines like the Linn were trying to sound like a real drummer playing real drums, the 808, intentionally or not, turned that idea on its head. It didn't sound like a real drummer playing real drums, or like a cheapo rhythm box churning out bossa novas – it sounded like ... well, it sounded like nothing else, really, and this is what made it so distinctive.
>
> (Carter 1997a)

For Carter, the TR-808's combination of warm, enveloping Bass Drum and tight, snappy Snare Drum made it an instrument of extremes, like 'sandpaper and velvet'. What is particularly notable, however, is that a decade before this, the TR-808 had already been re-evaluated and was proving itself to have a flexibility and staying-power unmatched by the instruments that had apparently stolen its thunder and rendered obsolete almost before it had been released.

> With the arrival of 'affordable' user-sampling machines like the Korg DDD1 and Casio RZ1, the curtain opens on another new era for the now ubiquitous drum machine. Roger Linn's revolutionary LinnDrum now lies subdued in the corner of recording studios, relegated to the reserves bench; the old Boss Dr Rhythm is quieter still ... Roland's curious TR808, however, continues its insistent rattle from transistor radios, and thumps its way across nightclub speaker systems the world over – three years after the last one rolled off the Hamamatsu production line.
>
> (Goodyer 1986: 70)

Making Beats (and New Genres)

For all the talk of lukewarm receptions, less than stellar reviews and the almost immediate impact of the LM-1, we might be forgiven for thinking that the launch of the TR-808 was a catastrophe with the instrument being instantly forgotten. There were, however, a number of high profile, mainstream uses of the instrument that spanned a range of genres. In fact, the instrument was first heard even before its commercial release as Roland supplied Yellow Magic Orchestra with a pre-production unit for use on their 1980 tour. In the following year, YMO would commit the instrument to record for the first time on their BGM album and the TR-808 can be heard prominently on the track '1000 Knives', for instance. LeRoy (2023) notes that, while this is typically cited as the first instance of the instrument on record, studio multitrack notes suggest that Suzanne Ciani's 'Seven Waves' might actually precede it, although the album was not commercially released until later in 1982.

The TR-808 found favour with a host of different artists across numerous genres including the upbeat poppy R&B of Whitney Houston's 'I Wanna Dance with Somebody (Who Loves Me)' where the iconic cowbell is one of the most prominent instruments in the track. Indeed, the TR-808 cowbell enjoyed a period of notoriety, throughout the 1980s across a host of pop and dance music tracks. As Goodyer (1986) notes, when Stock, Aitken and Waterman cheekily credited the drums on Mel & Kim's single 'Respectable' to 'A Linn', they were only partially correct as the track's cowbell part unmistakably came from the TR-808.

Perhaps the most well-known of the early uses of the TR-808, however, is Marvin Gaye's 'Sexual Healing'. Released in 1982, the recording saw Gaye take a self-consciously different direction from his previous work with Motown. Replacing lush string arrangements with the synthetic sounds of the Roland Jupiter-8 polyphonic synthesizer and live drums with the TR-808, this was a new form of stripped down, electronic soul. With these two instruments forming the backbone of the orchestration, their tones were unmistakable. Yet, there is more to the TR-808 in 'Sexual Healing' than can be explained just through its sounds. The programming of the two-bar pattern does not appear to try to replicate the kind of pattern that a drummer might play. Indeed, as Mike Butcher who was a recording engineer on the sessions suggests,

> As far as I know, he was the first person to program a TR-808 as its own instrument instead of trying to imitate a drum kit … He would do things like put the side-stick on a beat, which a drummer wouldn't do.
>
> (Smith 2020)

Even if this was not a conscious decision, it signals something important about the uses of the TR-808 and the drum machine more broadly. While the dominant discourse might have seen emulation take centerstage with realistic samples and programming that mimicked the playing style of a drummer, there was another way to approach the drum machine. The design of the machine meant that multiple sounds could be played at once, more than a single player could strike simultaneously. Moreover, while instruments like the TR-808 did not allow every sound to be played on the same beat, the Snare Drum and Rim Shot could sound together in a manner most 'unrealistic'.

Indeed, as Gaye demonstrated, 'side sticks' could be programmed to occur anywhere – or even everywhere – in the bar. Certainly, Gaye's use of the TR-808 in 'Sexual Healing' helped to transform soul music by introducing a new palette of electronic sounds, but it also speaks to a new approach to rhythm that arises from the programmatic possibilities of the drum machine. Perhaps the electronic sounds of the TR-808 helped as a guide, but essentially, 'Sexual Healing' stands as an eloquent demonstration of the ways in which the drum machine can be approached as an instrument in its own right, with its own set of affordances and musical capabilities that transcend the imitation of live human drumming performances.

While the design of the TR-808's circuitry does not allow every one of its sounds to be triggered simultaneously, a single sequencer Step can comprise Snare, Rim, all three

Toms, Hi Hat and Cymbal, for instance. Obviously, this ample polyphony makes it possible to programme rhythms that would be impossible for a human player with a single pair of sticks. In fact, we might argue that the design of the TR-808 sequencer and the arrangement of the sound generating circuits makes almost inevitable the creation of rhythms that could not be replicated by human players. Programmed combinations of continuous 16th note Hi Hats with a Snare on the backbeat, occasional Tom fills and some Ride and Crash Cymbals for good measure would challenge even the most adept of drummers yet such patterns effortlessly usher forth from the TR-808. Once we add percussion, such as Handclaps, Cowbells, Rimshots and an array of Accents, patterns soon lose any connection between the capabilities of the human player and are guided by the affordances of the TR-REC programming interface and polyphony of the analogue sound circuits.

As the programming interface separates out each sound and gives it its own independent sequencing lane, attention is shifted away from a drum 'kit' and towards the creation of individual rhythmic lines that weave in and out of one another and come together to form a cohesive performance. In this way, the TR-REC programming methodology is more akin to the performance of samba band in which repeating motifs played by each individual performer on their specific drum or percussion instrument combine to create a complex syncopated whole. The impact of this approach to drum programming is heard throughout the 1980s and beyond and is far from limited to just the TR-808. For instance, although it was produced with sample-based instruments such as the Fairlight CMI, Frankie Goes to Hollywood's cover of Edwin Star's 'War' from the band's 1984 'Welcome to the Pleasuredome' album is an object lesson in TR-style drum programming. Underpinned by a relentless four-on-the-floor Bass Drum, the song is based around a two-bar drum pattern that builds in complexity as individual cowbell, clap, tom and other percussion 'lines' are added and removed. As one listens to the evolving pattern, it is hard not to visualise the notes on each individual sequencer lane being programmed, muted and unmuted.

As such, just as its synthesized sounds were intended to capture the essence of drums rather than simply replicate them, the 16-Step TR-REC programming interface encouraged users of the TR-808, and those drum machines that followed it, to create a new aesthetics of rhythm that was unbound from the imitation, or perhaps limitation, of human performance. As we shall see in later chapters, for many artists, including the pioneers of Techno, the new approach afforded by instruments like the TR-808 and later TR-909 encouraged the deep investigation of relationships between humans and technology that was central to creative expression and around which sense of and community and personal identity was built.

Key to the continued development and visibility of the TR-808 throughout the 1980s was Afrika Bambaataa & The Soulsonic Force's 'Planet Rock'. In a time before 'mashups' or 'collabs', 'Planet Rock', produced by Arthur Baker, brought together an eclectic range of influences that straddled the electronic sounds of YMO, Gary Numan and Kraftwerk and the funk of James Brown, George Clinton and Parliament-Funkadelic. For Shapiro, 'Planet Rock' was, 'The song taught the world that machines were just as funky as James Brown' (Shapiro 2000: 107).

The genre-defying track mixes melodic elements from Kraftwerk's 'Trans-Europe Express' album with an extraordinarily punchy TR-808 pattern that draws heavily from the track 'Numbers', also by Kraftwerk. Even among the sampled orchestra hits courtesy of the Fairlight CMI sampler, the TR-808 bites through the mix and propels the track along adding an undeniably futuristic sound. As Iverson (2021) notes, part of the importance of the TR-808 goes beyond its sound.

> Here is where I think we can view the 808 most productively: as a hub that consolidated sonic signifiers. In 'Planet Rock,' the 808 beat is like the canvas for a visual artist, providing the groove above which the artist(s) can layer rich sonic signifiers – samples, new melodies, sound effects, quotations – that further involute and stretch racialized elements of style and genre. In this way, the 808 has double significance.
>
> (Iverson 2021)

Of course, while its $1000 retail price meant that the TR-808 cost considerably less than the Linn LM-1, this still represented a considerable barrier. However, the impact of the LM-1 and the new breed of sample-based instruments meant that a good number of TR-808s found their way onto the second-hand market and into pawn shops. As apparently undesirable, superseded pieces of outdated equipment, they were often available for a fraction of their original prices. It was this that brought them into the hands of a different generation of musicians, DJs and producers.

If any genre made the TR-808 its own, it must surely be Hip Hop. Artists such as LL Cool J, The Beastie Boys and Run-DMC each made the instrument part of their own signature sound and contributed to it quickly becoming a foundational element in Hip Hop. While most of the instrument's sounds were put to use with the snap Snare Drum and sizzling Hi Hats being particularly cutting, it was the TR-808 Bass Drum that came to dominate. Producer Rick Rubin, who worked with the Beastie Boys and would go on to co-found the Def Jam record label, was particularly instrumental in bringing the sound forward in the mix. The Beastie Boys' 'License to Ill' (1986), produced by Rubin, is a virtual paean to the TR-808 with tracks like Brass Monkey and Paul Revere putting it front and centre and, in the case of the latter track, even committing it to tape and reversing it. By extending the Bass Drum Decay and taking advantage of the 'unrealistic' flexibility that Kikumoto and Nakamura had built into the circuit, Rubin's sound took on a quality that transcended any emulation of a drum and became a vast, wide boom. In the appropriate environment such as a club sound system or a suitably-equipped car, this mighty, powerful bass tone is felt rather than heard.

While unaware of the rebirth of the TR-808 at the time, Kikumoto, the design team at Roland and even Kakehashi, eventually learned about 'Mid-O fever' (in reference to the numbering scheme for the instruments) in the early 1990s.

> I also remember a hip-hop producer – perhaps Rick Rubin – extending the decay of the 808 Bass Drum. Users like that took advantage of Mid-O according to their creative instincts. Those applications were sometimes different from my intention, but the expandable Decay knob of the TR-808 was my intention. I was very glad to hear that because the idea was basically not supported at the time, except by Mr. K [akehashi].
>
> (McCabe 2020)

The influence of Rubin's extended Bass Drum is felt not only through the records he produced but also in the recordings of countless others as this sound was endlessly sampled by other artists and producers. Speaking in Alexander Dunn's 2015 '808' documentary, Hank Shocklee, who worked with Public Enemy as part of the Bomb Squad production team, is clear, 'It's not hip-hop without that sound … If you don't have it, it's not authentic.' For McKee and Grierson (2017) just like Richards (2008), the centrality of the TR-808 to the sound of Hip Hop is equivalent to the Fender Stratocaster guitar and its influence on rock music. And, while the lack of realism had been the thing that had initially driven the TR-808 out of favour, and out of certain genres of music, it was precisely the electronic, futuristic tone that made it so appealing in this new context.

> The 808's electronically-generated sound was one that rap and hip-hop artists identified with: anti-establishment and more aggressive than regular percussive instruments.
>
> (Lewis in Hayes 2019)

So valuable was the TR-808 that an instrument often ended up being shared among a number of different artists in a studio. As DJ Pierre of Chicago-based Acid House group Phuture notes, there was a palpable respect and care for the TR-808 with one particular instrument moving around the city on a trust-loan system. "It was almost like we were joint parents of this child, and we all had to do our part to take care of it." (Sorcinelli 2022). It is important to note, however, that while the deep, booming sound of the TR-808 Bass Drum soon became a common touchpoint for DJs, artists and producers, its extremity was such that the meaning and importance of the sound was not always understood across the various institutions of music production. Recounting the difficulty of convincing studio engineers unused to the dominance of the bass frequencies he wanted to commit to tape that the TR-808 had to be the focal point of the recordings, Jazzy Jeff laments that the message was not communicated throughout the formal (and formalised) production process.

> Hip Hop is something that the drums have to drive the record. And I got him to allow me to do it to the point that I loved it, and what I never realised was I never told the mastering engineer that I wanted that. And he thought it was a mistake, and he took all of the 808 out of the album, and I don't think I've ever said this in public. I can't listen to He's the DJ, I'm the Rapper now. That is the biggest record we've ever done, and I absolutely hate the way that it sounds because they sucked all of the bottom end from the 808 out in mastering.
>
> (Jazzy Jeff 2024)

Becoming the 808: Thinking Outside and In the Box

Over time, the TR-808 has transformed, or rather has been transformed, into something surprisingly distinctive and, in some cases, disembodied from the original instrument. Kanye West's '808s and Heartbreaks' (2008), an album that uses the instrument on every track, is notable in speaking to and helping to popularise it both as a musical element and as a source of lyrical content. As such, the album pushes the TR-808 into new musical territory while helping to establish the '808' as a sonic and cultural artefact intimately connected with complex forms of musical, social and political identity. The continued transformation of Hip Hop and the emergence of (sub)genres such as Trap have similarly extended the uses, meanings and appreciation of the instrument with a particular focus on the Bass Drum and Hi Hat sounds.

Emerging from the Southern United States, Trap is a comparatively stripped back form that draws its influences from rap and electronic dance music. Although it evolved over many years, the modern Trap sound and formula can trace it roots back to Cool Breeze's

1999 'East Point's Greatest Hit' and Lex Luger's production on the Waka Flocka Flame album 'Flockaveli' in 2010 (with Luger going on to nod to the influence of the TR-808 in the name of the production team '808 Mafia'). It is interesting to note that the contribution of the TR-808 in Trap exists at the highest and lowest ranges of the frequency spectrum. At the high end, Trap Hi Hats celebrate their synthetic voicing and are programmed in highly syncopated double time and triplet patterns. The triplet pattern, in particular, has solidified as a key component of Trap and is very often echoed in the rhythm of the rapping itself which is typically delivered in what has become known as the 'Migos Flow' (named after the artist who popularised the triplet rhythm in the 2013 track 'Versace'). Similarly, the alternation between half and double time Hi Hat and Snare patterns allows rappers to shift between slow and fast styles. In terms of the lower end of the frequency range, Trap is rooted in the deep TR-808 Bass Drum with its extended Decay turning the sound into something far more akin to a synthesised bass tone than a percussion sound. The influence of Trap has been felt so strongly across a diverse forms of popular music, with tracks from artists such as Ariana Grande and Beyonce bearing the genre's characteristic hallmarks, that the TR-808 has been ensured a new lease of life and audience of producers and listeners.

Through its continued use by artists such as Original Concept and the formation of genres like Miami Bass, the TR-808 has been yet further transformed and not only fills out the low end but actually takes on the role of performing bass lines. With the booming sound ringing out for even longer than the original instrument allowed and playing distinctly pitched notes that underpin the chord progressions of the tracks, we might be forgiven for thinking that this genre was built around modifications to the circuits. After all, we noted earlier that the design of the TR-808 Bass Drum circuits affords no ability to change the tuning of the sound from the front panel, while Kikumoto had opined drum sequencers' and the later MIDI specification's absence of pitch programming for percussion tones. What, then, are these pitched 808 basslines that give Miami Bass not only its distinctive sound but even part of its name? What we hear in genres like these are actually processed and manipulated TR-808 samples which are triggered from an external sequencer rather than being the output of an original instrument or the product of inventive Step Button programming. The TR-808 Bass Drum can truly be said to have transcended the instrument to become, 'Less a product of engineering than a force of nature.' (Norris 2015).

In fact, so prevalent is this sound that the term '808' has moved into production parlance and is used today simply to refer to a deep sub bass part that delivers the weight and sustained Sine wave warmth of the original sound. And, as we can see form the myriad tutorials and guides that exist to create these sounds, contemporary music production does not require a TR-808 to make '808s'. As Reason Studios' introduction notes, 'There are many different ways to create an 808 bass line. You can use a bass synth, a sampler, or even a MIDI keyboard.' (Reason Studios 2022).

What is particularly fascinating about this situation is the way in which it reveals different attitudes towards sampling and the nature of the TR-808 as an instrument. Considering the question in the early 1990s, David Mellor draws attention to what is lost in the act of sampling.

> After all, a drum machine is easily sampled, and once installed in an S1000 or S750 the samples will sound pretty much like the real thing, with the advantages of MIDI programmability, lower cost, and a significant space saving. With any digital drum machine I would say this is mostly true. But with the TR808 and other analogue

> machines, however, there is a world of difference between using samples and having the real thing to hand. For one thing, many of the sounds can be modified in a way that can't accurately be imitated by a sampler's editing facilities. Also, a sampled 808 bass drum just isn't the real thing – it's like a Polaroid snap compared to real life.
>
> Mellor (1993: 43–44)

There is something in this sentiment that recognises the variability, dynamism and quirks of the analogue circuits along with the multifaceted real time controls that cannot be readily frozen in a single digital recording. Similarly, towards the end of the decade, Chris Carter suggests that attempts to freeze the sounds of the analogue circuits fall short.

> I've read countless interviews where producers, users, DJs and remixers have voiced opinions on whether the 808 should be sampled or not. Most agree that, no matter how good a sample you have, nothing sounds quite like the original, particularly when it comes to the 808 bass drum.
>
> (Carter 1997a)

Perhaps this reveals both a tendency towards the fetishisation of analogue circuits and the creative and technical limitations of sampling in the 1990s, yet it is fascinating to see how this argument is articulated around authenticity. There is certainly a delicious circularity in reading a discussion that ponders whether it is possible to accurately create digital samples of the living, breathing, sighing analogue circuits of the TR-808 and the dramatic echoes to the Linn LM-1 'real drums' advert from 1980 are surely unmissable. However, what is equally interesting is to note the way in which the practice of sampling the TR-808 continues to shift. Where Mellor and Carter proceed from a position that seeks authentic reproduction and nuance, and Shocklee was sampling a specific instance of the TR-808 on record, we see an alternative approach in more recent music production that actively attempts to explore, extend and transform the sound by pushing and pulling it in new directions.

That said, it is something of an irony to find that the name, or rather number, of the TR-808 has never been more widely known and spoken, yet this has become dissociated from the instrument itself. Similarly, accepting that it does not even require the instrument's presence, the '808' could be seen to have a reductive effect of boiling the TR-808 down to just a single sound. More positively, we might argue that the prevalence of the '808' to describe the booming bass tone speaks to a recognition of the solidity of the TR-808 and its continued role as the bedrock upon which a host of musical genres rest.

Afterlives

Given the prevalence and continued popularity of the TR-808 across multiple genres, it is unsurprising to find that the sounds of the instrument have become staples in the preset banks of countless keyboard synthesizers created from the 1990s to date. As those instruments shifted from analogue to digital synthesis in the late 1980s and then became increasingly based around the use of samples as their building blocks throughout the 1990s and onwards, a category of instrument called the 'Workstation' began to appear. With models such as Korg's M-1 sowing the seeds for later generations of Triton and Trinity along with Roland's own JV sound modules and Fantom series of keyboards, these multitimbral instruments (able to play multiple sounds simultaneously) typically included a multitrack sequencer and were sold as all-in-one music making studios. As such, a collection of drum samples became an essential ingredient with a mix of acoustic and electronic sounds often including various elements of the TR-808 kit. Of course, while Roland felt able to reference their sounds accordingly, other manufacturers often took a more circumspect approach to naming that strongly hinted at but ultimately pulled short of actually revealing their origin. Perhaps this was out of respect, on the advice of the legal department, or perhaps it was born of the fear of promoting another company's product inside one's own instrument. Similarly, the trade in libraries for hardware samplers soon saw floppy disks labelled 'Snappy Snare' or 'Mel&Cowbell'. But whatever they were called, the sounds were as immediately recognisable then as they are today.

Reflecting on some of the questions posed by Carter and Mellor above, it is fascinating to see the way in which these samples speak to broader trends in the use of the TR-808 with an increasing focus on the Bass Drum. Additionally, we see attempts to account for the range and impact of the tone controls by offering differently 'Snappy' Snare drums and short and long Cymbals, for instance. In this way, there is a crystallisation of particular combinations of settings into discrete recordings which reflect decisions about the usable or tasteful sounds of the original instrument.

In addition to their inclusion in the presets and sample banks of workstation style keyboards, the emergence of 'Groovebox' instruments such as the Roland MC-303 in 1996 explicitly recalled the x0x instruments. While its name actually references both the MC-202 Micro Composer (1983) the Roland TB-303 Bass Line (1981) explored elsewhere in this book, the MC-303 actually brought together sounds from the TR-808 along with selections from the CR-78, TR-606 and TR-909. In one sense, the MC-303 is an insight into Roland and perhaps reflects the company's realisation of the ongoing musical and cultural impact of its instruments and a desire to (re)claim them. Subsequent instruments in the MC series added more complex sequencers and a broader range of synthetic sounds and effects but left the TR-808 at the heart. Throughout the late 1990s and 2000s, Roland's MCs were joined by instruments such as the Novation Drum Station which, similarly, sought to provide a combination of TR-808 and TR-909 sounds. Here, the sounds were recreated using a mixture of digital synthesis and samples. Contemporary reviews point to the potential market and suggest that the cost of TR-808s (and TR-909s) were already rising on the second-hand market.

> And with advance orders for the DrumStation apparently numbering well into the thousands, it would seem that plenty of people are fed up with lining the pockets of those who were cunning (or fortunate) enough to pick up real 808s and 909s for a song when everyone was selling theirs!
>
> (Nagle 1996)

Setting aside any criticism of the individual sounds, one notable feature of the Drum Station is that, as a rackmountable module, it only provides a source of sounds and does away with any built-in sequencer whether 'TR-REC' or otherwise. The intention here is that the sounds in the instrument are triggered from a separate hardware sequencer or, more likely in the late 1990s, a computer-based sequencer such as Cubase. In one sense, the Drum Station predicts the disconnection of the 808 from the TR-808 that we see in modern genres like Miami Bass and Trap. In truth, what we see in the Drum Station's dissociation of the TR-808's sound generation from its sequencer is a continuation of the work of the workstation.

We should note that, as we have seen in our earlier discussions, it is often the interactions between the sequencer and sound circuits that give the TR-808 its specific nuances. While the front panel controls are the most obvious place where sound shaping might be seen to occur, there are often complex interactions and multiple changes that occur as a result of Accented or non-Accented sequencer step in the TR-808 beyond just an increase in volume. Replacing the instrument's sequencer with another, that may be altogether more powerful and flexible in some ways, cannot fail to have an impact both on the feel of using the instrument but also on the nature of its sounds.

Indeed, the separation of the instrument's sequencer and its sounds continues apace. The distillation of the TR-808 into the '808s' we noted above, has seen the creation not only of countless libraries of samples but also software instruments and plugins dedicated to providing the raw materials and means of processing booming sub basses. Taking the essence of the TR-808 Bass Drum circuit, for instance, this new breed of '808' plugins uses samples or virtual analogue sine waves and various forms of saturation and distortion to create the overtones that make these almost impossibly deep tones audible across a wide range of listening equipment. As such, many contemporary '808s' are created without the involvement of an original TR-808 even in sampled form, with software plugins expressly designed to operate within the context of a Digital Audio Workstation such as Ableton Live or FL Studio whose sequencers allow programming pitched 808 notes to create basslines and not just booms.

Not all recreations of the TR-808 have opted for this disintegrated approach, however. Propellerheads' Rebirth RB-338 software was notable in being one of the first 'virtual studios'. In its first iteration, released in 1996, the Windows and Mac-compatible software offered digital emulations of two TB-303s along with a TR-808 (a recreation of the TR-909 was added in Rebirth Version 2). Again, the software's focus on the Roland xOx instruments speaks to their centrality in electronic music making as well as their desirability and scarcity. Importantly, in addition to emulating the sounds of the instruments, Propellerheads recreated both the visual appearance and operation of the interfaces. For the TR-808, step recording still requires the use of the Instrument selector and row of 16 multicoloured buttons for inputting individual steps. Of course, while some familiar physical interactions can be neatly approximated via the computer interface, such as swiping the mouse pointer across all 16 step buttons to enter a pattern of Hi Hats as one might with a single finger, others become harder with just a single mouse pointer and click available. Since Rebirth, countless developers have offered software emulations of the TR-808 including instruments based around algorithmic modelling such as Nepheton by D16 Group or sample playback such as the Revolution Kontakt library by Wave Alchemy. Roland's own software emulation was released into their Roland Cloud service in 2018.

In the analogue realm, TipTop Audio produce a series of Eurorack compatible modules that each replicate many of the circuits form the TR-808 including Bass Drum, Snare Drum, Cowbell, Hi Hats among others while System80's 880 module mimics the look and feel along with the operation of the built-in sequencer. Behringer's RD-8 (and its mark 2 version) are standalone instruments that seek to emulate the analogue circuits of the Roland original but replace the sequencer with one they claim to be more flexible and intuitive. Roland, themselves, have also returned to the TR-808 in a number of pieces of hardware, although these have been resolutely digital rather than analogue recreations. Interestingly, with different instruments, the company has taken a different approach. With the TR-8 and TR-8S, released as part of the AIRA range that introduced the company's Analog Circuit Behavior (ACB) sound modelling, there are clear influences that draw from Roland's earlier drum machines. However, despite their names, these two instruments are probably better understood as culminations of Roland's ongoing development of the 'TR' series. As such, while the 'TR-REC' mode is at the centre of the programming method, the TR-8 (2014) and TR-8S (2018) instruments make a number of modifications to the workflow.

The TR-08 (2017), however, sees Roland similarly deploying its ACB technology but this time with an interface and workflow that more closely replicates that of the TR-808, albeit on a significantly smaller scale. As a member of Roland's Boutique series, the new instrument is a comparatively miniature hardware reconstruction of the original. While it is perhaps a consequence of the decision to standardise the size of the instruments in the Boutique range, it is notable that the TR-08 offers only a single stereo output with the separate outputs allowing processing of each sound available only over a USB connection. As such, the instrument sits in an interesting liminal space that straddles the affordances of both hardware and software. It does, however, offer a number of new features such as the ability to make further sonic modifications to many of the sounds. The Toms, Congas and even the Rim Shot are all tuneable as is the Bass Drum which is also switchable between Normal and Long Decay modes allowing for even lengthier booms that reflect the way the TR-808 has been used and its sound has been adapted over its lifetime. It is worth remembering that, even with all these new sound shaping features and lengthier booms, the TR-08 still falls short of the complex drum synthesizer that Kikumoto had originally envisioned.

It is also notable that the menus of the TR-08 offer some additional sound shaping functionality beyond the facsimile of the front panel. While the Bass Drum Decay knob is typically cited as the key to the reinvention of the TR-808, contemporary uses of the sound draw heavily on external effects and processing. In addition to overdriving the sound to add distortion, saturation and overtones, the heavy use of compression serves to draw out the Bass Drum even further as the quieter portions of its decay are effectively boosted in volume. The TR-08 adds a digital compressor to its Bass Drum and Snare Drum recreations precisely to bring into the box those previously outboard processing technologies and practices within the instrument. As such, we find an intriguing duality in the claims to authenticity in emulation combined with the recognition of particular artistic and technological practices that have shaped and reshaped the core sound of the instrument and its creative and cultural meaning. It follows that, either because particular uses of the instrument rely on extensive external effects and processing after the audio outputs or that the '808' sound does not actually emanate from the instrument at all, it is very often the case that the sound of the TR-808 is not the quite sound of the TR-808.

Indeed, while Roland's 2025 TR-1000 Rhythm Creator includes analogue sound circuits that are modelled after the TR-808, the new instrument includes a richly featured effects section that affords the ability to manipulate its sounds to replicate the tonality of the '808' aesthetic as it has developed over time. Moreover, the TR-1000 ships with a variety of preset 'kits' that process the output of its circuits to simulate the effect of aging components and thereby allowing a brand new instrument to sound 'vintage'. In addition to these fascinating negotiations around 'authenticity,' there is also a degree of curation at work as just some of the sounds of the original make it into the new instrument. Bass Drum, Snare, Hi Hats, Rim Shot, Claps and Cymbals all make the cut, but Congas, Claves, Maraccas are all absent. Roland do not offer any formal insight into their selection process and it is intriguing to observe that even those sounds that are present are sometimes offered in a markedly different fashion. The TR-1000 only offers Low and High Toms rather than the Low, Mid and High variants present in the TR-808, for instance. As Roland note, the tuning ranges of the new circuits have been extended so that the tonality of all three Toms can be replicated in the new instrument yet the design of the TR-1000 means that only two can be used simultaneously.

Although in some senses the palette may be reduced, in other areas the TR-1000 adds complexity and flexibility. Many parameters of the circuits drawn from the TR-808 are extended beyond the ranges of the original or offer control over aspects of the sound unavailable in the original instrument. Particularly notable is the TR-1000's implementation of the TR-808 Bass Drum which now affords user control of the sound's fundamental pitch. In combination with the new iteration of the TR-REC step sequencer and the instrument's snapshot facility, it becomes possible not only to tune a Bass Drum to match the key of a track but also to program pitched '808' basslines. In this way, even the analogue circuits in the TR-1000 are not so much recreations of the original designs but are better thought of as responses to the way in the sounds of the '808' have emerged and evolved over time.

Of course, while subsequent rhythm machines such as the TR-1000 have brought together analogue circuits and user sampling into a single instrument and, in doing so, have mixed synthetic and imitative approaches to rhythm sound design, 40 years earlier the emergence and rapid dominance of sample-based instruments left the TR-808 apparently swimming against the tide. The fetishisation of 'real drums' meant that that TR-808 was initially judged in terms of what it is not. As we have seen, however its success, influence and impact came after a generation of musicians, producers and DJs embraced its otherworldly, self-consciously electronic sounds. As Goodyer notes,

> the TR808 only began to approach its heyday when musicians and producers realised it was no use trying to turn the machine into something it wasn't – namely a facsimile of a human drummer hitting a traditional drum kit.
>
> (Goodyer 1986: 71)

In this new context, concern for realism and authenticity were irrelevant and the TR-808 was used, explored and cherished for what it was. As the decades have passed, the plethora of clones, remakes and emulations of the original circuits that have been released in analogue and digital hardware as well as increasingly software, ensure that there certainly are plenty instruments that offer a very convincing emulation of the TR-808. But for all that, perhaps the Beastie Boys put in best when they said that 'nothing sounds quite like an 808'.

4 TR-606 Drumatix

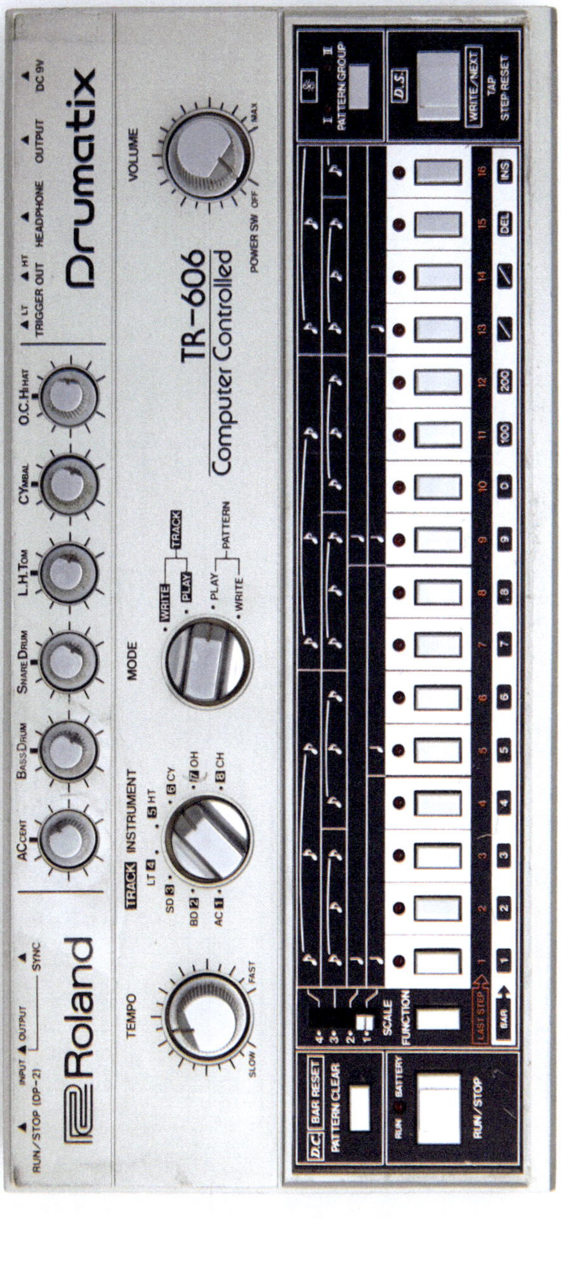

Figure 4.1 Roland TR-606 Rhythm Composer (image credit: James Newman).

DOI: 10.4324/9781003374510-4

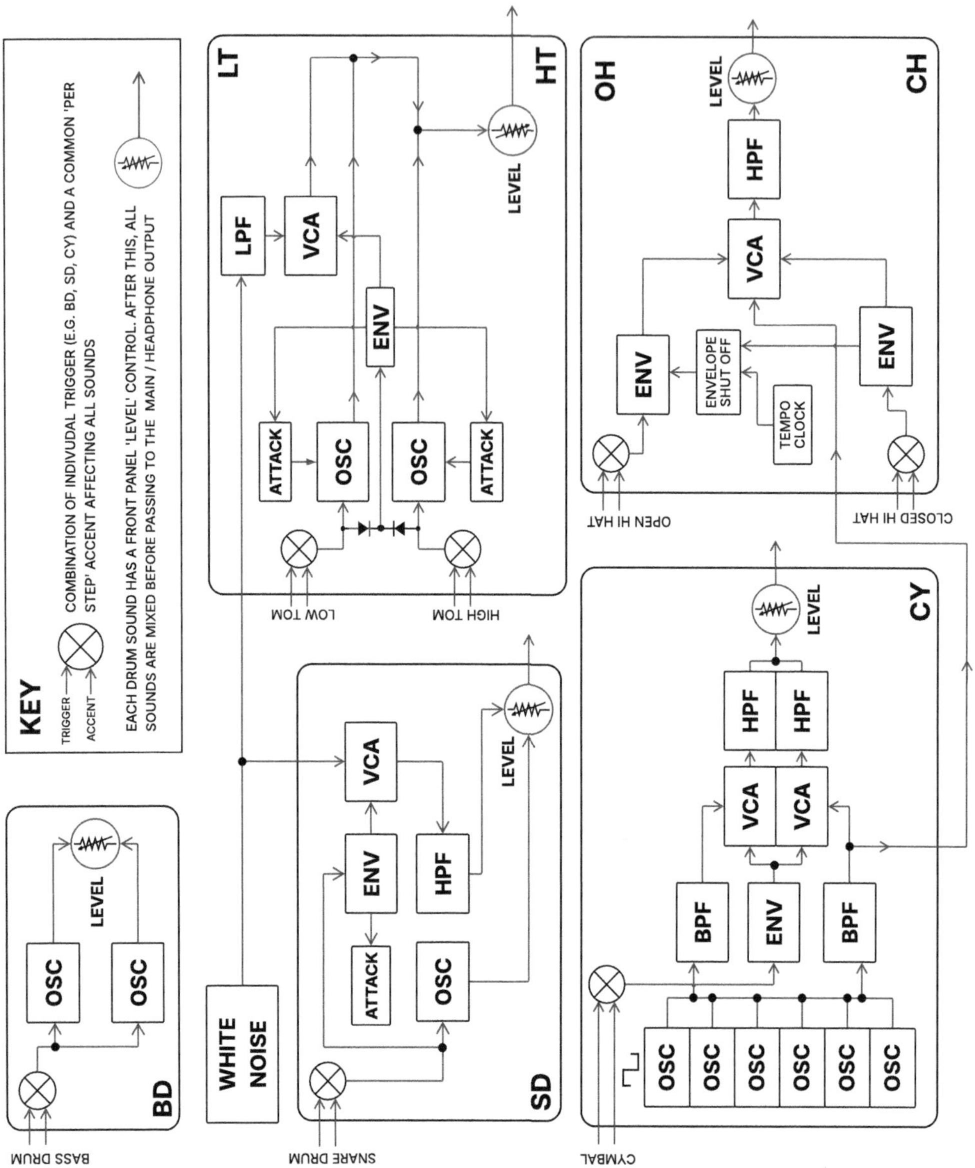

Figure 4.2 Roland TR-606 Simplified block Diagram (adapted from Roland 1982a TR-606 Service Notes).

TR-606 Drumatix

Released
1981 (discontinued 1984)

Sound source
All sounds generated by analogue circuitry
Bass Drum
Snare Drum
Low Tom
High Tom
Cymbal
Closed Hi Hat
Open Hi Hat

Real time controls
Individual Volume Controls for: Bass Drum' Snare Drum; Low and High Toms; Cymbal; Open and Closed Hi Hat
Overall Accent Level
No sound sculpting controls

Tempo
Approximately 40-300 BPM (no display)

Memory
32 Patterns (16 in Pattern Banks I and II)
8 Tracks (Songs)
Tracks 1-7 maximum 64 measures Track 8 maximum 256 measures

Connectivity
Line Level and Headphone audio output (monaural)
DIN Sync with Input and Output switch
Trigger Outputs echo programmed Low Tom and High Tom patterns
Run/Stop Pedal input

Power
9V power supply or 4 x C-type batteries

Dimensions
300 (W) x 146 (D) x 55 (H) mm
Weight
3.7 kg

Notes
The TR-606 is the only x0x rhythm instrument that can be switched between 'Play' and 'Write' modes without stopping the sequencer
It is the only 1980s TR-x0x instrument not to bear the name 'Rhythm Composer'
When the Open and Closed Hi Hats are triggered together, a third 'half open' sound is generated

Suggested listening
Sisters of Mercy 'Alice' (Merciful Release,1982)
Drexciya 'Rublick's Cube' (Underground Resistance, 1996)
Erika 'The Filtered Sun' (Interdimensional Transmissions, 2020)

Nothing Sounds Quite Like an 808

The TR-606 is perhaps the most enigmatic of the xOx rhythm instruments. Certainly, it is the most frequently overlooked. As a drum machine, it has always, in some way or other, been overshadowed the TR-808. When it was originally released in 1981, the TR-606 was seen as a low-cost derivative of the bigger, more powerful instrument whose analogue drum synthesis technology had been cut down and stripped back almost to the bone. Compared with the TR-808, the TR-606's diminutive size and obviously cheaper construction clearly marked it out as a different kind of instrument. Metal gave way to plastic, and the host of sound shaping controls were discarded leaving just a handful of volume knobs. Although, as we shall continue to see throughout this book, Roland's xOx numbering system is not as straightforward as it might at first seem and rarely indicates a neat relationship between instruments, for some the connection is clear. For instance, Dave Crombie's preview of the instrument from the Japanese Music Fair notes that, 'The TR-606 is, as the model number implies, a scaled down version of the TR-808' (Crombie 1982: 12). Even today, the TR-606 is frequently described as the 'little sister' (as Hub 2015: 191 put it), or as an affordable alternative to the ever more expensive TR-808, implying that, while it might do service, the instrument is not exactly what one would really want if sufficient finances were available.

Over the decades, the extraordinary success and continued valorisation of the TB-303 Bass Line synthesizer has also cast another long shadow over the TR-606. Despite an initially lukewarm reception, by the late 1980s, the TB-303 would find favour as it provided the archetypal sound of Acid House. Over the decades, it would go on to become a near-fetishised staple of electronic dance music as we shall explore in the next chapter. However, it is important to remember that, for all its fame, the TB-303 was originally designed and released as an accompaniment to the TR-606, built in its image and sharing its distinctive, albeit rather inexpensive and plasticky, look and feel. Today, as commentators seek to trace the history of the TB-303, the TR-606 is often recast as an interesting footnote or relegated to the role of a somewhat curious companion piece designed and released alongside the more influential and important instrument (Dias 2020). As such, when the TR-606 is spoken about, it is often defined in relation to another instrument whether the TR-808 or as '1981's TB-303 sibling' as journalist and critic Peter Kirn (2020) put it in features on the TR-606. To complicate matters, unlike the TR-808, TB-303 and many other instruments we will examine in this book, the TR-606 was not a high profile or spectacular failure on release as, in truth, the aspirations for it were never that it be a flagship instrument, nor did it ever enjoy a meteoric resurgence in popularity. As Kirn also notes,

> It didn't define a genre, the way the 303 did with acid or the 808 and 909 did with hip-hop and techno and house. It has star owners – Massive Attack, Autechre – but one wouldn't necessarily associate the box with them. The 606 is more like a chameleon. Its dry, wet, distorted, modded, circuit-bent sound shows up everywhere from punk to IDM.
>
> (Kirn 2020)

It would be easy to simply disregard the TR-606 as an unremarkable instrument, especially in the company of such illustrious and well-documented family members. Yet, while some aspects of its early years resonate with the taken-for-granted assumptions about the fates and reception of other Roland instruments, the TR-606 has an absolutely fascinating history that is anything but straightforward or unremarkable. Seemingly eschewing the limelight, the TR-606 bubbled along in the background while other x0x instruments fell and rose ... and rose. It is possible to argue that, although its overall sound is notably metallic and perhaps even harsh at times, it is precisely the lack of a singular, definitive 'signature' sound that has bestowed upon the TR-606 a flexibility and adaptability that has enabled it to find a place in a wide variety of musical contexts.

Of course, none of this should lead us to think that the instrument is not well-loved. As its history unfolds, we will find bands for whom the instrument is so central that it is given a name and treated like any other musician in the group. We will find it presaging a new era of portable electronic music making devices and, in the spirit of convenience, we will even encounter a future Nirvana producer wearing a TR-606 around their neck, programming and listening to it as though it were a Walkman. While this sight might be an unexpected one, the TR-606 is well-suited to the task. Not only is it the smallest of Roland's x0x drum machines, it can also be battery powered for portability. Most importantly, it is designed for performance unlike any other x0x instrument before or since, even the TR-808 and TR-909!

Despite finding an enthusiastic and loyal group of users, the TR-606 was not a mainstream success and Roland discontinued the instrument in 1984. The recent release of both analogue and digital recreations based around the TR-606's circuits in Eurorack-compatible sound modules, as well as hardware replicas, clones and emulations that capture the sound along with the look and feel of the original, has certainly increased the visibility of the instrument. However, even this has not seen the prices of 1980s TR-606s skyrocket in the same manner as TR-808s, TR-909s and TB-303s. We might even suggest that the popularity and recognition of those more celebrated instruments offers an opportunity to reappraise the TR-606 as 'now that you've heard way too many 808 and 909 hats, you might appreciate this just for some variety.' (Kirn 2019). And, where those other instruments might be almost too precious and valuable to use let alone tinker with, the comparative availability of the TR-606 has led it to become one of the most comprehensively modified drum machines. Some of these 'mods' seek to (re)instate features and functions apparently excised in the process of creating an affordable instrument. Others, however, add entirely new features and a level of sound sculpting control that exceeds the capability of many of the instruments that have overshadowed the TR-606 in the popular consciousness, and even hint at the level of drum synthesis control Kikumoto had originally envisioned for the TR-808.

Less is More

For anyone familiar with drum machines or even Rhythm Composers of the 1970s leading into 1980, the TR-606 immediately stands out. It is small. At just 300x146x55mm, it is a remarkably compact instrument. It is entirely plastic construction means that it is also extremely lightweight, tipping the scales at almost 1.5kg less than the far larger, metal-cased TR-808. That instrument had already proved that drum machines did not need to be clad in wood or require the skills of a cabinetmaker to assemble them. However, the TR-606's silver case and matching silver knobs, buttons and switches still manage to appear like a product of the space age and restrained at the same time, especially when compared with the rainbow colours of previous Roland instruments like the TR-808 and CR-78, for instance. Continuing the theme of understatement, all the legends and labels indicating the functions of the interface elements are written in black directly on the silver case or in silver on sections that are painted black. One breach of the monochrome design language is a single row of numbers underneath the 16 step/pattern buttons. These legends annotate one of the multiple functions performed by what are now known as the 'Selector Switches'. In this case, choosing the length of a pattern from 1–16 steps which are written in an orange text that directly references the palette of previous generations of Roland instruments, and is, ironically, rather difficult to read on the black painted background of the lower section of the TR-606.

Among the other text on the front panel, the design of the TR-606 continues the trend of communicating technological currency by proudly announcing that it is 'Computer Controlled'. Indeed, while the sound generation circuitry remains resolutely analogue, the TR-606's sequencer, programming and memory management functions are all handled digitally by a NEC D650C-128 integrated circuit. Altogether more unusually, however, in addition to the TR-606 model number, this instrument is given a name. Drumatix, the derivation of which is explained nowhere in any of Roland's literature, must be assumed to be a pun or portmanteau of drum and dramatic. Certainly, the name elicited a less than positive reaction in the introduction of Tony Bacon's (1982a) review of the instrument in *Electronics & Music Maker* magazine which acknowledges 'the so-called "Drumatix" (groan)' and then proceeds to refer to it as the TR-606 from then on. Indeed, the TR-606 is unusual in the entire history of Roland drum machines in having a name in addition to the more familiar Letter-Number identifier. If Boss' earlier 'DR-55 Dr Rhythm' had started a trend, it was one that did not last for long and no other instruments in this book were given such an honour. Even Roland's other instrument ranges were only sometimes named. Synthesizers like those in the Juno and Jupiter ranges sit alongside SHs and MCs. Perhaps we might consider the TB-303 'Bass Line' to be a name rather than a functional descriptor but, even then, it has none of the futuristic whimsy of Drumatix.

Roland's US marketing and promotion gave a sense of the instrument's essence describing it as 'a totally programmable drum synthesizer/sequencer.' The Owner's Manual goes further in capturing some more detail of the instrument.

> The Roland TR-606 is a portable, economical, automatic and fully programmable rhythm device, which can memorise your own rhythms with exceptional flexibility. With it you can write and record different percussive rhythms into non-volatile memory (battery back-up), with up to seven different drum sounds plus accent. Then you can adjust the volume of each percussion sound with individual level controls for full flexibility over the final mix.
>
> (Roland 1981b: 3)

In short, it is programmable, portable and cheap. In a world that had been dominated for so long by preset instruments, being able to program original patterns was still something of a novelty. Roland's US advertising copy in its 1982 'Rhythm Machines' brochure was keen to emphasise just this point and separate the TR-606 from the previous generation of preset instruments. 'Unlike conventional rhythm machines with preprogrammed patterns, this unit is capable of memorizing any rhythm the user wishes to create.' (Roland 1982b: 3).

Perhaps 'any rhythm' is stretching the point a little, especially given the reduced set of sounds available compared with the TR-808 and even the CR-78, but it is certainly the case that the TR-606 was still one of only a handful of Roland/Boss instruments to offer programmability. What is missing from the advertising copy and, indeed, from the front panel of the TR-606 itself, however, is the phrase 'Rhythm Composer'. As we have seen, this soubriquet was a key means by which Roland communicated not only its shift from preset to programmable instruments but also, crucially, from drum machines based around programming single patterns like the DR-55 and CR-78, to one that could chain together strings of different patterns in order to define the structure of a complete song or performance. In addition to its Intro and Fill-In functionality, the TR-808's Track Write and Track Play modes in which songs could be assembled, edited and recalled, helped to transform the instrument into a complex compositional tool. We could be forgiven for thinking that the lack of a 'Rhythm Composer' designation of the TR-606's case is a consequence of the instrument being a cut-down version of the TR-808. Perhaps in the desire to produce a lower-cost version of the instrument, this is a feature that was too costly to implement and did not make the final cut. And yet, looking at the instrument's front panel, we find a four position 'Mode' switch with settings for 'Pattern Play', 'Pattern Write' as well as 'Track Play' and 'Track Write'. The TR-606 is clearly more than it at first seems.

In fact, delving deeper into the US Owner's Manual, we find that there is very little consistency in the naming of the instrument with 'TR-606', 'Drumatix', and the altogether more generic 'rhythm device', used almost interchangeably. Intriguingly, this English language version of the Owner's Manual even occasionally refers to the TR-606 as a 'Rhythm Composer' despite that designation appearing nowhere else in Roland's literature or on the instrument itself. For the TR-606, only the model number and Drumatix name appear on the front panel making this unique among x0x drum instruments. As we shall see, TR-505, TR-707, TR-808 and TR-909, as well as the many variations thereof, not only have no other name beyond their x0x designation, but also all proudly bear the 'Rhythm Composer' descriptor on the instruments themselves.

Regardless of what it was officially (or even mistakenly) called, it is clear that TR-606's functionality was understood and welcomed by contemporary reviewers. Indeed, Bacon's comments are particularly revealing in this regard.

> Why are there so many pattern locations, you ask? Well, the whole point of the 606 is its ability to use the patterns linked together, forming complex rhythm compositions to fire your music and making the machine a composing tool rather than merely a rhythm box.
>
> (Bacon 1982a)

It is worth noting here that, while the TR-808 offered 12 Basic Rhythm patterns, each of which could have two Parts and completely separate A and B variations (along with four separate Fills/Intro patterns), the memory of the TR-606 allowed storage of 32 separate patterns. This might not seem like 'so many' to modern players or even to those familiar with some of the more capacious instruments of the time like the LM-1 which offered 100 memories, it is evident that, upon its release, the TR-606's storage provision was considered sufficiently impressive to warrant comment. This was especially the case in an instrument so comparatively inexpensive. Where the LM-1 had initially retailed at $5000, and Kakehashi and the engineering team at Roland had tried to keep the TR-808 below $1000, the TR-606 was listed by UK retailers at £199 or less on its release.

Writing and Playing

Although its physical appearance, form factor and even its name make it feel like a very different instrument, the TR-606 inherits a great deal from the TR-808. The 16-button interface is translated almost wholesale and programming will be immediately familiar to anyone used to working with the TR-808 (or the Eko Computerhythm, for that matter). Unlike the LM-1 which offered multiple pads that could be played simultaneously in real time in a manner that directly connects to what we might now know as 'finger drumming', TR-606 patterns are built up one sound at a time. With the instrument in Pattern Write mode, each of the 16 Selector Switches represent a single subdivision of a bar. A four-position switch allows a different 'Scale' to be set for each Pattern thereby allowing individual steps to represent semiquavers or triplets, for instance. In combination with a setting to indicate the 'Last Step' of a Pattern (taking advantage of those hard-to-read orange legends), sequences can be written in different time signatures such as 3/4 and 6/8. The default Scale and Last Step settings give a bar of 4/4 with each of the 16 Selector Switches representing a semiquaver.

With only a single row of Selector Switches, the eight-position 'Instrument Select' dial is used to select a specific sound whereupon the LEDs above each Selector Switch illuminate to show the arrangement of notes and rests already programmed. Through the combination of the Instrument Selection control and Selector Switches, notes can be added and removed while the sequencer is running making it possible to build up complex patterns. In addition to the step programming method, a TAP button located to the right of the row of individual step buttons allows rhythmic sequences to be played in real time. With the appropriate sound selected, pressing the TAP button adds steps to the grid with any imprecision in timing seeing notes quantised to the nearest subdivision (based on the Scale setting). Roland's method for tapping in rhythms bears something in common with the LM-1 although in the world of the TR-606, live performances are built up one sound at a time. However, unlike the Linn method, notes programmed in real time immediately appear illuminated in the step interface and, therefore, may be moved or removed just as if they had been entered directly onto the timeline. So far, so TR-808.

In truth, there are some important differences. Although the hard plastic TAP button of the TR-606 does not offer quite the tactile feel or travel of the one on the TR-808 or the pads of the LM-1, it is serviceable. It does, however, speak to the rather less luxurious design of the TR-606. The economy of the hardware interface becomes further evident in use. There are comparatively few controls on the front panel. One positive result is that, despite the size of the case, the layout of the interface does not feel cramped. That said, the reduced number of hardware controls might give rise to an instrument that is more efficient and cheaper to produce, and might allow all this functionality to be packed into a considerably smaller case, but it also means that some features become less immediately intelligible from the front panel alone.

The presence of a Function button which acts as a 'Shift' key provides the means of unlocking some of the interface's secondary functions but even here, there are some interesting quirks of the operating system as functions are almost always dependent on the state and status of the instrument. For instance, setting the Last Step of a pattern involves ensuring the instrument is in Pattern Write mode, pressing and holding the Function button and selecting the appropriate Selector Switches. While this might just about be intuited from the colour coding of the 'Last Step' and the orange 1–16 text labels under the Selector Switches, it is not clear that this action can only be performed while the sequencer is running. Similarly, clearing a pattern is achieved by holding the relevant Selector Switch and pressing the 'Pattern Clear'. This, however, can only be undertaken when the sequencer is not running. With the sequencer running, though, holding the Clear button can be used to erase steps from the currently selected sound's sequence as they play as a kind of real time eraser. It might be unfair to call the interface confusing and, obviously, continued use brings familiarity with its vagaries. However, the key combinations, dual functions and the impact of the different Modes of operation means that a number of operations appear hidden or hard to unravel without recourse to the Owner's Manual.

Like many of the Owner's Manuals that accompany Roland's x0x instruments, it is fair to say that the English language version of the TR-606 guide is detailed and thorough. Whether it is the most accessible document is rather more debatable, however. Aside from the brief initial introductory notes, the manual is split into two sections. Or rather, it is split into two 'Courses'. If this terminology conjures images of lessons, homework and the possibility of an examination at the end, reading the manual does little to dispel the impression.

> This Basic Course will provide essential information needed to satisfactorily operate the TR-606 Drumatix. For complete information and more complex applications, refer to the Advanced Course Manual.
>
> (Roland 1981b: 4)

True to its name, the Basic Course does, indeed, start with the fundamentals. Labelled images aid identification of the various sections and controls and describe the variety of ways that the instrument can be powered and connected to external amplification, speakers or headphones. Things soon get considerably more complex, however and just a few pages into the discussion of Pattern Writing in the Basic Course, we find diagrams illustrating the translation of musical notation into the programming language of the instrument. Much of the initial description in the manual is oriented around explaining a cascading process of translating a Rhythm Score into a Transcribed Score. Both Score are represented in Western musical notation but differ in terms of the way notes and rests are visualised. A further translation turns this into what Roland's calls a Program Score which notates the rhythmic information as a series of step settings ready for inputting via the 16 Selector Switches on the instrument's front panel. If this method sounds familiar, many of the instructions in the TR-606 Owner's Manual are duplicated verbatim from the TR-808 documentation including the somewhat convoluted explanation of the translation of musical notation into the instrument's programming procedures. Of course, while Roland's multi-stage translation process is already conceptually complex, it is important to note that deciphering the operation of the instrument from the Owner's Manual assumes considerable a priori knowledge of the concepts and orthographic conventions of Western musical notation, and stave-based percussion notation in particular.

As the TR-606 is a 'Rhythm Composer' in all but name, the discussion in the Owner's Manual quickly moves beyond the creation of discrete bar-long patterns to cover the process of assembling them into full Tracks. The TR-606 offers eight Track memories for assembling collections of patterns. With the instrument in either Track Write or Play mode, the eight-position rotary control shifts mode from Instrument to Track Selection. Tracks are not identical, however. The first seven allow combinations of up to 64 Patterns while Track 8 has four times the storage and can store up to 256. In fact, the first seven Tracks can also be combined in different ways to create fewer but longer sequences if required. Patterns are selected from either of the two Pattern Groups (I and II) and can be memorised in any order to build up the composition. While individual Patterns can be arbitrarily triggered using the Selector Switches while the instrument is in Pattern Play mode, the ability to create, edit and play back long-form compositions with form and structure are clearly significant features of the instrument. This flexibility, which extends to being able to store and play up to eight 'songs' without requiring any further human input, enables the TR-606 to be used in a variety of rehearsal, accompaniment, live and studio contexts. However, while this functionality is undoubtedly powerful, its implementation, or at least the description of it in the manual, is not exactly straightforward.

> Start the rhythm by pressing RUN/STOP
>
> The rhythm previously recorded in the 1st bar of the Track will be heard.
>
> Choose a new 1st bar Rhythm Pattern by pressing a SELECTOR switch, in this case #1.
>
> The new Rhythm Pattern will play at the end of the old rhythm, and will repeat.
>
> The running Rhythm Pattern will be recorded in the Track Memory as the 1st bar, and the Track Memory is advanced to the 2nd bar, and it will play the previously recorded old rhythm.
>
> You can re-select a new 2nd bar Rhythm Pattern and then press the TAP button to record it in Track Memory.
>
> Continue switching Rhythm Patterns and logging them into Memory by pressing the TAP button.
>
> (Roland 1981b: 14)

It is worth bearing in mind that this excerpt is taken from the Basic Course and there is significantly more complexity added in the procedures and features set out in the Advanced Course. It is also notable that, just as we saw in the overview of Pattern creation, the description of Track Writing in the Basic Course assumes familiarity with concepts and symbols of modern Western musical notation including *da capo* and *dal segno*. In fact, the abbreviations D.C., D.S. along with the segno symbol actually appear on the front panel of the TR-606 itself as secondary functions of the Pattern Clear, Tap and Pattern Group buttons respectively. *Da capo*, usually translated to mean 'from the beginning' and literally translated as 'from the head' is a term directing a musician to repeat the previous section of a printed score, usually from the beginning. *Dal segno*, literally 'from the sign' is another navigation maker used in Western scores that indicates a passage of music should be repeated from the point marked by the sign. The reason for the creation and use of these navigation markers and symbols in sheet music is to save space by avoiding the need to reproduce identical sections of a printed score. The reason for their appearance in the Owner's Manual and on the front panel of a drum machine is rather similar although, instead of saving paper, here they save computer memory.

> The D.S. function is applied by designating one bar in the Track as the sign Bar. This sign Bar is the bar that the Track goes to when the TAP button is pressed during TRACK PLAY. The sign Bar is usually the first bar of the verse or chorus that is to be repeated.
>
> Once the Track returns to the sign Bar and repeats the verse, it continues on through the end of the song unless the TAP button is pressed again. The D.S. Function can be activated at any time during Track play, and can be activated either by the TAP button.

> At the end of the song, there is another designated bar, the D.C. Bar The D.C. Bar should be the last bar in any Track. After the D.C. Bar plays, the Track automatically resets to the first bar of the Track, and plays all over again. There can be only one sign Bar, and one D.C. Bar for each Track.
>
> (Roland 1981b: 30)

It may not come as a surprise to learn that we are firmly in the Advanced Course here, but complicated though the procedure might seem, essentially both the D.S. and D.C. features are describing different kinds of looping. D.C. allows the entire Track to be looped or replayed from the beginning while D.S. allows a specific set of Patterns to be identified for looping during Track Playback starting from the measure designated with the 'Sign' and initiated by pressing the Tap button. For all its complexity, the benefits are considerable. With seven of the eight Tracks comprising a maximum of only 64 measures, D.C. and D.S. allow sections of music to be repeated while minimising additional storage requirements. What is so fascinating about this implementation and the description of it in the Owner's Manual is that it represents a collision of memory saving pragmatism, Western sheet music notation and the conventional structures of pop/rock songs. As the manual suggests, *dal segno* can be put to use to repeat the verse/chorus section of a song or perhaps even the ABACAB structure of 12-bar Blues (where the capital letters describe song sections A=Verse, B=Chorus, C=Bridge). In this way, the operation and design of the TR-606 not only requires a priori knowledge of aspects of classical music theory and orthographic convention but also speaks to an implied use of the instrument in the context of popular music genres.

The flexibility of Track Writing and the ability afforded by the D.S. and D.C. features that eke out lengthier song structures from the comparatively limited memory makes it even more surprising that Roland denied the TR-606 the title of Rhythm Composer. Certainly, these capabilities were noted when the instrument was released. However, while their usefulness was recognised, contemporary reviewers also suggested that the complexity of the functionality and the sparseness of the interface required a systematic and decidedly non-computerised approach.

> to get the best from this lovely new machine, you really have to use the old, old method of pen and paper to write down rhythms. To get the most from it you'll have to commit thoughts to paper at an early stage, otherwise you'll get into an awful mess. Whether you go the whole hog and use proper drum notation or your own code is up to you. But write you must.
>
> (Bacon 1982a: 19)

Ironically, what Bacon suggests here is that to fully harness the power of the instrument, it is advisable to map out Patterns and Tracks on paper in advance of entering the data into the TR-606. In other words, the user should use pen and paper to write out Rhythm and Transcribed Scores in order that the Programe Score information can be entered into the TR-606 via the Selector Switches. Just like Roland suggested! Perhaps this is why it is sometimes claimed that 'Most TR-606 users stay in Pattern mode and never venture into Track Mode' (Social Entropy 2013: 2). However, for those who craved more than a single Pattern but did not want to venture into the depths of the Track Writing process, the TR-606 had another trick up its sleeve.

With the instrument in Pattern Play mode, it is possible to play up to four consecutive Patterns in a continually looping sequence. Referred to by Roland as 'Chaining the Rhythm Patterns', this opens up a range of creative possibilities. In combination with the Scale feature (which alters the number of crotchets [quarter notes] in a Pattern), this allows adjacent patterns to be joined to create a single bar with step resolution of a semidemisemiquaver (64th note). In combination with the Last Step feature, chained Patterns can be crafted to create a single measure in complex time signatures such as 5/4 (one Pattern of 4/4 followed by a Pattern of 1/4, for instance). Alternatively, four 16-beat Patterns can be chained together to create a mini version of a Track (with its own built in, de facto *da capo* looping feature). Compared with the complexity of the Rhythm Track procedures, chaining Patterns is remarkably simple. While holding down the Selector Switch of the first Pattern and pressing another to set the end, a chain is set that continuously loops between the two including any in between. As a live performance feature, it is extremely useful and allows the user to escape the tyranny of the single Pattern accompaniment and incorporating fills or variations.

There is a noteworthy limitation in the Pattern Chaining feature, however. A consequence of the memory architecture of the TR-606 means that Patterns are actually split into four 'Blocks'. The Owner's Manual describes the arrangement with Patterns 1–4 in Block I, 5–8 in Block II, 9–12 in Block III, and 13–16 in Block IV. Importantly, only Patterns from within a single Block may be chained. Thus, Patterns 1–4 may be chained but not 3–6 meaning that some planning is required. Nonetheless, this restriction notwithstanding, Pattern Chaining is important in adding flexibility to the instrument through a remarkably easy to use interface. It also means that, while the TR-606 appears, at first blush, to be missing some of the complexity of the TR-808, with its A-B Variations and two-Part Patterns, this is actually not the case and precisely the same four-bar looping patterns are perfectly possible. What is missing, however, is any indication that this is possible on the front panel of the TR-606. Where the TR-808 had dedicated hardware controls to switch between or set Variations and Parts, the TR-606 front manual is less forthcoming, with the existence and arrangement of Banks only listed in the Owner's Manual.

Not everything that the TR-606 inherited from the TR-808 was as positive. Continuing the transfer of DNA between the two instruments, the TR-606 sequencer remained resolutely and metronomically locked to the grid and did not include a shuffle or swing feature. With Shuffle being the feature so often attributed to bestowing the celebrated 'human feel' to the Linn LM-1, its continued omission in Roland's instruments is surprising. However, although it may not have been able to simulate the looseness of the Linn, the TR-606 did include an Accent feature. As with the TR-808 and CR-78, the TR-606 Accent is applied globally rather than to individual sounds and so when an Accent is set on a particular step, its effect is applied to all of the sounds that play. Moreover, there are no degrees of Accent and the feature is either active or inactive per step.

Despite the apparent limitations, however, and although it does nothing to affect the timing of the strictly quantised notes, the Accent can break the overtly machine-like (or perhaps even machine gun-like) monotony of consecutively triggered notes. Through judicious use of Accented and non-Accented-steps, a considerable amount of variation of intensity can be introduced. Kakehashi has spoken about the importance of the Accent feature in discussing its inclusion in the CR-78 stating that, 'I built many rhythm machines previously, but none of them had accents. Rhythm with no accents? Unbelievable! Nobody would buy it.' (Kakehashi in Vail 2000: 282).

Clearly, the importance of the Accent was such that it made the jump to the TR-606 even where other features had been stripped away in the service, reducing the cost and size of the instrument. Of course, for those favouring the machine (gun) like quality of the instrument, the Accent can either be ignored or set for every step. And because these were analogue circuits built from components triggered by voltage, applying the Accent did not simply increase the volume of sounds, it also changed their timbre in subtle, and sometimes not so subtle, ways.

Nothing Sounds Quite Like a 606

On first inspection, the TR-606 might seem like something of a disappointment. The instrument offers fewer than half the number of sounds than were available on the TR-808. There is a Bass Drum, Snare, High and Low Tom, Open and Closed Hi Hat and a Cymbal but gone are the Congas, Maracas, Cowbell, Rim Shot and even the Hand Clap so beloved of Don Lewis and so central to the development and distinctiveness of the TR-808. As such, the TR-606 palette is stripped back to the barest fundamentals of a four-piece pop/rock drum kit with none of the additional percussion embellishments or even a Mid-pitched Tom. Interestingly, though they both offer more sounds, Roland's later TR-707 and TR-909 Rhythm Composers continue the trend set by the TR-606 in focusing on the core rock pop/drum kit at the expense of other percussion tones.

Coupled with its diminutive size and plastic housing (and name and number), it is surely the case that the comparatively limited range of sounds must have been the cause of some of the unfavourable comparisons with the TR-808 and even the CR-78 both of which had larger and more varied arrays of sounds in their arsenal. However, few though they may be, the sounds of the TR-606 are unique and utterly unmistakable. Like its predecessors, the sound generation circuits in the TR-606 are entirely analogue but, unlike the smoothness and sheen emanating from those previous instruments, there is an aggressive, abrasive quality to the tonality of the silver box that is quite unlike that of any other Roland drum machine. It is somewhat ironic given the instrument's plastic construction that the overall sound of the TR-606 has an undeniably metallic quality to it. Sonically, at least, this is far from simply a cut-down TR-808.

Much of this abrasiveness of the TR-606 comes from the sound of its Hi Hats and Cymbal. Perhaps this should not surprise us too much. After all, these are emulations of percussion instrument made from metal. However, there is a high frequency sizzle in these TR-606 sounds that some commentators (including Roland) refer to as 'crisp' (Kirn 2020) but which could be said to approach harshness. As we noted in the previous chapter, recreating the sound of cymbals with analogue synthesis circuits is notoriously challenging. Real cymbals sit somewhere between pure noise and harmonically pitched sound with wide variations over time and as the result of different sticking materials and techniques. As such, many early approaches to synthesising cymbals started with white noise as it comprises a random signal containing different frequencies at equal intensities. By contouring the amplitude (and sometimes the frequency range through filtering) to simulate the natural decay of the cymbal as it rings out to silence, an approximation can be achieved though the effect tends to exhibit a whooshing or hissing sound.

To create something more clangourous and metallic, Roland's circuit eschews white noise in favour of a mix of six high pitched square waves generated by a Hex Schmitt trigger. As can be seen in Figure 4.2, this circuit is clearly inherited from the TR-808 with the six square wave oscillators each individually tuned and mixed to create a highly dissonant sound that is altogether more metallic than a whoosh of white noise. Where this atonal cluster did triple duty in the TR-808 by providing the sonic foundation for Cymbals, Hi Hats and, in part, the infamous Cowbell, the reduced palette of the TR-606 places less burden on the circuit (see the CY, CH and OH sections in Figure 4.2). However, efficiency is clearly still the watchword and the 'metallic noise' circuit still forms the basis of the instrument's Hi Hats and Cymbal with each individual sound being created through further filtering and contouring to accentuate certain frequencies and varying the duration or decay to silence.

By passing the metallic noise through resonant bandpass filters, specific frequency ranges can be made more or less prominent with distortion adding further emphasis by adding overtones. While there is a fundamental similarity in conceptual approach and circuit design of the Hi Hats and Cymbals in the TR-606 and TR-808, the reduction of components and variations in calibration and user control of tone mean that the resulting sounds are markedly different. As we can see from the TR-606 'CY' circuit, there is considerably less complexity in the processing of the square waves than in the TR-808 design. In the TR-808, the circuit creates a complex chain of processing that initially splits the cluster into two discrete frequency ranges and then feeds them in parallel through three separate VCAs with their own envelopes so that different aspects of the sound decay at different rates. The TR-606 takes a similar but more economical approach that still uses Band Pass Filters to split the square wave tone into two frequency ranges but significantly reduces the complexity of the contouring by using just a single envelope to control the level of the sound as it decays to silence. The sonic impact of the differences in these circuits is palpable and where the TR-808's Hi Hats have a full-bodied smoothness to them, the TR-606's are thinner, more strident and raspy, and perhaps even harsh. As the Cymbals decay, the complexity of the TR-808 circuit is evident in the interplay between the various frequency bands which is absent in the TR-606. There is also a noticeable bite to the attack transient of the TR-606 Hi Hats which, again, contrasts with the comparative softness of the TR-808 equivalent.

While listening to the sounds in isolation is telling, the TR-606 Hi Hats and Cymbal really reveal their unique character when they are sequenced and begin to interact with one another. As with the TR-808, though there are Open and Closed Hi Hat sounds which are primarily differentiated by the length of their decay into silence, both sounds share a single note of polyphony. This means that only one sound can play at a time. Thus, if an Open Hi Hat is triggered on Step 1 of a sequence, a Closed Hi Hat triggered on Step 2 will immediately cut off and replace the Open Hi Hat sound thereby interrupting its decay. This emulates the effect of a real Hi Hat and has become a standard feature of drum machines which now refer to such behaviour as 'choking' and often allow any combination of sounds to be placed in 'choke groups' to enable a host creative effects. In the case of the TR-606, as there are actually two distinct circuits processing the metallic noise, each of which have their own separate triggers as well as their own imprecision, the interaction of the Open and Closed Hi Hat results in an even more raucous noise with successive triggers appearing to generate yet more atonal harmonics that build upon and fight with one another. The interplay between these two sounds as they are repeatedly triggered and choke one another greatly adds to the abrasiveness of the TR-606 and, importantly, reminds us of the dynamism and inherent non-linearities and instabilities of analogue circuits. Indeed, pushed to extremes, the high end frequencies take on an almost digital quality that resembles clipping, aliasing or Bit reduction adding a flavour quite unlike the smooth, softness of the TR-808.

It should also be noted that, while there are only two Hi Hat sounds listed on the front panel of the TR-606 and they do form what we might now call a 'choke group', it is actually possible to trigger both sounds at the same time. This situation arises because, while it is true that the Closed Hi Hat will interrupt and replace the decay of the Open Hi Hat if triggered on the following step, each sound actually has its own entirely independent programming lane and separate audio signal path. This means that an Open Hi Hat and Closed Hi Hat can be placed on the same step. Indeed, this means it is possible to program a full 16 steps of Open Hi Hat and 16 steps of Closed Hi Hat to play simultaneously. Perhaps we might expect some logical operation that would give the Closed Hi Hat preference meaning that only the Closed Hi Hat would be heard but, in fact, the result is actually a third Hi Hat sound. Setting aside this philosophical debate as to whether this Hi Hat is half open or half closed, it is nonetheless a different tonality, that actually has appreciably more punch than either the Open or Closed Hi Hats from which it is derived. This is likely an adopted design quirk rather than a specifically implemented feature and is the result of the combined trigger voltages as the Owner's Manual intimates (albeit in cryptic terms and tucked away in the corner of a page). 'If you write OH upon CH, CH will gain Accent.' (Roland 1981b: 9).

With appropriate programming, the combination of the three Hi Hat sounds, each also differently reacting to the presence and absence of Accented steps, can create sequences possessing a degree of liveliness, dynamic range and energy that exceed what might be expected from the specification of the instrument.

Although the metallic percussion sounds use of swarms of carefully-pitched square waves, the TR-606, like the TR-808, does also contain a white noise generator and it is this that forms part of the basis of the Snare sound. In fact, as the SD section of Figure 4.2 shows, the Snare Drum comprises two elements that broadly simulate the body of the drum and the rattle of the snares themselves. The tonal element of the sound is created by another Bridged-T network circuit like that used in the TR-808 Snare. The design of this circuit gives a natural decay, the duration of which is governed by the values of the components and is not directly adjustable by the user. An Accented trigger, however, causes the circuit to resonate at a higher amplitude which slightly lengthens its decay again adding dynamism to the sounds beyond just changes in level. The second element of the Snare Drum sound is generated by applying a volume contour to the output of the instrument's common white noise circuit and a High Pass Filter to remove unwanted low frequencies and emphasise the highs. When triggered, this creates a burst of 'snappy' noise that, when summed with the Bridged-T component, creates a composite sound with a defined attack and release. The TR-606 Snare Drum is extremely punchy with a far shorter decay than that found on the TR-808. As such, like the Hi Hats, the TR-606 Snare Drum has a bite and aggressiveness that, if not quite unrefined, certainly has an untamed edginess to its hard-hitting transient.

In a similar fashion, the TR-606 Toms are composite sounds making use of different circuits. The tonal element gives the sounds their body with a variation in the fundamental pitch differentiating the Hi and Low sounds. Again, the white noise circuit is put to work here but, rather than contributing to the attack of the sound or even to its body, here it is used in an attempt to simulate the reverberant effect of a room so as to give the Toms a sense of location in space. There are some parallels here with the TR-808 Hand Clap which employed a 'reverb' envelope to add sustain to its ratcheting attack. For the TR-606 Toms, this is an interesting aesthetic decision. First, the effect is not wholly convincing and adds a detectable hiss even with a Low Pass Filter shaving off some of the high end.

Second, the other sounds in the TR-606 are conspicuously dry and are synthesised without even the suggestion of such nods to realism or spatial context. Like the Hi Hats, the two Tom sounds have their own independent sequencer lanes and can, therefore, be programmed to play separately and together. However, with just two Toms and no ability to alter their tuning, there is less opportunity for complex patterns of sequences than with the three sounds offered in the TR-808.

It is the Bass Drum, however, where the difference between TR-606 and TR-808 is most keenly felt. Again, the circuits of the two instruments are somewhat similar although the implementation in the TR-606 is subtly differently as can be seen in the BD section of Figure 4.2. As with the TR-808, the TR-606 uses the qualities of the Bridged-T network that, when excited by the trigger pulse from an active sequencer step, creates a resonant sound that rings out to silence. However, where the TR-808 used a single Bridged-T network with a circuit that generated a swooping pitch that fell from high to low in order to simulate the effect of an initial strike and decaying body of the drum, the TR-606 uses two. By tuning these independent Bridged-T networks to different frequencies and allowing them to contribute to different elements of the sound, the TR-606 circuit appears to offer an extremely elegant and flexible solution that combines both a solid, defined attack and a natural decay. Indeed, the design is reflective of Kikumoto's ongoing focus on synthesising the attack or transient phases of percussion sounds and it is certainly the case that the TR-606 has a punch that is absent in the TR-808. However, for all the initial kick, the tuning of the TR-606 Bass Drum oscillators and the comparatively short decay time defined by the amount of feedback in the Bridged-T networks create a sound that has little of the weightiness of its genre-defining counterpart in the TR-808. Where the TR-808 Bass Drum is boomy with the potential for an unrealistic degree of decay, the TR-606 Bass Drum is notably high pitched and truncated. In fact, it occupies a very similar frequency range to the Low Tom rather than underpinning it.

To exacerbate the problem, the TR-606 offers no sound shaping controls for any of its sounds. The Toms cannot be pitched to more clearly differentiate them from the Bass Drum or to tune them to the key of music they may be accompanying. The Bass Drum's comparatively high pitch is similarly fixed as is its decay while the snappiness and brittle quality of the Snare are calibrated in the factory and not available to be tweaked from the front panel. There is a row of small rotary dials along the top edge of the instrument in exactly the place where we might expect to find such controls, but these are, in fact, the 'Instrument Mix' controls that affect the level of the individual sounds. Compared with the TR-808, the lack of sound shaping controls clearly speaks to the comparative simplicity and economy of the TR-606 and contributes to its framing as a cut down version of a larger machine. However, it should also be noted that even the volume mixing capabilities of the TR-606 are more limited than the TR-808 which provided a dedicated volume control for each sound. In fact, even counting the number of 'Instrument Mix' controls on the TR-606 is potentially misleading as one actually affects the level or intensity of the Accent which is applied to all sounds simultaneously, and there are just two controls which affect the level of the combined High and Low Toms and Open and Closed Hi Hats respectively.

The lack of sound shaping controls and somewhat limited mixing is compounded by the fact that the TR-606 has just a single monaural audio output through which the combined output of all sounds is routed. In contrast with the TR-808 whose 'Multi Outs' provided external audio routing possibilities, the TR-606 offers no opportunity to further process its sounds individually or to externally adjust them through a mixing desk with outboard effects units. As such, there are no straightforward ways to compensate for the absence of sound shaping controls and mixing configuration. The TR-606 is, in this way, a far more self-contained instrument than the TR-808. The lack of low end in the Bass Drum might have found favour with mixing and mastering engineers reluctant to cut the booming decays to vinyl, yet the limited controls certainly create a far less flexible instrument that could be seen to take a step back towards the the CR-78 in offering programmable patterns but preset sounds.

In fact, it is not entirely true to say that there are no sound shaping controls as there are yet more unexpected features lurking in interaction between the Hi Hat and Cymbal circuits and the TR-606 sequencer. First, the duration of the Open Hi Hat and Cymbal are related to the sequencer's Tempo setting. By connecting the values of the envelope that fades the amplitude to silence, the duration of the two sounds is reduced as the tempo is lowered and vice versa. This has the effect of filling out or cleaning up the overall sonic space as the tempo is varied. However, while this might not be noticed in the usual course of play back, when the TR-606 is synchronised to an external clock source (e.g. to a sequencer or another drum machine), an unexpected level of control is revealed. This is because the envelope control of the Open Hi Hat and Cymbal decay is not connected to the sequencer tempo per se but rather to the instrument's Tempo controller. As such, when the sequencer is externally clocked, the Tempo knob of the TR-606 transforms into a dynamic Hi Hat and Cymbal decay control.

Less is Less: The Sound of One Hand Clapping

Regardless of these hidden affordances, contemporary reviews suggest that the TR-606's sounds were not an unequivocal hit with comments ranging from observations that the instrument 'can sound thin and weedy' (Mills 1985: 8) to the inevitable comparisons, 'The Drumatix sounds are better than average but not quite as powerful as those on the TR-808. The Bass Drum is a little weak but the Snare is good, the cymbals are reasonably metallic and the Toms are passable.' (Beat on the Cheap 1984: 34). For some, the particular atonality of the metallic percussion circuits were a little too abrasive. 'My only real criticism of the 606 is a personal one – I didn't like the cymbal and open hi-hat impersonations at all, and kept their "chinging" sound down low with the mix rotaries when playing back patterns and tracks.' (Bacon 1982a: 19).

In addition to the critique of the sounds that were present, attention is drawn to a sound notable by its absence. As the overview of the TR-606 in the January 1983 issue of *One Two Testing* magazine opines, 'I missed the 808's handclap sound (apparently it's very expensive to produce)'. (Roland Beatboxes 1983: 38–39). Given the lack of controls on the TR-606, we might think that little could be done to incorporate this sound so central to the TR-808. Roland had a solution, however, albeit one that relied on an additional purchase, but which made use of some powerful functionality of the TR-606.

As well as sporting one of the best product names, the Boss HC-2 Hand Clapper is a remarkably focused instrument. It is a single function Percussion Synthesizer dedicated to making the sound of a Hand Clap. With an analogue circuit at its heart, the sound of the Hand Clapper's Hand Clap sits somewhere between a TR-808 and TR-909. Either way, it is a classic Roland x0x-style tone that captures the nuances of the sound that Kikumoto had so disparaged. Housed in a metal chassis only a little larger than a Boss compact effects pedal but with a larger rubber covered section, the instrument is designed to be struck with a drumstick. It is essentially an electronic drum pad with a single built-in sound. In addition to being hit with stick or hand, the sound of the Hand Clapper can also be triggered by a voltage pulse sent from a connected sequencer … or drum machine. And, as luck would have it, the TR-606 is equipped with not one but two independent Trigger Outputs. Echoing the sequences of notes programmed into the Low Tom and Hi Tom lanes, these signals can be used to activate the HC-2 as well as the PC-2 Percussion Synthesizer which was housed in the same casing but produced altogether more space-age electronic sounds and disco-stye toms.

> Boss Hand Clapper and Percussion Synthesizer at £63 each. These two units can stand alone or be triggered from the pulse outputs of a Drumatix, for instance, which while it has some new friends hasn't been superceded.
>
> ('Roland Revelations' 1984: 13)

With 'Dry' and 'Hall' controls broadly affecting the tone of the Hand Clap sound and the intensity of a simulated reverberation effect created by extending the decay envelope of some of the noise generators in the circuit in much the same way as the TR-808 Hand Clap and TR-606 Tom sounds, the HC-2 was also available in kit form through Roland's Amdek division as the HCK 100 Hand Clapper Kit. Interestingly, although perhaps inevitably given its availability as a DIY project, a review of the HCK 100 in the February 1984 issue of *Electronic & Music Maker* details a number of modifications that could be made to further alter the pitch and tonality of the sound. This willingness to look inside the instrument and make adjustments to its circuits in order to explore the range of latent sounds within dramatically echoes one of the key uses of the TR-606 itself as we shall see below.

That Roland made an instrument dedicated just to the sound of the Hand Clap goes some way to underscoring just how significant a part of contemporary popular music that sound had become. It also makes the decision to omit the sound from the palette of the TR-606 an even more baffling one and causes it to hold the dubious honour of being the Roland xOx drum machine not to feature a Hand Clap. However, it also reminds us that the TR-606 was peculiarly well-equipped with interfacing capabilities. And so, while its rear panel may not have been bristling with separate outputs for the individual sounds, the instrument was equipped with the two Trigger Outputs we noted above as well as Roland's own 'DIN Sync' connection. This analogue communications protocol predates MIDI and allows compatible instruments such as drum machines, synthesizers and sequencers to play in sync with one another. Essentially, DIN Sync connectivity allows the TR-606 to be integrated into a complex electronic music studio, perhaps even operating as its beating heart by sending timing information around the system and triggering other instruments. Indeed, the Advanced Course in the Owner's Manual dedicates many pages to detailing a number of setups including synthesizers and sequencers. For a little box, the TR-606 packs a considerable punch in both its sounds and its functionality. However, even the implementation of DIN Sync betrays the influence of economic considerations on the design of the instrument as, rather than including the costly hardware required for separate In and Out sockets, there is just one that operates in combination with a switch that alters its function. Nonetheless, with or without a Hand Clap and notwithstanding the lukewarm reception of some contemporary reviewers and commentators, the TR-606 soon found a fanbase.

The Sound of the Underground

With its resolutely synthetic sounds and a sequencer whose dedication to machine-like precision makes absolutely no concessions to shuffle, swing or human timing, it will come as no surprise to learn that the TR-606 became popular with electronic music artists. It features prominently on Orchestral Manoeuvres in the Dark's 1983 album 'Dazzle Ships' where it is even listed in the liner notes and provides a characteristically brittle backbeat to Italian duo Victrola's 'Maritime Tatami' EP of the same year. Since these early outings, the instrument has continued to underpin the work of countless pioneering electronic artists. It can be heard on Platikman's 1993 'Helikopter' track and across almost all of 'Incunabula', Autechre's influential first album from 1993 where the noisy Snare and metallic Hi Hats lend a crispness to tracks like 'Bike' as well as 'Autriche' (see Bein 2020). Similarly, the TR-606 is near-constant in Massive Attack's 1994 'Protection' album where its sharply truncated sounds punctuate the open spaces in the arrangements, while Drexciya's 1996 'Rublick's Cube' is practically a paean to the instrument. More recently, performers and DJs such as Erika have not only made extensive use of the TR-606 in live and studio contexts but have also spoken of their continued attachment to the instrument having learned to program with it (see Kirn 2020). Speaking to the instrument's influence and the depth of affection felt towards it, we find it providing inspiration for artist names such as Kid606 as well as track titles such as '606' from 808 State's 2002 'Outpost Transmission' album (a group who had already named themselves after a Roland Rhythm Composer).

One of the most celebrated examples of the TR-606 in dance music tracks must surely be 'Flat Beat' by Mr Oizo. Based around a pulsating bass line showcasing the characterful filters of the 1978 Korg MS-20 synthesizer, the track accompanied a series of 1999 TV adverts promoting Levi's Sta-Prest range which ensured it went viral long before that term existed. In truth, the extraordinary level of international success was thanks in no small part to the presence of Flat Eric, a small yellow puppet who starred in the advert and subsequently in the video promo for the track itself. While an undeniably catchy track, 'Flat Beat' is based around an extremely stripped-down arrangement. The all-important Bass Drum is provided by the TR-606. In fact, it is a sample and is the only sound from the instrument that appears on the record. While the retrospective of 'Flat Beat' published on Roland's corporate website recognises that the TR-606 Bass Drum might not have had the heft of some of its competition, it unsurprisingly paints a picture of the sound as the foundation upon which the success of the track is built. 'Its ultra-slim kick wasn't beefy like the 808, or sharp like the 909. However, in the context of this lo-fi electroclash beat, it was the perfect addition.' (Dias 2020).

Interestingly, Mr Oizo (aka Quentin Dupieux) has a slightly different take on the sound.

> The finishing touch was the kick from a Roland TR-606: "Maybe the weakest in the industry," he smirks. "When I play in clubs, the difference between "Flat Beat's" kick and some new-school song is crazy – but it's part of the charm of this tune. It was really homemade: I never got round to doing a proper mixdown!"
>
> (Ravens 2019)

In addition to fuelling some key electronic music tracks, albeit without the depth of low end power of other instruments, the TR-606 actually found favour across a wide range of genres and while industrial rock might not necessarily be the first to come to mind, the abrasive, aggressive sound of the instrument makes it extremely well-suited. Nine Inch Nails' provocative track 'Closer' from the 1994 album, *The Downward Spiral* for instance, features the instrument in its densely layered, distorted soundscape. However, this was far from the first time the TR-606 had been used in such a context. When Steve Albini, who would later go on to produce the Pixie's 'Surfer Rosa' (1988) and Nirvana's 'In Utero' (1993) albums, was working on his earliest material under the name Big Black, he was so enamoured of the TR-606 that it was treated as a member of the band and given the nickname 'Roland' (Gordon 2023).

In fact, albums like 'Lungs' (1982) truly reveal the inherent brutality of the sounds of the TR-606. This is a collection of tracks not merely possessed of the TR-606's aggressive sounds but positively revelling in their harshness. Albini's relationship with the TR-606 is a longstanding one:

> When I first got the 606, I carried it around and listened to it like a Walkman. Over the course of a day I would gradually build and re-build a rhythm until it was satisfying to listen to on its own.
>
> (Albini in Heywood 2020)

Albini is not the only artist to incorporate the TR-606 into a more distorted, rockier sonic landscape. Nor is he the only one to name his TR-606. Post-punk rock band The Sisters of Mercy similarly made use of the instrument and its characteristic tonality in much of their early 1980s work. Their instrument, however, was given the altogether more theatrical moniker of 'Doktor Avalanche'. Subsequently, the figure of Doktor Avalanche has gone on to become a fully-fledged member of the band although, in reality, a number of instruments have taken on the role with the TR-606 replacing the earlier Boss DR-55 Doctor Rhythm.

It is fascinating to note such anthropomorphism in the naming of the instrument especially given the relatively inorganic quality of its sounds and notable lack of any 'humanisation' features that could move its beats off the regimented, computer-controlled grid (and let us not forget that the Drumatix already has a name). Perhaps the desire to humanise and elevate to the status of band member arises from the instrument's palpable underdog status.

Ultimately, for The Sisters of Mercy founder, even though its sound and programmability meant that the TR-606 incarnation of Doktor Avalanche represented an upgrade from the DR-55, there were yet more improvements to be made. As the page dedicated to the history of Doktor Avalanche on the band's official notes,

> The Roland TR606 came next. A few more sounds, a bit more memory, and individual volume controls. We liked it. Eventually someone worked out that it was possible to drill holes in the back and gain individual outputs.
>
> (Doktor Avalanche n.d.)

Rockers and Mods

There were no official upgrades to add features to the instrument. Rather, such modifications or 'mods' were developed by industrious engineers and circuit benders with information and guidance circulating in popular music magazines. Reviews such as that in *Electronic Soundmaker & Computer Music* (1984) hint at just such alterations to the instrument, 'Apparently it's simple to modify the unit for separate audio outputs, which would give the potential of improving the sounds even further.' ('Beat on the Cheap' 1984: 35). Brian Jones' 1984 'Roland Drumatix Modifications' feature in *Electronics & Music Maker* provides just one example.

> The Roland TR606 has proved to be one of the most popular drum machines. It is relatively inexpensive considering the features it offers, and is ideally suited to the electronic music-based musician. Here is a simple and extremely cheap modification which can be made to the TR606 to provide individual instrument outputs like its big brother, the TR808.
>
> (Jones 1984: 42)

A number of points are notable here. First, the TR-606 is reported as a popular instrument rather than the embarrassing commercial flop that it is very often presented as being in more recent histories. Second, the design of the instrument makes such modifications comparatively straightforward. Indeed, as Kirn (2020) notes, opening the case of the TR-606 reveals a suite of invitingly accessible and clearly marked circuits, traces and test points ripe for experimentation and play. In fact, that Jones' article is so brief at just a single page including some large photographs, speaks to the comparative ease with which audio signals can be intercepted and routed to a new set of sockets mounted on the rear panel – 'drill holes in the back and gain individual outputs'. Third, it is notable that this modification does not seek to extend the instrument into new and unexpected territory but rather to (re)instate functionality apparently lost as a consequence of Roland's creation of a less expensive instrument. Then and now, TR-606 is appealing to modders precisely because it is cheaper than a TR-808 but can, with some ingenuity, have added to it at least some of what makes that a more expensive product. The importance of separate outputs is considerable as it potentially allows sounds to be processed with external effects and EQ (thereby giving a degree of sound shaping not possible in the box perhaps adding some heft to the Bass Drum or removing (or adding!) harshness in the metallic percussion, for instance). Similarly, individual audio outputs allow sounds to be panned across the stereo field rather than summed together through the instrument's single monoaural output. Additionally, if the sounds are run through an external mixing desk, they can be brought under the control of larger, more tactile faders instead of the small, recessed dials whose awkwardness makes them best suited to creating static mixes rather than dynamic movements. We will explore the affordances of these controls in the next chapter as they also appear on the TB-303 synthesizer and similarly reveal interplay between design intention and creative practice.

Finally, it is worth remembering that the article detailing this mod dates from 1984 when the TR-606 was still in production. Typically, we might expect a modding scene to develop later in an instrument's life or even after it has been discontinued by its manufacturer, but here we see contemporary documentation that tackles and overcomes one of the instrument's key hardware limitations and potentially even encourages new purchasers to the take the plunge in the knowledge that it can be transformed into a more capable device.

Importantly, the individual audio outputs mod is far from the only one that has been developed for the TR-606. The April and May 1998 issues of 'The Mix' magazine ran DIY guides authored by Geoff Waterston that dig deeper into the circuits and alter the sounds of the instrument. Among the many modifications, changes to the capacitors in the Bass

Drum circuit allow its pitch to be lowered, moving out of the mid-range, and its decay lengthened. As Waterston notes of the original Bass Drum,

> I also thought the decay of the drum was far too short, so I used a bit of 808 technology. By doing this you can quite easily get decay times of up to five seconds. and added an extra filter network into the drumoscillator ... It will boom even longer and deeper than an 808.
>
> (Waterston 1998: 98)

It is worth pausing to consider the implications of this statement. When we talk of the TR-606, it is important to remember that, for all the flexibility of its sequencer, its sounds are presets. They are defined by calibration settings applied in the factory or set by the inherent properties, specifications and tolerances of the components in the circuits and, above all, shaped by the aesthetic decisions made by the designers of those circuits. As such, when Heywood discusses the role of the TR-606 in Albini's Big Black oeuvre, it makes perfect sense to talk of the, 'Punchy, mid-range kicks, popping toms, the flat-backed smack of the snare on the staccato beat. The sounds that introduce "I'm a Mess" are the signature tones of this classic drum machine.' (Heywood 2020). The mid-range Bass Drum and short Snare are defined by the instrument just as they define the instrument. As such, we might argue that they are not really the 'signature' tones of the instrument, so much as they are the only ones.

It is for this reason that modifications are so important. They unlock a level of customisation over the sounds that the design of the TR-606 simply does not. Waterson's articles document further mods that alter the decay of the Snare Drum to draw out the sound beyond its abrupt default settings. Interestingly, Waterston's mods do not only reference the TR-808. Alongside the ability to alter the level of the Noise circuit that feeds the Snare Drum, there are modifications that add a pronounced 'pitch sweep' to the Bass Drum and that give the Low Tom the sharp-edged thump of the TR-909. Crucially, these mods do not merely aim to reinstate features apparently missing from the TR-808 but, in explicitly simulating the sounds of the TR-909, they transform the TR-606 into the instrument that replaced it.

Inventive as they are, the Waterston mods are somewhat hamstrung by the physical constraints of the TR-606 itself. With such a small case and existing circuitry already comparatively tightly packed, it is a challenge to add significant quantities of new components. This lack of space also impacts the ability to add new front panel hardware for dynamically controlling the sounds. As a result, while many of the Waterston mods reveal a wide range of tonalities lurking in the circuits of the TR-606, each sound effectively becomes a preset once again as soon as the settings and components have been altered and the plastic case reassembled.

However, the simplicity of the TR-606's analogue circuits and the comparatively low cost of the instrument (certainly when compared with the astronomical prices that the TR-808 and TR-909 have reached on the second-hand market), have ensured that the instrument has remained a focus of modding activity. It would be unfair to suggest that the TR-606 is expendable, but it is certainly the case that the increasing scarcity and replacement cost of the TR-808 and TR-909 would give many owners pause before opening the instrument let alone tinkering with the circuits or drilling holes in the case to accommodate new controls. As such, the TR-606 is the most modified of the analogue Roland xOx instruments and has essentially become its own platform for exploration and experimentation.

The range of mods available today are extensive and make it possible to retune the six square waves of the metallic noise circuit, for example, or even to inject an entirely new external audio signal into the circuit to create altogether new tones. For all the individual tweaks and alterations, perhaps the biggest shift in TR-606 modding in recent years has been the addition of dynamic control over the circuits. Sometimes, as is the case with the 'TR-608' modification, this involves bringing together a suite of mods and adding new knobs to the exterior of the case itself, often in a slightly awkward or even suboptimal manner as dictated by available space. Even the naming of 'TR-608' mod once again speaks to the idea that this is not simply an exploration of sound but also an upgrade of status to rival that of a more widely revered instrument.

As the scope and number of modifications have increased, however, the limitations of the form factor have encouraged yet more inventive solutions that literally think outside the box. There are numerous examples of mods housed in completely separate breakout boxes that bristle with knobs, switches, inputs and outputs. Often connected by ribbon cables to the original instrument which services almost as an external sound source or sonic donor, these external peripherals allow dynamic control of a wide range of synthesis parameters and bring an unprecedented level of sonic control over the circuits unhindered by the real estate on the instrument's font panel. These mods and their associated hardware are demonstrably less inspired by a desire to replicate other x0x machines. Rather, they are dedicated to exploring the potential of the TR-606 in and of itself. In this way, such mods help to reconceive the TR-606 as a drum synthesizer that recalls Kikumoto's original design intentions for the x0x project.

It should be noted that, while most of the mods developed for the TR-606 focus on its analogue circuits, US company Social Entropy developed a now-discontinued replacement for its CPU. As well as adding USB and MIDI interfaces to allow straightforward connection to modern computer-based studio, the 'Quicksilver 606' adds new features to the computer-controlled sequencer. The upgrade extends patterns from 16 to 32 steps and expands on the instrument's global Accent so that each sound can be Accented individually. Marketed with the strapline 'Your 808 May Be Getting Jealous', the Quicksilver CPU upgrade is another way in which the TR-606 is positioned as an instrument brimming with untapped potential.

Remaking and Reimagining the TR-606

Given that it has spent much of its life in the long shadow of the TR-808 and TR-909, it might be surprising to learn that there have been a number of recreations of the TR-606 and instruments inspired by its circuits. Companies such as NoisyFruitsLab and Erica Synths have created Eurorack-compatible modules based around the Hi Hats, for instance, helping cement their status as iconic sounds in their own right and bringing some degree of awareness of the distinctive approach Roland took to its metallic percussion when it replaced the more common white noise with a swarm of atonal square waves. Reflecting the approachability of the TR-606 sequencer, German manufacturer Tubbutec's '6equencer' is a Eurorack module inspired by the instrument's workflow albeit with some significant additions drawn from modern instrument design such as variable play directions and per-step probability that can introduce degrees of randomness.

As well as breaking out the individual components into separate Eurorack modules, there are some more complete recreations of the entire instrument. In the analogue realm, the Cyclone Electronics TT-606 is described as 'analog drum machines with timeless electronic sounds and modern professional features' while the Behringer RD-6 is similarly marketed with a mix of reverence 'a timeless design from one of the most classic drum machines of yesteryear' and updated features to create 'a fresh and modern approach on a classic drum machine'. Roland, themselves, have also added a TR-606 reissue to their Boutique range of hardware. Like the other instruments in the series, the TR-06 is based around digital emulation and includes no analogue sound generation circuits, though it does broadly recreate the physical interface of the original.

What is particularly notable about these instruments is that this 'modernisation' not only includes adding connectivity such as MIDI and USB but also incorporates a number of the mods that have been developed, documented and codified by DIY circuit benders over the past decades. As such, the RD-6 includes separate audio outputs for each sound while the Boutique TR-06 offers control over tuning of each sound, the level of the noise introduced into the Tom circuits accentuating the simulated reverb, along with decay settings that extend the sounds, the attack of the Bass Drum and the transient noise burst to create an even sharper Snare sound. Interestingly, the TR-06 manual refers to this control as the 'Snappy' setting, explicitly referencing the front panel control of the TR-808 Snare. In fact, the TR-06 does not simply incorporate the functionality of the mods, it explicitly references the practice and materiality of modding and the struggles to find space in the front panel to place new controls and connections. As Hiromu Nagasawa, one of the engineers who worked on the Boutique recreation notes, 'On the TR-06, the trigger outs are on the line on the top panel. That is a nod to the geeks who loved and modified the TR-606 by adding individual outputs by themselves.' (Rosenschein and Ryan 2022).

As well as reflecting the aftermarket modifications, these new interpretations of the TR-606 also incorporate features that speak to the ways the instrument has been used in production contexts. The TR-606's use in experimental and industrial music means that distortion has become a key means of processing the instrument's sounds. The TR-06 includes controllable Saturation and Distortion effects while the Behringer RD-6 goes even further in its dedication to emulation by recreating the circuit of the Boss DS-1 distortion pedal. Importantly, each of these instruments, along with software recreations such as Roland's 'Cloud' TR-606, also include additional sounds. The TR-06 offers the most with a selection of Bass Drums and Snares that can be switched out to offer more TR-808 or TR-909 inspired flavours. As with some of the analogue circuit mods we noted above, this could be argued to connect the TR-606 to its x0x siblings whilst simultaneously drawing attention to a sense of lack in its palette whether this is measured with reference to its predecessor or successor. Perhaps to cement this feeling, both Behringer's hardware and Roland's hardware and software recreations now offer at least one Hand Clap. Clearly, this harks back to the absence of the sound in the original instrument and the potential for expansion with the HC-2. However, in these new instruments, everything is self-contained and does not require any additional hardware or the use of the Trigger Outputs which are retained for other duties. It is notable, however, that the TR-06 Hand Clap does not merely augment the existing TR-606 sounds but rather is offered as a replacement. In this way, the inclusion of the Hand Clap disrupts the integrity of the original instrument's palette by recognises its potential shortcomings, but perhaps does not quite apologise for them.

Interestingly, Roland's TR-06 and TR-606 plugin offer a degree of interconnectivity with the controls of the hardware instrument pre-mapped to the appropriate settings in the software to create a hybrid workflow. However, while the TR-06 continues to be referred to as the Drumatix, it is notable that the plugin's expanded sequencing features seem to finally give Roland the confidence to name it 'TR-606 Software Rhythm Composer'. A perfect recreation of an instrument that never was.

The Little Drum Machine That Could

Ultimately, the influence of the original TR-606 is felt in a number of ways. As an instrument, its sounds are simultaneously celebrated for delivering a tonality that is uncharacteristic in Roland's drum machine lexicon, while also being implicitly and explicitly criticised in terms of how they differ from the aesthetic set by those other instruments. In some senses, the emergence of the TR-606 as a modding platform similarly demonstrates this ambivalence being sometimes motivated by a desire to explore the latent sonic potential of the instrument on its own terms and sometimes to repair its deficiencies and recreate it in the image of the imaginary instrument it might have been.

Perhaps the true impact of the TR-606 is felt in the way it signalled a new direction for Roland and Boss to create comparatively inexpensive, portable instruments. While the DIN Sync and Trigger Outputs allow connectivity within a broader electronic music making context, features such as battery power and built-in headphone amplifier mean the instrument can be fully self-contained (as Steve Albini demonstrated by using it as a programmable Walkman). Of course, the Boss DR-55 had already pointed in this direction, and the TR-606 clearly charts a path to future x0x including SH-101, MC-202 and TB-303 (along with the Boss DR-110, the next Dr Rhythm) all of which are as at home in the studio or on the move.

To underscore its portability, the TR-606 even came with a matching silver (or rarer black) carrying case complete with shoulder strap. The bag is a complicated affair with pop studs and removable flaps and panels that allow access to the connectors so the instrument may be used in the case for those impromptu sessions where there simply is not even time to fully unpack.

Finally, it is important to note that, for all the comparisons between the sounds of the TR-606, TR-808 and TR-909, Roland's diminutive instrument offers one feature that sets it out from its relations. The TR-606 is the only x0x instrument to allow seamless switching between Pattern Play and Write Modes without stopping the sequencer. This means that Patterns can be selected at will in Play mode and edited on the fly by simply turning the Selector knob to Pattern Write. This one feature, unavailable on the earlier TR-808 as well as the TR-505, TR-707 and TR-909 Rhythm Composers that succeeded it, means the TR-606 is uniquely well-suited for live performance. The ability to freely switch between playback and editing opens up an endlessly mutable, improvisational space in which new Patterns are not simply selected or chained, but can be programmed and reprogrammed in step time or performed live with the Tap button. It is no wonder the 808 might be getting jealous.

5 TB-303 Bass Line

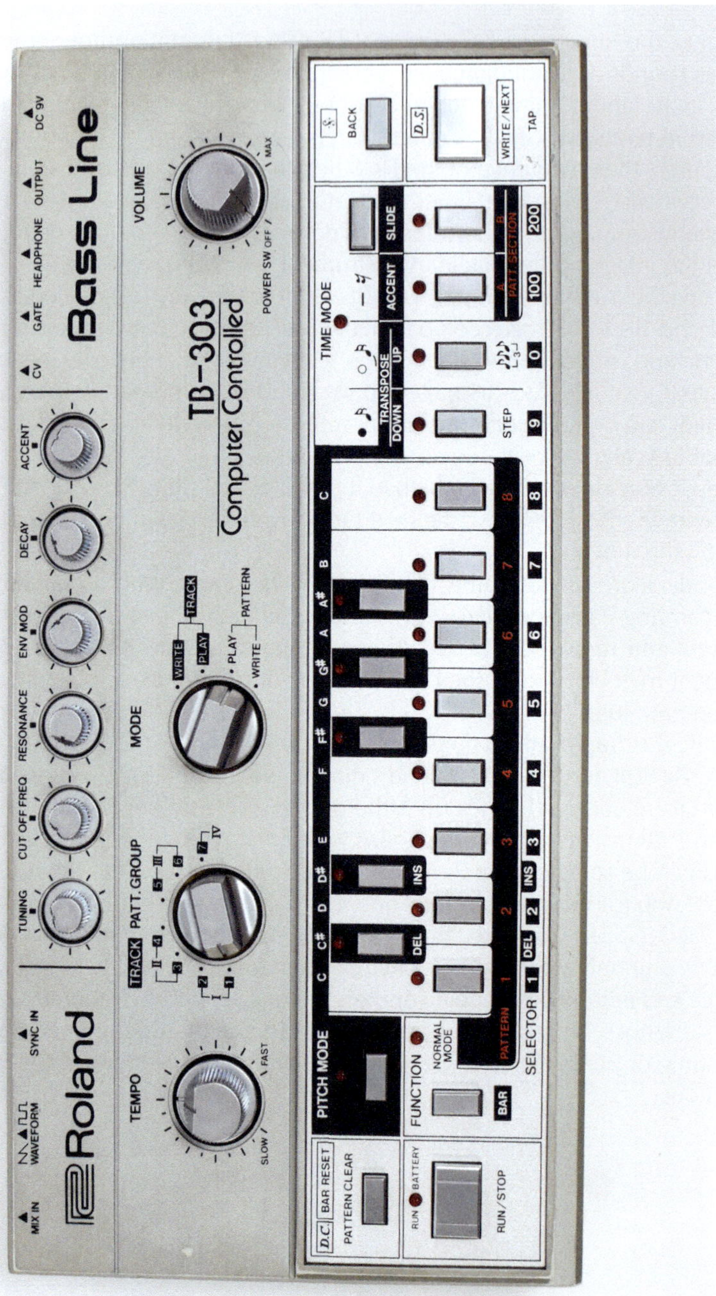

Figure 5.1 Roland TB-303 Bass Line (image credit: implant4).

DOI: 10.4324/9781003374510-5

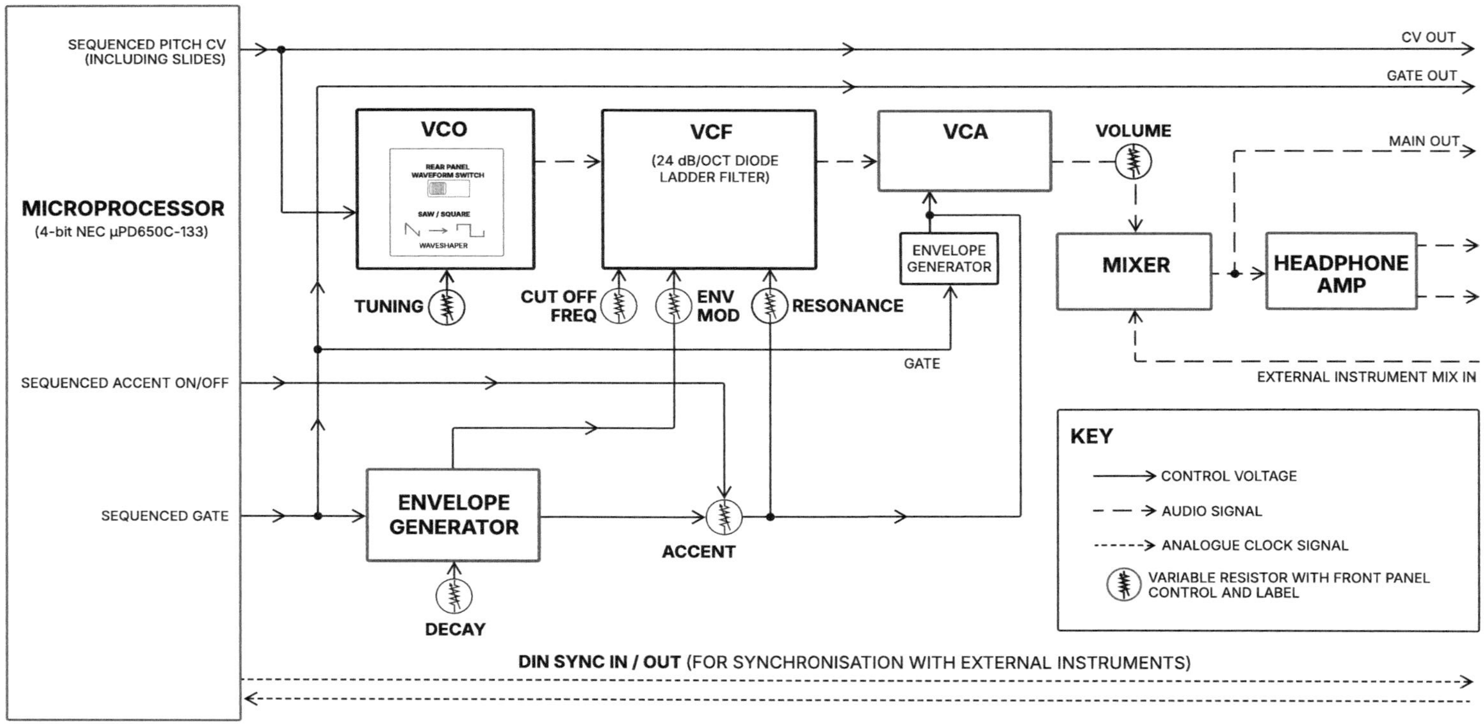

Figure 5.2 Roland TB-303 Simplified Block Diagram (adapted from Roland 1982c TB-303 Service Notes).

TB-303 Bass Line

Released
1981 (discontinued 1984)

Sound source
Analogue oscillator with selectable Sawtooth or Square waveforms
3-octave range (4 octaves in Track using Shift parameter)

Filter
24 dB/Oct Low Pass Diode ladder filter with resonance
The Filter does not self-oscillate

Modulation
2 Envelopes control Amplitude and Filter Cutoff

Keyboard
1-octave switches for sequencer step input
(Octave Up and Down buttons allow programming 3-octave range)

Sequencer
Parameters: Pitch, Duration, Accent, Slide
64 Patterns
7 Tracks (up to 256 measures in total)
Step and Tap input methods
Tempo approx. 40-300 BPM

Connectivity
Line Level and Headphone audio outputs (monaural)
CV/Gate Output
DIN Sync switchable between Input and Output
Mix In for routing external audio (does not pass though Filter)

Power
9V (centre negative) power supply
4 x C-type batteries

Dimensions
300 (W) x 55 (H) x 146 (D) mm

Weight
1.35kg / 3lb (without batteries)

Notes
Carrying Case (Black or Silver) with shoulder strap. The TB-303 user interface comprises just 10 knobs and 24 buttons. Only the 6 knobs at the top of the instrument provide control over the sound. Unlike the MC-202 with its Tape Save feature, the TB-303 offers no way to backup or save sequences

Suggested listening
Orange Juice 'Rip It Up' (Polydor, 1983)
Charadit Singh 'Synthesizing: Ten Ragas to a Disco Beat' (The Gramophone Company of India, 1982)
Phuture 'Acid Tracks' (Trax, 1987)
Josh Wink 'Higher State of Consciousness' (Strictly Rhythm, 1995)

The Most Successful Failure?

Of all the instruments explored in this book, the TB-303 Bass Line is the one that seems to most readily conform to the oft-repeated x0x narrative of the commercial flop, unloved and unwanted in its time that is later discovered by a generation of musicians who unleash its previously hidden sonic potential to create new forms of musical expression.

We have already seen that this overarching sequence of events, though frequently told, is prone to being overstated and is perhaps even untrue for some instruments. In the case of the TB-303, however, the story seems to ring true. Initially presented in marketing materials as a convenient, portable instrument providing an automatic accompaniment for the solo pianist or guitarist to play along with in live performances, the TB-303 was marketed as a replacement for the electronic bass guitar (and bass guitarist, for that matter). In combination with the TR-606 Drumatix, with which it was almost always pictured in Roland's promotional literature and with which it shared a suspiciously similar industrial design, the TB-303 and TR-606 combo promised the gigging musician a complete rhythm section. Indeed, Roland briefly referred to the two instruments as the first of the what was promised to be the 'ProForm' family. And in case the idea alone was not sufficiently appealing, Roland's 1982 print advertisement of the instruments infamously pictured the TB-303 and TR-606 perfectly angled on the music stand of a Roland Piano Plus 70 electric piano played by jazz virtuoso Oscar Peterson.

For all this promise, however, the TB-303, just like the TR-606, fell rather flat. As we shall see, unlike its comparatively accessible percussive counterpart, the Bass Line's interface and sequencer operation created an extremely complex and unintuitive workflow. To compound matters, while the TB-303's sound is certainly recognisable, the consensus is that the instrument does not sound much like a bass guitar with some poetic descriptions suggesting its sonic fingerprint is 'more reminiscent of a psychedelic mouth harp than a stringed instrument' (Hamill 2014).

As such, unable to neatly fulfil the role that Roland's marketing had attempted to carve out for it, the TB-303's origin story usually continues to note that, following a manufacturing run of 10,000 units (Roland 2019a 'The 303 Story'), production was halted with otherwise unsaleable units being heavily discounted and many eventually finding their way into thrift stores (Brotman 2022). Rather than becoming an intriguing footnote in corporate and technological history, however, it was precisely its newfound availability and (relative) affordability that meant that TB-303s were snapped up for a fraction of their original retail prices by a generation of young North American musicians. Through a process of creative experimentation and improvisation, these new musicians and DJs coaxed from this unassuming plastic box a sound that would go on to define an entire genre of electronic dance music. The characteristic sound of the TB-303's filter along with the esoteric features of its almost inscrutably enigmatic sequencer gave rise to dramatically, and aggressively, accented notes and sinewy portamento slides, which soon became the underpinning sound of Acid House.

In fact, although it remains inexorably linked to that genre, the tone and esoteric sequences of the TB-303 have found a home in myriad musical contexts and the instrument has been remade in hardware and software for many decades. The ongoing appeal and centrality of its sound has led the 303 to be considered as an '… iconic piece of electronic music culture, analogous to the Stratocaster for rock and the MPC for hip-hop.' (Kalepu 2015). As Hamill (2014) notes, 'It is small wonder the TB-303 has been referred to as one of the most successful failures in tech history'. The irony of this situation is that the combination of the desirability and scarcity of original units means that, as Davis (2022: 8) notes 'the TB-303 has achieved a legendary status and is now too expensive for the very musicians that used it to create Acid House' as well as other forms of music in the mid-late 1980s. It will come as little surprise, then, to find that the TB-303 is among the most emulated and recreated instruments with hardware and software versions first emerging in the 1990s and continuing to be released apace today by Roland and a plethora of other manufacturers. For Jenkins, the instrument is so significant that it is 'almost solely responsible for the "analog revival" some years later' and the renewed interest in 1980s instruments and synthesis techniques that emerged throughout the 1990s-present (Jenkins 2020: 99).

However, while much of this well-documented origin story and history does bear scrutiny, there remain many more twists and turns to be revealed and what follows in this chapter draws out, among other things, an album inspired by classical Indian music that puts the pulsating TB-303 front and centre many years before the invention of 'Acid'. In addition, more recent interviews with the TB-303's designers suggest that they had more than just the emulation of 'realistic' bass guitar tones in mind when they voiced the instrument suggesting a slippage between design intent and marketing that may have significantly contributed to its misunderstanding among musicians and initial commercial failure.

Central to the dominant narrative of the TB-303 is the concept of the 'happy accident' (Smith 2022; Rosenschein 2020) or as Vine (2011) puts it, 'once it got into the right (wrong) hands … its thrilling, squelchy, endlessly tweakable sound was perfect for the emerging house and techno scene.' However, while the instrument may not have been designed with Acid House in mind, that the distinctive sounds of the genre are available through the front panel controls without requiring any internal modifications or adjustments to the instrument's circuits, must surely invite us to revisit the straightforward notion of the TB-303's rise to fame being inexorably connected with the revelation of new sounds through creative 'misuse', 'abuse' or being used in the 'wrong way' (Sky Speir 2019). Instead, we shall see that it is as much a recasting of the musical context in which it is used as the nature of its underlying sound that is responsible for its transformation into a vital and vibrant instrument.

Of course, none of this should detract from the rich and diverse creative contributions of the influential actors that helped shape the TB-303. Rather, as is the case with so many of the instruments encountered in this book, by augmenting the dominant discourses around the TB-303's creation and use, we are reminded that the instrument sits within a web of complex, overlapping and parallel histories and that the journey to becoming 'The world's most famous electronic instrument' (Hamill 2014) is, inevitably, far from a simple one.

The Case for the TB-303

Despite its now legendary status, one could easily be forgiven for mistaking the TB-303 for the TR-606 drum machine. The two instruments share the same all-plastic silver casing with battery compartment for portability and the same complement of sockets dedicated to interfacing with the world beyond their rear panels. The top halves of the TB-303 and TR-606 front panels are identical in layout with identical rows of identical knobs for affecting the sound, setting tempo and changing the operating mode of the instrument. While the lower halves of the front panels differ in layout, each arrangement is created with the same switches and buttons. The results are so similar that even Roland managed to confuse the two instruments when they mislabelled them in their 1982 SH-101 advertising brochure!

Interestingly, the reason for this striking similarity is more than just a desire to create a coherent design language across two related products. As project lead Tadao Kikumoto has recently indicated, 'The TB-303 was developed to depreciate the molds of the body case and knobs for the TR-606' (Kikumoto 2020). Given that it was designed to fit into exactly the same box, it might be tempting to imagine that the TB-303 was a reaction to commercial shortcomings of another product. Certainly, such an apparently inauspicious beginning goes some way to explaining the persistence of the TB-303's origin story and its apparently meteoric rise from a literal space-filler to genre-defining icon. For all this, however, Kikumoto is clear that the TB-303 was conceived not simply as a way to recoup investment by salvaging and reusing parts from the TR-606 but rather was intended to operate as an accompaniment to it. Perhaps it could even be seen as an attempt to reignite interest in the earlier machine by providing it with a companion that created a new musical and commercial context.

It would not, however, dazzle with its specification as, at its heart, the TB-303 appears to be a very simple synthesizer that offers a single oscillator with a choice of a buzzy sawtooth or notably hollow-sounding square wave. There are just six knobs providing control over the shape and tonality of the sound. One is dedicated to the overall pitch/finetuning of the instrument (+/- one octave), two adjust the Cutoff Frequency and Resonance of the 24dB/Oct filter circuit respectively, with a further two altering the Envelope Generator Depth and Decay. Finally, the sixth control changes the depth of the Accent which is activated by the sequencer and is broadly analogous with the feature of the TR-606. The oscillator has a range of three octaves and, as one might expect of an instrument dedicated to creating bass lines, is tuned to cover the lower frequency range. As we shall see, there is far more to the TB-303 than these bare specifications or the apparent simplicity of the circuit shown Figure 5.2 might suggest. Indeed, the design of the filter and the interaction between the sequencer, Envelope and Accent circuits are actually extraordinarily complex and widely misunderstood aspects of the instrument.

For now, however, let us note that, notwithstanding any marketing hyperbole, the idea of a synthesizer providing bass tones was hardly an innovation. In addition to providing lead lines and melodies, synthesizers of all kinds had been used for low end duties with modular systems and compact instruments like the Minimoog and ARP Odyssey being employed in this wat by players since the early 1970s. By 1975, Moog went further still with the Taurus series which were dedicated to bass and controlled via organ-style pedals rather than a keyboard. The intention here was that the pedals be played with the feet while two hands played either one or two keyboard instruments. The almost mythical Moog 'Constellation' system brought together the monophonic 'Lyra' and polyphonic 'Apollo' keyboards and augmented them with the Taurus pedals. The astronomical cost and unwieldiness of the Constellation meant that, although it was used extensively on Emerson, Lake & Palmer's 1973 'Brain Salad Surgery' album, it failed to gain much mainstream traction and the bass pedal synthesizer remained largely in the domain of prog rock.

Aside from the cost and size, however, there were other obvious differences between the TB-303 and the dedicated bass of the Taurus pedals and even the Minimoog and other keyboard instruments. Where those instruments were intended to be played live with the performer using the appropriate limb to trigger notes via keys or pedals, perhaps embellishing their riffs with modulation or pitch bend applied via real time controllers such as wheels, ribbons or rocker switches, the TB-303 categorically could not be played even by the most virtuoso of performers. Quite simply, it was an instrument designed to be pre-programmed with sequences of notes that would be played back either in the studio or in a live setting (Bacon 1982b: 20). In this way, the TB-303 functions much more like a drum machine than a traditional performance-oriented (bass) synthesizer and its built-in sequencer was not simply a neat addition to facilitate automated playback but was an utterly essential part of the instrument without which it did not emit a sound.

It is worth noting here that, even though the TB-303's centring on creating a deep connection between a built-in sequencer and sound source might seem like an innovation in design given the plethora of earlier performance synthesizers, there is a precedent, albeit one that is largely forgotten and rarely discussed. Prior to the TB-303's release, a Japanese instrument manufacturer had launched the 'Firstman SQ-01' (as Jenkins [2020: 114] notes, this device was subsequently sold in the UK and US under a number of names including Hillwood in 1980 and Multivox from 1981). Although housed in a considerably larger and sturdier metal casing, the SQ-01 is built around a remarkably similar design ethos and functionality to the TB-303. As the Multivox iteration of the Owner's Manual puts it,

> The MULTIVOX SQ-01 Mini Music Lab SEQUENCE SYNTHESIZER is a monophonic multi-sound producing source (a synthesizer) and a digital recording device (a computer memory storage sequencer) in one compact control box.
>
> (Multivox 1981: 3)

With its integrated sequencer, built-in one octave touch plate keyboard for entering notes and a handful of sound sculpting controls giving access to the Cutoff Frequency and Resonance of the 24db/Oct filter, the SQ-01 bears much in common with Roland's later instrument including its positioning as a programmable device rather than one for live playing. Recognising the 'increasing acceptance' of sequencers and their value in freeing the performer from having to play repetitive riffs, Ellis' review (1982: 12) cites the Owner's Manual and Multivox's claim the 'Mini Music Lab is to music what the calculator is to math'. Roland's promotional materials similarly emphasise the innovation of the TB-303 and draw attention to the functionality of its sequencer in allowing bass parts to be programmed in advance and saved for later playback.

> Roland introduces the world's first computerized bass machine, the great new Bass Line TB-303. This remarkable little unit covers three full octaves. ... Up to 64 patterns can be created for use anytime.
>
> (Roland 1982b)

Roland's vision is clear here and perhaps deviates slightly from Multivox's in an important way. Where Multivox evoked the technical underpinnings of the SQ-01 through a connection with the pocket calculator, Roland's marketing discourse suggested a different relationship. Where the allusion to the calculator clearly speaks to the commercial and cultural currency of new technologies such as the circuitry used to create such a complex sequencer, it also has the effect of portraying the device as a tool rather than an instrument. In this way, its function appears to centre on the mechanical play back of riffs that were either too repetitious or too challenging for the human player to perform rather than functioning as an accompanist, per se. Roland, on the other hand, foregrounded the ways in which their new computerised technology created instruments that enabled the solo musician to play in different ways and which created new contexts for their creative practice.

One such context was live performance and it was a design decision inherited from the TR-606 that made this possible. While the SQ-01 could be run on batteries, the compartment and wiring were rather fragile and the device consumed a lot of power, 'A 12V AC adaptor obviously makes everything much more secure, and, as I found out, it's also rather essential if you're after a fairly consistent performance from the synthesizer as far as tuning is concerned.' (Ellis 1982: 13). Running the TB-303 on batteries was an altogether more straightforward proposition (although we shall see later in the chapter some unexpected creative practices emerge that take advantage of some quirks of the battery operation). Coupled with the lightweight, plastic construction of the case, this made Roland's instrument an inherently portable one.

> Completely portable, the TB-303 runs on either battery or AC line voltage and comes completely equipped with its own carrying case. A headphones jack is also provided as standard.
>
> (Roland 1982b)

Pre-programming sequences to be played back in a live performance setting clearly shifts the TB-303 into a different category of device and amplifies its role as a musical accompanist rather than a studio-based technological tool. That the instrument also shipped as standard with a soft carrying case complete with shoulder strap only added to the sense of this being a mobile instrument equally at home at a gig, in a rehearsal room or somewhere on the journey in between.

Moreover, where the SQ-01 was a single device, Roland's marketing almost always positioned the TB-303 alongside the TR-606 which afforded the company an opportunity to create a somewhat different musical proposition. Taken as a pair, these two instruments could be seen to comprise the constituent parts of a complete, portable and fully programmable rhythm section. In order to communicate this point, Roland's promotional materials centred on describing the ways in which the TB-303 and TR-606 not only complemented one another but effectively operated as one. Surely, the enforced similarity of the case design lent a degree of coherence to the ensemble and seeing them propped up on the music stand of Oscar Peterson's (Roland) electric piano in the infamous promotional shot, they almost dissolve into one double-length instrument. The neatness of the familial resemblance very nearly, but not quite, overcomes the awkwardness and unlikeliness of the scene. Nonetheless, as Raval suggests, this picture of the TB-303 speaks to Roland's musical vision in which 'They imagined a future where virtual musicians were playing in harmony alongside real musicians.' (Raval 2019).

While even the most committed fans of the silver boxes might struggle to make a case for them replacing Niels-Henning Ørsted Pedersen and Ed Thigpen (among Peterson's most celebrated bass and drum collaborators, respectively), it is clear that Roland wished musicians to see the connections between the TB-303 and TR-606. Perhaps because it was conceived of as a companion device to an existing instrument, the TB-303 was launched with a remarkable clarity of purpose. It may have been to the bass what the TR-606 was to the drum kit, but it was together that Roland hoped the instruments would really shine.

In fact, many of the technological and functional capabilities of the TB-303 speak to ways in which it was designed with easy connection to the TR-606 in mind. Chief among these was the implementation of straightforward synchronisation. By making using of Roland's own analogue DIN Sync protocol, the two units could be run in time with one another with their sequencers also starting and stopping in unison. A rather less widely discussed feature that further locks the two instruments together is found in the TB-303's 'Mix In' capability. Repurposing the TR-606's 'Run/Stop' socket that allowed the connection of a foot pedal to offer hands-free control over starting and stopping a drum pattern (already a nod towards the use of the instrument in broader live performance contexts), the 'Mix In' allowed external audio to be merged with the internal bass synthesizer sound with the combined signal sent to the instrument's main audio Output. Although the signal added through the 'Mix In' does not pass through the synthesis engine and so cannot be filtered or otherwise effected, this simple reuse of the TR-606's hardware, allowed the TB-303 to act as a rudimentary mixer. Crucially, taken together, the provision of DIN Sync and Mix In meant that with just two cables, the TB-303 and TR-606 combo could be kept in perfect sync with their audio signals mixed and ready to send to an amplifier or PA. Illustrations in Roland's promotional materials for the instruments map out just such connectivity, making great play of the ease with which they can be combined.

Indeed, so keen were Roland to cement the technical and musical connections, their marketing and promotion would go on to describe the TB-303 and TR-606 as the first two instruments in their 'ProForm' series. These PROgrammable PerFORMance instruments were marketed as,

> A group of inter-related products, each providing a specific musical function. Like individual members of a band, each of the ProForm products can sync with others in the series to produce a totally programmable musical performance.
>
> (ProForm series 1982)

Here, then, the programmability of the ProForm instruments was not presented as offering the player a way of generating musical phrases of such complexity that they could not be otherwise played but rather as offering an automated rhythm section of bass and drums that did not need to be played. Or, more precisely, a rhythm section that did not require a drummer or bassist. The key selling point of the ProForm series, then, is not found in its ability to create an electronic orchestra or to provide a host of new, never-before-heard sounds, but rather to support the solo musician in their rehearsal and live performance. By providing the 'individual members of a band', the marketing message here echoes the image of Oscar Peterson and makes clear that the ProForm series is as much aimed at the gigging musician as at the synthesist wishing to explore new sonic potentials. By way of example, while the description of the TB-303 in the company's 1982b 'Roland Rhythm Machines brochure', acknowledges the 'full control over resonance, envelope modulation and other important factors influencing sound quality', it quickly pivots to note that 'Instead of strings, you use keys to program bass patterns.' And so, while slightly awkwardly recognising that the instrument was emblazoned with the controls and terminology of the synthesizer (with its knobs marked 'Cut Off Frequency', 'Resonance', 'Mod Depth' and the like), Roland's European and US promotion and marketing of the TB-303 made it clear that the instrument was intended as more of a substitute for the bass guitar and guitarist than as an alternative to the Taurus or Minimoog for the synthesist. As the

1982 ProForm advert notes, 'The programmable Accent and Slide functions bring true bass technique capabilities to the TB-303'. Similarly, the Owner's Manual makes frequent reference to the bass guitar in its explanation of the uses of the TB-303. Moreover, of the four 'Sample Sounds' that give example settings of the tone controls, one is labelled as a 'Synthesized Bass' with the others being an 'Acoustic Bass' and two variations of 'Electric Bass' (Roland 1982d: 34–35).

These marketing materials and the framing of the TB-303 as a companion for the gigging musician form the basis of the usual TB-303 origin story (Roland 'The 303 Story'; Saxelby 2012; Stadtfeld 2021). Some commentaries go even further in suggesting that Roland had intended that their instrument directly compete with Fender's bass guitar (e.g.Raval 2020), or that the instrument was intended for use by guitarists (mn2s 2016). Certainly, Roland had a number of guitar-based products and had earlier enlisted the support of Led Zeppelin's Jimmy Page in endorsing a suite of their guitar synthesizers and effects as well as studio equipment including the TR-808 and System 100M and 700 modular synthesizers. The TB-303 described in Roland's marketing is not nearly so targeted, however.

Whether or not it had been intended to replace the Fender Precision Bass in the marketplace, the TB-303's apparent inability to produce a convincing emulation of a bass guitar tone is understandably given as a key reason for its initial commercial failure (Raval 2020; Vine 2011). For Jenkins (2020: 99), 'The "flat" bassline sound was a rather uninspiring "plonk" not particularly reminiscent of a real bass guitar', while others suggest that the instrument's 'incredibly tacky synthetic sound [meant that] no respectable guitarist could justify the crude concession over a real bassist' (JaegerOslo 2017). We will revisit the idea of the TB-303 as a bass guitar substitute later in the chapter when we explore the design and voicing of the synthesis circuits, but first it is useful to explore the other key reason usually given for the TB-303's initial failure to launch. Quite simply, it is almost unfathomably complex to use.

A Three Course Manual

There are no two ways about it, the TB-303 is not an intuitive instrument. Sure enough, once programmed, the large Run/Stop button makes it comparatively easy to start and stop playback, while the provision of the Mix In and DIN Sync connections mean that sequences can play back in perfect synchronisation with a TR-606 without fuss or extraneous cabling. In this way, the 'Form' of ProForm is relatively straightforward. 'Pro'gramming sequences, however, is an altogether more onerous and potentially frustrating undertaking.

For those musicians outside Japan who initially sat down with the new TB-303 upon its release in 1981, we might think that some of the confusion arose from the fact that the instrument shipped without a manual. However, even the arrival of an English language translation of the previously missing tome a year later would do little to dispel any conception that this was a complex and unforgiving device. Indeed, the English translation of the Owner's Manual makes no bones about it.

> The switches and control knobs have several functions, so operating the TB-303 may seem a little complicated. You may find difficulty using the TB-303 at first, because it is so different from a Bass Guitar or a keyboard instrument. Therefore, this manual includes a Basic, Intermediate and Advanced Course to help you understand these operations step by step.
>
> (Roland 1982d)

So much for an immediate, musician-friendly instrument for live accompaniment. Spread over more than 90 pages, completing all three courses is quite the undertaking and immediately underlines the difference between the comparatively accessible TR-606 and its rather more inscrutable relation. The complexity of the TB-303 is multifaceted but is the result of two major factors. Most obviously, while the reuse of the casing and button layout from the TR-606 might have conceptually connected the instruments and made sound financial sense, it significantly impacted on the freedom afforded the designers of an altogether more complex instrument and, in particular, a more complex sequencer. As we have seen, the TR-606 inherited the methodology and interface for drum sequence programming that had been pioneered in the TR-808. With each of the 16 buttons dividing a musical bar into 16 steps and an accompanying LED indicating whether a step would trigger a hit for the selected drum sound, the interface was, and remains, as effective as it is efficient.

There was no other visual display imparting information as there was no real need for one. The array of Selector Switches provided everything required to program and edit Patterns and as there was just a small selection of preset sounds, only six rotary controls were required to mix their outputs. Even with such a finely-tuned interface, most of the controls had multiple functions, and while the more esoteric button combinations were reserved for secondary rather than core features, it was clear that the TR-606 had been produced to minimise hardware costs. For the TB-303, the task was to create a means of controlling a synthesizer by using the array of buttons, switches that had been specified and physically arranged to create an already pared back interface for a completely different type of instrument.

Transforming a hardware interface built around sixteenth note on/off triggers into one that could support the entry of musical notes each with pitch, timing and duration cried out for a more sophisticated visual display, yet none was available. The multiplicity of functions and somewhat cryptic combinations of key presses highlighted in the opening pages of the TB-303 Owner's Manual self-evidently arise from the result of this recycling and inheritance of drum machine hardware. Indeed, as Kikumoto has noted in more recent reflections on the design challenge, 'The only available low-cost display was the LED, hence unlike the visual and friendly programming on the 606, the 303 system became quite difficult and tedious.' (Kikumoto 2020).

In such a situation, usability was always going to present a challenge. However, there is another facet to the TB-303's complexity and it is also, to some extent, the consequence of a Roland family inheritance. Musicians today might be familiar with musical step sequencers in which the workflow involves entering notes by selecting a step on which to work, setting the pitch of the note to sound, defining its duration (in musical time or a number of arbitrary ticks) and then applying any other modifiers (e.g. their loudness, or whether they are staccato or overlap with the following note in a legato fashion), before moving onto the next step and repeating the process. Essentially, each step is edited individually with all of the musical parameters pertaining to it being definable and modifiable at once. Typically, one might expect to be able to arbitrarily return to a step in order to adjust parameters or to change the pitch or duration of a note. Those familiar with the Elektron series of devices such as the Octatrack or Roland's more recent instruments such as the MC-101 will recognise the fundamental principles here, though there is considerably more additional complexity available in these instruments.

The TB-303 does things rather differently, however. Pitches, timing and other musical embellishments are each treated as different entities that must be entered as discrete streams of data in distinct modes of operation. In order to write a single 1-bar Pattern, the instrument must be switched to Pattern Write mode using the four-position knob (unlike the TR-606 which can switch between Pattern Write and Play modes without stopping the sequencer, the TB-303 must be stopped to create or edit a Pattern). Once in Pattern Write mode, the interface must then be set to 'Pitch Mode' using the dedicated button above the Function key. At this point, each of the pitches that comprise the desired sequence are entered in turn using the Selector Switches which are arranged as one-octave musical keyboard. The pitches can be confirmed and edited using the Write/Next and Back buttons. Once this process is complete, pressing the Function button returns the instrument to 'Normal' mode. To enter timing information and note durations for the sequence of pitches, Time Mode must be activated using a button on the right of the front panel (initially, it appears this button is labelled 'Slide' but this is actually a secondary function). Timing information is programmed using the group of buttons to the right of the Selector Switches. These are used to designate whether a step should sound a note (a 'Step'), should create a Tie from the previous note, or should be a Rest. With timing information entered and checked by using the Write/Next and Back buttons, the instrument can be set back to Pitch Mode in order to move through each step again and apply further embellishments such as Accents or Slides.

With each of these processes complete, the Pattern is written and can either be re-edited or played by setting the instrument back to Pattern Play mode and pressing Run/Stop. If that sounds complicated, that is because it is and, among other things it means contending with the fact that, even if 16 pitches are entered in Pitch Mode, the addition of any rests or ties in Time Mode means that not all of the programmed pitches will play. For instance, entering 16 ascending pitch values beginning with a low C along with timing information mapping out four rests, a note and 11 rests will result in a sequence comprising just a single low C on the second beat of an otherwise silent bar. Essentially, for every non-sounding step entered, a pitch from the end of the inputted stream of data is rendered superfluous.

The separation of musical phrases into separate streams of pitch and timing data is actually inherited from Roland's earlier MC-4 (1981) and MC-8 (1977) microprocessor-controlled sequencers and carried forward to the MC-202 MicroComposer (1983). We will reserve a more detailed exploration of the rationale and impact of the disintegrated 'Pitch' and 'Time' sequencing paradigm for our discussion of the MC-202 in Chapter 7. For now, suffice it to say that, if the TB-303 programming procedures seem unintuitive and longwinded on the page, the actuality is somehow even more awkward with an already cluttered front panel further complicated by multiple labels and legends as well as the need to switch between different modes of operation in strict order. As the TB-303 manual intimates, to use the instrument requires climbing inside the mind of the designers in order to ascertain how the process of music-making and composition have been translated into a series of codes, tables and programming procedures.

Trying to fathom all of this without a manual is a pretty tall order. In truth, trying to fathom it *with* the manual is not necessarily that much easier. Like other x0x manuals, examples lean very heavily on Western musical notation, the conventions of popular music structure and require a good amount of theoretical fluency on the part of the user. As we saw with the TR-606, the section on looping will make much more sense to readers already familiar with the notational conventions of the *dal segno* instruction, for instance. Similarly, the 'Shift' (transpose) feature, is discussed in the context of the rock/pop song in which a single pattern or bass riff might continue throughout an entire track with transpositions in different sections. In exploring the uses of the Track of chained Patterns, the manual sets out musical notation and the corresponding TB-303 procedures to create the I-IV-V chord progression commonplace in a host of popular music genres including 12-bar blues.

The esoteric nature of the TB-303's modes and its unintuitive procedures led many contemporary reviewers to categorise this as an instrument best approached with some advance planning.

> As with the 606 drum machine, you really have to write things down at an early stage to get the best from the 303 – you can if you like use it for off-the-cuff bass lines in a sequencer-like fashion, but for anything beyond very basic use you have to plan out what you're going to do before even switching the machine on. Playing with the 303 for a few seemingly wasteful hours – making a few mistakes, losing a few programs, swearing once or twice, etc – helps you understand that things which seemed like stupid idiosyncrasies at first are merely the machine's inherent insistence on mathematical order.
>
> (Bacon 1982b: 20)

Jamming on it, Ripping It Up, and Synthesizing Ragas

Just as we will see in our analysis of the MC-202, the TB-303's manual and its musical reference points speak eloquently to the contexts in which at least some folk at Roland conceived of the instrument and imagined it might be used. Certainly, a number of artists did put the instrument to use in more traditional pop and rock contexts. In the UK, Heaven 17's 'Let Me Go', released towards the end of 1982, credits Ian Craig Marsh with 'Roland Bass Line Programming' and, sure enough the track is underpinned by a one-note ostinato that is unmistakably the product of a TB-303 filter continually bubbling and percolating away as it moves around to mark out the root note of the chord progression. Likewise, 'Rip it Up', released in early 1983 by Scottish Indie Pop band Orange Juice features a prominent use of the TB-303. Here, though, the programming is more adventurous with five patterns underscoring the song's key sections. Like the Heaven 17 track, however, the bassline in 'Rip It Up' finds the TB-303 in a particularly resonant mode that eschews any pretence of realistic bass guitar emulation and, to some extent, mimics the unashamedly synthetic sounds of funk recordings such as Herbie Hancock's Arp Odyssey on 'Chameleon' (1973) and Bernie Worrel's Minimoogs on tracks such as Parliament's 'Flash Light' (1978). Indeed, the Resonance and Envelope settings are so extreme in 'Rip it Up' that the TB-303 is the most pronounced instrument on the track.

Elsewhere, the TB-303 was put to use in a variety of different ways. The 1984 single 'Jam On It' by New York-based Hip-Hop group Newcleus, is a case in point. Listening to the record, one could be forgiven for thinking the weighty bass tone that combines with a thundering kick drum and punchy snare courtesy of the TR-808, might emanate from Roland's silver box. After all, the diminutive instrument is proudly emblazoned with the phrase 'Bass Line' on its case. However, the bass on 'Jam On It' is actually provided by a Sequential Circuits Pro-1 mono synth and it is the comparatively high-pitched, fluttery, staccato sequence of sixteenth notes that marks the contribution of the TB-303. Taking advantage of the transposition and tuning capabilities and the synthesizer's three-octave range, it would be hard to argue that this sequenced line reveals a hidden versatility in the TB-303, but it certainly demonstrates – and presages – uses of the instrument beyond (or perhaps above) the lower frequency ranges. And, despite being drenched in echo, the interplay between the TB-303 and TR-808 is key in driving forward and giving energy to what is a remarkably stripped down but undeniably influential rap track.

Perhaps the most intriguing and important of the early uses and recordings of the TB-303, however, is Charanjit Singh's 'Synthesizing: Ten Ragas to a Disco Beat'. Originally released in 1982, the album brings together a triumvirate of Roland instruments with the Jupiter-8 analogue polyphonic synthesizer joining the TR-808 and TB-303 as the foundation of the recording. By 1982, Singh was already a well-practiced live and session musician and was enjoying a successful career as an arranger working with acclaimed film makers and music directors such as R.D. Burman. Reflecting on his contributions to countless songs and movie soundtracks and describing him as 'The Most Influential Musician You May Have Never Heard Of', Suprateek Chatterjee (2015) observes that 'Millions of Indians have heard him play key parts on iconic Bollywood songs for decades'. As Geeta Dayal (2010) writes, Singh was no stranger to experimentation and made use of a 'panoply of interesting instrumentation throughout his career – steel guitars, Farfisas, electric violins, Transicord electric accordions' which were soon augmented with synthesizers and drum machines. Indeed, throughout the 1970s, the landscape of music in Hindi film had been transformed through the uses of electronic instruments with pioneering musicians such as Kersi Lord introducing Moog instruments shortly after their release, along with drum machines and a multitude of effects units. Alongside this, the 1970s had seen a growing influence of dance music genres such as Disco on both film and popular music in India. Whilst travelling in Chicago, musician Bappi Lahiri had been introduced to Disco through the Bee Gees' 'Saturday Night Fever' and when asked by Director Ravikant Nagaich to produce something new for a forthcoming film, 'the beat of the disco came back to me, leading to the making of "Mausam Hai Gaane Ka, Gaane Ka Bajane" Ka from Surakksha (1979) … From thereon, I made folk to the beats of disco.' (Team FC 2022).

Given this context, we can see that 'Ten Ragas to a Disco Beat' did not emerge entirely from nowhere. Indeed, as Singh notes in a 2011 interview,

> There was lots of disco music in films back in 1982 … So I thought why not do something different using disco music only. I got an idea to play all the Indian ragas and give the beat a disco beat – and turn off the tabla. And I did it. And it turned out good.
> (Aitken 2011)

In addition to the overarching creative inspiration to meld musical styles, the TB-303 offered specific aesthetic and technological opportunities. Singh recounts that he acquired the instrument from a music store in Singapore shortly after its release in part because it was new and promised a novel sound but also because of its ability to synchronise with the TR-808 as well as the Jupiter-8's built-in arpeggiator. Indeed, it was this aesthetic and literal connection between the instruments that really set apart Singh's work from what had preceded it. As Pandey (2019) notes, the synthesizer had been integral to the Hindi and Tamil film industries since the early 1970s, but 'it had previously been used as a standalone instrument like any other but nobody was clocking a drum machine to a bassline synthesizer or sequencer to get electronic grooves'. Singh dedicated time to understanding the potential of these new instruments, 'At home I practised with the combination and I thought "It sounds good – why not record it."' (Aitken 2011). And so, while Singh was well-known for session and film soundtrack work, 'Ten Ragas to a Disco Beat' was conceived as a standalone album from the outset and, as its name suggests, is a collection of electronic arrangements of traditional Indian classical music modes.

The role of the TB-303 in the recordings is particularly notable not merely because of the centrality of its unmistakable sound but also because its ability to create sinuous, gliding, sliding portamento is expertly harnessed by Singh to map out the ragas. Appearing to bend the strings of an instrument that Roland assured us was stringless, Dayal (2010) notes that, 'The 303 ends up being perfect for the slithery glissandos in Indian music, the sliding between one note and the next.' Perfect it may have been, but the album did not make a significant impact on the record-buying public. As Singh notes, 'It didn't click ... It didn't have any publicity. Only sometimes you could hear it on the All Indian radio station, filling the gaps.' (Aitken 2011). As such, and dramatically echoing the journey of the TB-303 itself, the lack of commercial success of 'Ten Ragas to a Disco Beat' meant that it slipped into relative obscurity. However, in 2002, record collector Edo Bouman happened upon a copy of the disc in a shop in Delhi and subsequently set about tracking down Singh. Eight years later, 'Ten Ragas to a Disco Beat' was re-released on Bouman's 'Bombay Connection' label and is now rightly recognised as being crucial to our understanding of the histories of electronic music in India. Moreover, the album is an influential and prescient recording that, as we shall see later in this chapter, appears to predict an altogether more widely-documented US and European compositional and artistic practice by many years.

Notwithstanding these notable early uses of the instrument, the TB-303 was ultimately sidelined. Even in the context of pop compositions with comparatively simple harmonic and chordal structures, the complexity of the instrument's programming procedures made using the TB-303 a laborious exercise.

The x0x That Wasn't

Although there is little to substantiate the Museum of Failure's curatorial note that, 'It was rumored that the company dumped thousands of unsold synthesizers in the Tokyo bay.' (Museum of Failure 2020), with approximately 10,000 instruments sold worldwide (Brotman 2022; Roland 2019a 'The 303 Story'), the TB-303 has been described by none other than Kikumoto (2020) as an 'utter failure in terms of sales'.

As such, it might seem poetic, if not a little sad, to note that an instrument born as an attempt to breathe life into the TR-606 and the operational need to make use of an inventory of available parts, ended up significantly underperforming. And so, while there had been some initial enthusiasm in the specialist press and some intriguing uses of the instrument by artists across the world, at least some of whom seemed to have risen to the challenge set out in the promotional material and manual, Roland ceased production of the TB-303 in 1984. In their attempt to amplify the dramatic arc of the instrument's history, retrospectives of the instrument have a noticeable tendency to focus on the negative response to the TB-303's release, and while it was clearly not the success that Roland hoped for, whether it is fair to call it a 'toy' is debatable (Reid 2004b). Similarly, there was clearly a mismatch between the hyperbolic marketing discourses, the actuality of the instrument and the desires of the market, it is perhaps stretching the point to suggest that 'It was heralded as one of the worst sounding synthesizers ever produced.' (Heermans 2016). However, although the TB-303 initially failed to capture the imaginations of musicians, it dramatically changed the future of Roland product development.

Given the less than stellar sales, it might come as little surprise to learn that the ProForm series that Roland had been so keen to promote was never mentioned again after the discontinuation of the TR-606 and TB-303. Despite what their 1982 marketing suggested, there were not 'more to come'. In fact, rather more prophetically, the same advert proclaimed that 'as the Drumatix is to the drums, the TB-303 is to the Bass.' Unfortunately for the accountants at Roland, in the short term at least, that would turn out to be true.

In case we are tempted to dismiss the ProForm monicker and any notion of a family of interconnected instruments as over-enthusiastic marketing, it has recently been revealed that at least one further instrument had been planned as part of the series. Reflecting on his work in Research and Development at Roland nearly four decades after the release of the TB-303, Kikumoto made a tantalising revelation. 'As a matter of fact, we were planning on creating a Roland TC-404 chord machine ... We gave up on developing the TC-404 due to the TB-303's low sales and commercial failure.' (Kikumoto in McCabe 2020).

It is clear then that there was at least one further x0x instrument planned by Kikumoto and the engineers at Roland and that it would interface with, and form an integral part of, a modular family of instruments. Part of the inspiration for the TC-404, and perhaps even part of the reason for its failure to materialise, appears to arise from the drafting of the MIDI specification which, along with Sequential Circuits, saw Roland and Kikumoto in particular, play a leading role. "I proposed the use of Chord Melody Tracks for rhythm machines, but there was not so much interest in this idea." (Kikumoto in McCabe 2020).

More than four decades later, it is impossible to avoid speculating about the relationship between the TC-404 and Roland's J-6 Chord Synthesizer released in 2022. Centring on easing the creation of complex progressions by drawing from a library of 'chord sets' that can trigger either an internal synthesizer or control external instruments over MIDI, the J-6 seems to fit the bill of a 'Chord Machine'. Similarly, like the fabled TC-404, the J-6 is a member of a family of small, portable, accessible instruments – the Aira Compact series – including drum machines and synthesizers with sequencers, which can be easily synchronised with one another and which are equipped with Mix Inputs for chaining together their audio without need for an external mixer. The T-8 Beat Machine even offers emulations of the sounds of the TB-303 and TR-606. Even if we cannot be certain that the Chord Machine lives on in the J-6, Kikumoto's revelation at least helps explain the notable absence of a '404' from Roland's 1980s x0x lineup!

Phuture Bass

The failure to connect with a wide enough pool of musicians meant that the TB-303 was soon heavily discounted by musical instrument retailers with even the units that had been sold often traded in for other more interesting and usable equipment. As such, the instrument soon became available for a fraction of its original cost with Raval (2010) noting that, in the US, 'many of the unwanted units were being sold second hand for as little as $50'. Even as late as 1990, DJ Josh Wink bought a TB-303 in 'a pawn-shop in Philadelphia for $70 with the original carrying case' (Rosenschein 2021).

Jenkins similarly suggests that, within a few years of its release, 'prices of Basslines had fallen almost to nothing, certainly to £50 or less in the UK.' (2020: 100). Indeed, UK musician and DJ A Guy Called Gerald confirms this in noting that, 'I bought my first one in 1986 for £50 from a secondhand shop called Johnny Roadhouse on Oxford Street in Manchester.' (Sparks 2020). All of this could well have marked the end of the line for the

TB-303. However, the instrument would soon go on to enjoy an extraordinary rebirth that would see an entirely new genre of electronic music, not merely centred around it, but irrevocably defined by its characteristic sonic fingerprint. And crucially, much like the TR-808, this resurgence did not come despite its initial commercial failure but because of it.

It is rare for a genre or any field of music to be linked so unequivocally with a single instrument and yet it is utterly uncontroversial to say that Acid House could not exist without the TB-303. The sound of Acid is the sound of erratic sequences of no more than one bar of sixteenth notes endlessly repeated to the point where they become absolutely coherent. It is a sound that is simultaneously made of slithering, gliding portamento and punctured, staccato bites, and it is the sound of a filter that, moment-by-moment, ominously bubbles before erupting into an almost animalistic, visceral scream. For Kodwo Eshun, the 'Inhuman or post-human' sound creates a 'tense, frightening and orgiastic music' (2000: 76–77).

Underpinned by a solid four on the floor kick pattern and assorted percussion elements typically supplied by Roland drum machines like the TR-707, TR-727 and TR-909 with their clipped punchiness, Acid House is stripped-back with an almost sparse sound that places the 'squelchy' TB-303 front and centre to an almost provocative extent. It is poignant to note that, even though it was not the most commonly-used device, Acid House producers' occasional use of the TR-606 even reunited the members of the ProForm family, albeit it in a shockingly new context.

Central to the birth of Acid House is the Chicago-based trio Phuture, a three-piece electronic music group comprising Nathan Pierre Jones (better known as DJ Pierre), Earl Smith Jr (known as 'Spanky'), and Herbert Jackson. While accepting that musical origins are invariably contentious, and there is debate about whether Sleezy D's 'I've Lost Control' is actually the first Acid House track to make its way to vinyl, it is ultimately Phuture's 'Acid Tracks' that set out the sonic blueprint and gave name to the genre. The story of the genesis of the track has been extensively chronicled (DJ Pierre 2012; Collin 1998; Rietveld 1998), perhaps even to the point where it has become as mythologised as the creation of the TB-303 itself, and it is worth attempting to disentange as many of the creative and musical intentions behind it are frequently underemphasised.

It is often taken that the TB-303 came into Phuture's possession simply as a result of it being a cheap, relatively readily-available piece of equipment that was left languishing and otherwise unwanted in the bargain bin. Certainly, the comparative affordability of the instrument played a crucial important role in making it available to a different audience both in Chicago and later in the UK. Had the instrument not been an initial commercial flop, it may never have entered the possession of many of the young musicians and DJs that subsequently transformed its musical and cultural value. However, it is important to note that there was more at work in rendering the TB-303 potentially appealing than cost and availability alone. As DJ Pierre notes,

> We were trying to find a sound, and we bought keyboards and everything, and nothing kind of clicked with us. We would come up with basslines, we would try out different bass sounds, and I don't know why but we just weren't satisfied with any sound that we were working with. This guy named Jasper G, who was just a friend of ours, made a track, and when I heard his track, I was like 'Yo, what did you use for the bassline?' And he said, 'A Roland TB-303.' We didn't know what that was. I went by his house and I saw it, I was like, 'Ah, that's nice right there.'
>
> (DJ Pierre 2012)

With the group subsequently seeking out a TB-303 to use on their own recordings and purchasing one for either $40 or $200 depending on the account, what we begin to see is an altogether more measured and deliberate approach to finding, honing and owning a new sound. This is not a simple tale of making do with dusty, discarded instruments but rather one predicated on actively seeking out stimulating new possibilities and unexplored affordances. Identifying the TB-303 as offering 'The most unique sound in the history of the world [that] doesn't sound like any sound you've ever heard before', DJ Pierre (2019) makes it clear the instrument played a critical role in defining the group's musical and cultural identity. Continuing, he notes, 'We got our sound. All of a sudden, we became confident. All that uncertainty just was gone.' Similarly, while the inspiration for Phuture's acquisition of a TB-303 came from hearing its use on a previous record, the distinctive sound that would go on to define Acid Tracks, the genre of Acid House and the new conceptualisation of the instrument itself was not immediately evident. Many commentaries suggest that the revelation of the characteristic 'Acid' tonality came by chance but, as DJ Pierre notes, this downplays more active and attentive creative and critical processes.

> People say it was made by mistake. No, we just listened to what we were doing and decided that it sounded good, and then recorded it, as we knew people would like it. It sounded like good music that people could dance to.
>
> (Ransom Note 2021)

As predicted, people did like it and did dance to it, although not immediately. Indeed, the first night influential DJ and tastemaker Ron Hardy played a homemade tape of the track at the Chicago Music Box club, it was a slow burn with the first airing reportedly emptying the dance floor. By the end of the night, however, it was a hit.

> 'Ron Hardy had the heart and the guts to play something like that, and to play it first,' enthuses Spanky. 'He trained the people in the club that night to like that song, and it worked. The fourth time he played it, they lost their minds.'
>
> (Arnold 2015)

Bootlegs of what was commonly referred to as 'Ron Hardy's Acid Tracks' proved so popular that it was inevitable that Phuture would go into the studio to commit the track to vinyl. Produced by Marshall Jefferson, a 12-minute cut of 'Acid Tracks' would be released on Trax Records in 1987.

It is impossible to ignore the gatekeeping role played by Ron Hardy here. However, there are other crucial factors at work and it is similarly hard to overlook the impact of the institutional power structures of the clubs and record labels that ultimately monetised this creative output. Indeed, echoing the situation documented by Negus (1999) in relation to Hip Hop, Harkins and Prior (2021: 11) note, 'very few of the black musicians associated with acid house were particularly well compensated for their pioneering work.' Discourses of discarded and rediscovered instruments like the TB-303 often centre on the enabling qualities that arise from their creative reinvention and the opportunities their comparatively low cost offers musicians otherwise unable to participate. Indeed, for some, the design of the TB-303 plays a crucial role in this regard. Berk (2000: 193), for instance, suggests that, 'The difficulty of programming the TB-303 accurately made it a very democratic instrument – a novice had just as good a chance of producing a usable bassline as did an expert musician.' We must, however, be mindful of the realities and how 'institutionalized and systemic forms of racism, including the exclusion of black personnel from positions of power, temper any benefits afforded by "cheap" and "democratic" technology.' (Harkins and Prior 2021: 11). This is especially significant given the international impact that Acid House would have with the UK, in particular, enthusiastically embracing the new sound and eventually transforming it into the underpinnings of a mainstream youth culture movement that would eventually become known as the 'Second Summer of Love'.

The journey of Acid House from Chicago to the UK was not a direct one, however, and made an important layover at the Amnesia club in Ibiza. For a number of years, resident DJ Alfredo had built a reputation for eclectic and diverse sets that spanned numerous genres along with various flavours of House music from the US and latterly Acid House from Chicago. As Stephen Armstrong's history (Armstrong 2012) of the Ibiza club scene notes, it was on visiting Amnesia in 1987 that DJs Danny Rampling, Paul Oakenfold, Nicky Holloway and Johnny Walker were exposed to these new sounds and set about bringing them back to the UK. By the end of 1987 Rampling's Shoom club had put on the first night that sought to explicitly replicate the Balearic experience. It was at Shoom that the 'Smiley Face' logo was recycled to become the emblem of Acid House and UK rave culture. The success of Shoom led Holloway, Oakenfold and others including Pete Tong and Carl Cox, to found other clubs and Acid House nights across central London including The Trip, Future and Spectrum. Elsewhere in the country, other clubs such as The Hacienda in Manchester, increasingly dedicated more of their time and floorspace to Acid House. As it increasingly took center-stage and gained both popularity and notoriety, the connections between Acid House and the Class A drug MDMA (commonly known as Ecstasy) become the subject of considerable media reporting. The subsequent reaction has been characterised as a moral panic (after Cohen 1972) with popular tabloid newspapers including the Daily Mail and The Sun routinely running stories on the indivisibility of the music, club culture and drug use.

Commentaries on UK rave culture do not generally downplay the presence of Ecstasy although many have a tendency to point to a more complicated situation than presented in the popular press. Simon Reynolds, for instance, notes that, 'generally speaking, Ecstasy seems to promote tolerance. One of the delights of the rave scene at its height was the way it allowed for mingling across lines of class, race and sexual preference.' (Reynolds 1999: 238). Nonetheless, as Passie (2023) notes, reports of deaths attributed to MDMA use at parties ensured the continued demonisation of the drug and the scene with which it was connected as a whole. Speaking to the tension between populism and panic, page 15 of the 12 October 1988 issue of 'The Sun' newspaper contains both an article titled '10 Reasons to Say No to LSD' detailing the 'daft drugs that plague the scene' and an offer to purchase for just £5.50 a 'groovy and cool' Acid House T-Shirt along with 'essential Acid House gear like smiley-face headscarves (£3.99), bandanas (£2.99) and cycling shorts (£25.99)'. Later, the Smiley Face logo would go on to be vilified, becoming a shorthand for the apparent excesses and endemic nature of Ecstasy use and banned from appearing on the screen of the BBC TV chart show 'Top of the Pops'.

The popularity of Acid House saw a number of tracks enter the UK pop charts with titles such as A Guy Called Gerald's 'Voodoo Ray' and The KLF's 'What Time is Love?' both released in the middle of 1988 and prominently featuring the instrument spitting out relentlessly repeating one-bar sequences and the characteristic squelch of the filter. Towards the end of the same year, D-Mob's anthem 'We Call It Acieed' had become a virtual anthem with its rallying call yet the track did not feature a TB-303.

As Hill (2003) notes, the Acid House panic in the UK was motivated as much by the perceived threat to the Thatcher government's bureaucratic authority as much as the music itself particularly as Acid House gatherings, often known as 'raves', moved out of city centres and clubs and into abandoned warehouses on the outskirts or even in rural areas. The incursion of the mob into the English countryside along with the lack of control over the ability to generate revenue from this shadow nighttime economy is at least part of what informed the parliamentary debate and legislation. Eventually passed in 1990, following some resistance including a 'Freedom to Party' rally in Trafalgar Square, the Entertainment (increased Penalties) Act, set out fines of up to £20,000 for large-scale unlicensed parties (mcmahonp 2023). The threat of financial penalty did not put an end to the rave scene in the UK which continued to operate and was publicised through more clandestine word-of-mouth networks or telephone answering machine messages posted at very short notice in attempt to skirt around disruption.

As the 1980s gave way to the 1990s, the sound of Acid House fell from the mainstream charts. However, Phoenix-like, the TB-303 was reborn in the middle of the decade although its sound was transformed as musicians and producers began processing the instrument through various combinations of guitar effects pedals to add distortion and overdrive. Among the first to bring this new saturated sound to the fore were German group Hardfloor whose 1992 release, 'Acperience 1' brings together an edgier, more menacing TB-303 with the punch and weight of the TR-909 to create a harder, more Techno inspired form of Acid. Moving from overdriving the instrument to applying out-and-out distortion, the processed TB-303 could be seen to have reached its apotheosis in Josh Wink's 'Higher State of Consciousness' (1995). Here, the sound is raw and sharp with none of the warmth that can be coaxed from the instrument. The combination of extreme distortion and high filter resonance creates a sound that is largely bereft of bass frequencies but which is almost eye-wateringly piercing in the upper register. In a similar vein, Daft Punk's 'Da Funk' (1995) stands as another influential example of this heavily processed, overdriven TB-303 sound. There is certainly something of the tone of the distorted lead guitar in this reinvention of the instrument with its mix of sustain and biting attack, yet the sequence of notes is undeniably simple and it is really in the manipulation of the filter that the track finds its defining feature. With resonance at maximum, the movement of the Filter Cutoff picks out harmonics that are almost brutally accentuated by the distortion unit. In places, the sound shifts between breaking up and reducing to a pure sine wave and, while not always matching the ear-bleeding extremity, the sound of the distorted and overdriven TB-303 has become a mainstay of artists such as The Chemical Brothers and The Prodigy. As we shall see when we explore the myriad remakes and clones of the instrument, this sound has become so much a part of the character and flavour of the TB-303 that circuitry to recreate the effect of external processing units and effects pedals is often built into the unit itself.

The reinvention of the uses and sound palette of the TB-303 has ensured that it has remained a staple of electronic music and, while it came to define the sound of Acid House, its contribution is now far more widely-felt, appearing in tracks from S'Express ('Hey Music Lover' and 'Coma AM/OK' (1989) the latter reprising the use of the instrument in its highest register) through to Massive Attack and Madonna where the use of the instrument in 'Protection' (1994) and 'Ray of Light' (1998), respectively, comprises a repeating single-note pattern that both harks back to Heaven 17 and perhaps speaks to the difficulty of programming the sequencer.

With its sounds endlessly emulated and found in almost every list of digital synthesizer presets from the 1990s onwards, perhaps Fat Boy Slim said it best in the title of his 1996 track 'Everybody Needs a 303'. However, wherever it is used, whether in Techno or mainstream pop, its presence is always accompanied with the word 'Acid'.

The Voice of the Machine

Looking at the simplicity of the block diagram in Figure 5.2 while considering Hamill's not unreasonable assessment of the TB-303 as 'a simple analog instrument which produces just two different waves', we might wonder what possible sonic richness could be coaxed from the device and whether this talk of never-before-heard sounds is just hyperbole. And yet these simple circuits interact with the esoteric sequencer in ways that create extraordinarily visceral sounds. By discovering and harnessing these qualities and by placing them centerstage, the pioneers of Acid House not only created a new genre but, in doing so, they transformed the TB-303 into one of the most immediately identifiable electronic musical instruments ever created. Key to this sound, and perhaps the most widely discussed aspect of the instrument, is its filter.

> The core of the acid house sound is the TB-303's low-pass filter. With the cutoff frequency and resonance turned up high, the sound veers between angry wasp and robotic gurgle, with a rawness that other synths, such as the MiniMoog, do not have.
>
> (Edwards and Wilson 2009)

But, while the filter may be the most immediately recognisable feature of the of the TB-303, it is also the one that is most shrouded in mystery and misunderstanding. As Tim Stinchcombe (2010) notes, 'Start reading about the Roland TB-303 and it will not be long before you come across mention of its "18dB 3-pole lowpass filter"'. In this context, the number of 'poles' and dB/Oct specification refers to the frequency response or 'steepness' of the filter. In musical terms, these abstract specifications have a profound effect on the sonic quality and, given the importance of the filter in the design of a subtractive, analogue synthesizer, they contribute a large part of the character of an instrument as a whole. With most synthesizers, including all of Roland's previous instruments, offering either 24dB/Oct (Minimoog, Sequential Circuits Prophet-5) or 12dB/Oct (Oberheim-style) filters, the TB-303's apparently singular implementation of the unusual 18dB/Oct topology has become the stuff of legend and is typically understood to be what gives the instrument its unique character (see Jenkins 2020: 199). So revered is the design that 18dB/Oct filters are sold by a number of companies promising to bring the sound of the TB-303 into Eurorack modular systems with Doepfer's A-103 module being one of the first to capitalise.

If we dig deeper, however, we find that this lauded filter design is not quite what it seems. In fact, as Stinchcome's remarkably detailed analysis unequivocally demonstrates, the TB-303's filter is not an 18dB/Oct, 3-pole design at all. By exploring both the schematics and the frequency responses, Stinchcome's investigation is so thorough and conclusive that there is absolutely no doubt that the TB-303's filter is, in fact, a 24dB/Oct 4-pole design. In the face of such certainty, we might ask why the instrument has been the subject of so much mystery and misinformation. Part of the reason might be found when we realise that even Roland, themselves, have made reference to the distinctive qualities of the instrument's '18dB per octave ladder low-pass filter' in promotional literature celebrating the TB-303's ongoing influence and accounting for its 'magic' (Roland 2014).

Without doubt, the TB-303's filter has a distinctive sound and the idea that there might be some 'secret sauce' or unusual topology certainly adds to the mystique of the instrument. However, the truth is, in many ways, more interesting. From a technical perspective, while the filter is a 4-pole ladder-based design that draws more than a little inspiration from Bob Moog's influential creations perfected in the Modular and Minimoog instruments, the TB-303 uses an array of diodes in place of Moog's transistors. As Riesterer (2017) suggests, this might be, in part, born of an attempt to avoid infringing Moog's design as protected in Patent No. 3,475,623 'Electronic High-Pass and Low-Pass Filters Employing the Base to Emitter Diode Resistance of Bi-Polar Transistors'. One consequence of the use of diodes is that the design creates considerable interaction between the poles which gives the filter an unpredictable and raw sound. However, even this does not tell the full story as Kikumoto's team put much energy into honing the filter and crafting its characteristic sound. Noting that the synthesizer engine and filter design were led by Yoshiro O-e, Kikumoto notes that,

> The VCF developed here had a cleverly designed stable feedback circuit so that it will not self-oscillate, and perhaps this led to the bold sound designing in the later Acid music etc.
>
> (Kikumoto 2021)

Self-oscillation here refers to the quality of a filter in which, regardless of the incoming signal, setting high levels of resonance, results in a sine wave (Goodwin 2019: 228). In many implementations, such as the Minimoog, this is the only way to produce a sine wave and setting up the filter to be controlled (or 'tracked') by the keyboard makes this filter and its pure sine wave output playable as a 'hidden' oscillator. As such, unlike the majority of other filter designs including those at the heart of other Roland instruments (and other x0x instruments such as the SH-101 and MC-202), the TB-303's filter retains the diode 'squelch' with the foundational sawtooth or square wave oscillator always present even at high levels of resonance.

Further adding to the filter's characteristic sound is the apparently innocuous Accent control. Set on a per note basis in the sequencer with an overall intensity amount continuously variable via the sixth rotary control on the front panel, users familiar with the TR-606 might expect the Accent to provide a momentary spike in the amplitude or volume of an affected note. However, the Accent actually affects both the amplitude and filter creating a distinctive 'wow' perhaps in an attempt to mimic the percussive sound of a plucked or 'popped' bass guitar string. More than this, however, a feature of the design of the Accent circuit means that successive Accented notes actually increase in the intensity of their pluck. This occurs because a capacitor in the Accent circuit does not fully discharge and return to zero before it is reactivated thereby leading to a steadily accumulating control signal. At high tempos, this effect is particularly pronounced and so, as music genres get faster, the bite intensifies and grows yet further as the components fail to keep up.

Regardless of whether the effect of the Accent increasing as it is repeatedly triggered is an intentional design choice or an unintended consequence of the interplay of users and circuits in action, it is so distinctive and unusual that many attribute it with the long-lasting appeal of the TB-303 and the connections musicians and listeners develop with it. Among the most poetic, Robin Whittle suggests that,

> Human listeners – and probably a few furry quadrupeds – interpret this output of increasingly high pitched audio signals as sounding like the cry of a living creature becoming increasingly distressed, apparently in response to quickly repeated bursts of stimulus.
> (Whittle 1999)

As so much attention has been given to the filter it has overshadowed a number of similarly important facets of the TB-303. The sequencer's Slide feature, for instance, that allows adjacent notes to smoothly glide in pitch between one another without retriggering the attack of the amplitude and filter envelopes is one example. This smooth transition between pitches is more usually known as portamento and, as a real time performance feature, has been available in synthesizers almost since their inception. However, it is the implementation in the TB-303 and, in particular, the way the sequencer makes it possible to incorporate portamento into sequences in ways that categorically do not emulate the performance of a bass guitarist, that makes its presence so characteristic. Because a Slide can be programmed between any two notes in the sequence, the pitch swoops can vary widely in range. Moreover, Slides can be applied to more than two consecutive notes to create a swirling effect that is wholly 'unrealistic' but utterly captivating especially when combined with movements of the Filter.

In truth, it is impossible to pick any single facet or feature of the TB-303 and claim that it defines the instrument. Rather, it is the coming together of all of these features that truly gives the TB-303 its uniqueness. As with other instruments we will see in this book, including the MC-202, to fully understand the TB-303 we must move beyond thinking of it as a synthesizer or a sequencer, or even as a synthesizer with a built-in sequencer as the two elements are, in reality, indivisible. Pragmatically, this is partly because there is no other way to play the TB-303's sounds, but it is also because the sound of the instrument is so profoundly affected by the specific features and functions of the sequencer. The combination of Accents and Slides means that beyond merely defining the timing and triggering of pitches, the sequencer allows the sound to twist and turn. The sequencer does not merely trigger the sound circuits, but manipulates them. It is for these reasons that the TB-303 is celebrated for its malleable and 'liquid' qualities (Van Der Vaart 2012).

That the sequencer is so unintuitive even for those with sufficient knowledge of music theory to interpret the manual, means that the practice of creating patterns remains, for a great many musicians, a matter of exploration and discovery rather than deliberate transcription. Given that 'the sequencer would shun the familiarity of all previously known forms of music.' (Raval 2010), it is small wonder that experimentation and attentiveness become key. As A Guy Called Gerald notes,

> In those days I used the 303 in a very unique way. I didn't have the manual initially, so I used it with a more serendipitous approach. I would take the batteries out and put them back in again to reset the memory. In 'Write' mode, I would tap out a rhythm, write in a few notes, and press play. There you would have the beginning of a new track. All of the writing I did on the SH-101 and 808 would be direct and conscious – exacting and purposeful. The 303 was the improvisational conductor of the orchestra.
> (A Guy Called Gerald in Sparks 2020)

And so, the compositional act here is driven by the technical complexity and affordances of the instrument which are not limited to the features documented in the operating manual (even if one is available). The practice of entering data in Pitch and Time Modes, saving what is desirable and discarding what is not, clearly arises from the impenetrability of the instrument's workflow. However, that the instrument has been designed to temporarily retain patterns while its backup batteries are replaced gives rise to yet more opportunities for experimentation. By rapidly turning off and on the power, or by using batteries that are nearly depleted, the unit's memory can be deliberately corrupted to create new patterns based on glitches and the manipulation and warping of previously saved data. By 'power starving' the instrument in this way and creatively embracing the effects of its malfunction, users can essentially add randomisation and generative features to the sequencer.

As such, despite early reviews suggesting that the key to the instrument was writing down musical ideas and transcribing them into the instrument, the improvised and experimental workflows that emerge through subsequent use in the hands of musicians must, at least in part, be responsible for the characterisation of the TB-303 as a 'happy accident' machine. Yet, this designation underplays the role of the musician and the complex relationship with the instrument. The careful act of listening to what the instrument is doing, while guiding and influencing it, might be better characterised as co-creation. Indeed, DJ Pierre often refers to working with the TB-303 not in terms of programming but rather as a process of discovery.

The Origins of Acid

Since its re-release and 'rediscovery' in 2010, there has been considerable discussion about the role of Charanjit's Singh's 'Ten Ragas to a Disco Beat' in the origin of Acid House with some commentators declaring that 'Acid House Was Born in a Mumbai Basement' (Wong 2014) and that the 'Pioneer of Acid House' (Weiss 2015) invented the genre 'accidentally' (Chatterjee 2015). However, as Dayal (2010) notes, the album's title and its reference to a 'Disco Beat' might be slightly misleading for contemporary listeners as what Singh produces is more closely connected to the expressly synthetic 'minimal, skeletal Moroder-style disco' than other expressions of the genre. Indeed, we might argue that the drum and percussion elements in 'Ten Ragas to a Disco Beat' sound far more like modern forms of electronic music like House or Techno than 1970s Disco. This is due, in no small part, to the continued and unabashedly electronic contribution of the TR-808 which, necessarily, makes no concessions to acoustic drums. Rauscher (2010), however, seeks to explain the reason for the potential misnomer of Disco in the album's title by noting that, 'there simply wasn't a genre called techno or house that could be invoked.' Certainly, it is unarguable that Charanjit Singh's work appears ahead of its time in this regard.

Of course, the strongest connection between 'Ten Ragas to a Disco Beat' and Acid House centres on the foregrounding of the TB-303 and its sound which appears prized not for its ability to emulate an electric or acoustic bass, but rather for its ability to sound like precisely nothing else. It is certainly the case that Singh's album and 'Acid Tracks' are both celebrations of the electronic sound of the TB-303. However, what distinguishes the two is the role and function of the TB-303 as a musical instrument. Rather than using it as a pre-programmed 'set and forget' accompanist that could provide an additional set of virtual hands while a soloist busied themselves with other instruments, 'Acid Tracks' transformed the TB-303 into a tactile, hands-on performance instrument. The TB-303 was the instrument the soloist busied themselves with. In this way, part of the transformation of the

TB-303 in its second life from the mid-late 1980s sees it take centrestage rather than being propped up on a music stand providing unacknowledged accompaniment. In its new guise, it is the row of six rotary dials that truly unlock the TB-303 and turn it into a playable instrument.

Reflecting on the emergence of Acid House, the influence of Phuture's 'Acid Tracks' and the role of the TB-303, Trask noted as early as 1989 that Phuture's record was important in,

> (re)introducing the idea of timbral modulation as a significant (indeed, prominent) musical feature. Much of the interest in 'Acid Tracks' springs from its syncopated cross-rhythm within the bass riff, which tugs against the 4/4 beat with varying strength depending on the timbral changes and associated degree of rhythmic reinforcement from the 'external' resonance. I can't help feeling this is something not many people picked up on in the subsequent rush to cash in on the acid – and TB303 – craze. In fact, the possibilities thrown up by 'Acid Tracks' have for the most part been sadly under-explored. As house producer Bam Bam (Chris Westbrook) has said: 'It's one thing having a 303, it's another thing knowing how to use it.'
>
> (Trask 1989: 26)

Some commentators have sought to connect the emergence of this focus and the real time manipulation of the instrument's tone controls with the esoteric sequencer and, in particular, its modal design that disallows real time editing or modification of notes. As Olney (2022) notes, 'The combined difficulty of sequencing and the ease of using parameter knobs to sculpt timbres led early acid house pioneers to focus on the latter. Thus, repeating sequences with shifting timbres became a defining feature of acid house.'

For contemporary musicians, the idea of real time control over tone parameters might be taken for granted as the raison d'etre of analogue synthesizers and their tactile control surfaces. Yet, it is fascinating to remind ourselves that the Moog Taurus bass pedals placed a perspex cover over its tone controls to protect them from being accidentally adjusted during performance. Seen in this light, DJ Pierre's subsequent discussion of the approach to the TB-303 during the writing of 'Acid Tracks' is revealing.

> It wasn't a mistake to twist the knobs. I didn't trip and fall onto the 303 and hit a knob then go, 'Wow, what's that?'!
>
> (Ransom Note 2021)

Perhaps unencumbered by the manual and its sense of how the instrument might be used in the specific rock and pop music contexts it sets out, this process of discovery encounters the instrument on its own terms and explores what it is capable of doing and saying. By eschewing a priori assumptions, the instrument can be embraced for its potentiality rather than being seen as a series of complex and esoteric procedures and limitations. With the instrument and performer locked together, Heermans (2016) suggests, the creative work of DJ Pierre and others created a context in which 'the machine spoke to us in its own language' and revealed itself to be a 'rhetorical actor' with a 'unique sonic character and strange mode of programming were persuasive and affective on those who heard it or used it to produce music.' In conceiving both the instrument and user as actors, there are parallels with improvisational performance as the sequencer exerts its agency by generating unusual combinations of notes and Slides which are sculpted, contoured and given life and movement through the manipulation of the synthesis controls.

> By manipulating knobs, it creates sounds but it stretches sound. It'll twist it, it'll pull it. It's like messing with a rubber band – you can stretch it in all kinds of different ways. No instrument does that. You're able to do things that can't exactly be repeated. Every twist and turn is never to be heard exactly the same way.
>
> (DJ Pierre 2019)

It is perhaps in the dynamism and real time manipulation of the tone controls that Acid might be differentiated from other (earlier) uses of the TB-303. In particular it is in the interaction of sequencer and synthesis, between the two actors, that the characteristics of Acid are found. Sequenced notes and Slides appear and disappear as Envelope settings are increased and decreased, Accented steps set in the sequencer differently punch through as the real time Accent and Envelope amount controls are twisted. And, of course, the filter Cut Off Freq and Resonance are continually kept in motion alternately adding bite, squelch and liquid sinuosity. And it is in these interactions that the truly distinctive features of Acid and the new voice of the TB-303 are unveiled.

By contrast, Charanjit's Singh's TB-303 is more tonally static. The sound has an elegant, sinuous quality, without doubt, but it does not continually writhe as if struggling to be contained. In many ways, Singh's is a more polished sound and while the reinterpretation of motifs from Indian classical music is as innovative as it is unmistakably electronic, the TB-303 is more controlled and less sonically shocking. The musical material in 'Ten Ragas', however, has considerably more movement and variation than the audaciously repetitious and spartan sequence in 'Acid Tracks'. Where Acid privileges repeating sequences and timbral movement, Singh's melodically and harmonically complex music is built around a TB-303 that is comparatively tonally restrained, yet whose sequences continually twist and turn.

None of this can in any way detract from the importance of 'Ten Ragas to a Disco Beat' which remains a foundational recording. Indeed, the renewed interest in the album and its creator gives us a richer understanding of the musical application of the TB-303 at the time of its release as well offering considerable insight into the histories, cultures and practices of electronic instrumentation and composition in Indian film and popular music genres. In later interviews, even Singh seemed somewhat perplexed at the suggestion that his work might have been an originating force. Reflecting on the connection while listening to excerpts of Acid House tracks, Singh suggests that 'It's very simple this music ... What I played are ragas – there's a lot of variation' (Aitken 2011) while DJ Pierre similarly notes, 'The acid sound isn't the programming of the notes, it's the twisting of the knobs that made the sound become something different' (Cameron 2022). Ultimately, whatever the lineage between 'Ten Ragas to a Disco Beat' and what became known as Acid House, it is certainly the case as Wong (2014) elegantly puts it, Charanjit Singh 'peered into the future of electronic music.'

Using and Misusing: The Machine Speaks

It also is notable that many commentaries suggest that the re-imagination of the TB-303 through Acid House and beyond occurs as a consequence of the instrument being 'misused'. The implication here is that the dynamic, highly resonant, sinuous sound had not been anticipated by the original designers who were solely occupied by their attempts to recreate the sound of a bass guitar. Accounts of Acid typically point to artists who 'pushed the 303 beyond what it was meant to do' (Sky Speir 2019) with the instrument having 'unconventional settings applied' (John 2016). As Cameron (2022) suggests, 'The resulting squelch when specific knobs are turned farther than the synthesizer's manufacturers likely intended was what elicited a visceral reaction on the dance floor.'

Plainly, it is unarguable that all the parameters and controls needed to create these sounds are available directly by manipulating the knobs on the instrument's front panel and there is no need to open the unit to make any modifications or tweaks to the circuitry. With this in mind, we might reasonably ask whether it is plausible that Kikumoto and the development team at Roland were unaware that the instrument was capable of making such sounds. 'Someone must have twisted the knobs before us though, surely. They must have just thought it sounded like crap. We were the ones to say that it sounded good and thought that it could be something.' (Ransom Note 2021). Continuing the more recent interrogation of the TB-303's origin story, Brotman (2022) notes that,

> If it was truly intended to purely replicate a bass guitar synthetically, then wouldn't they have limited the 303's sonic signature to just the low frequencies? Or made other choices to restrict the sound to be more, well bass like?

As such, while Kikumoto has been reasonably silent on the development of this and other instruments during most of his career at Roland, there have been recent glimpses that suggest that, just as with the TR-808 and its elongated Bass Drum, there was a conscious decision to provide latitude and flexibility for sound design beyond the emulation of electric bass tones.

> The TB-303 was meant to emulate the electric bass guitar sound. However, my team and I did allow the sound to expand with the filter. The one aspect that we consciously limited was self-oscillation with the filter resonance. In fact, if you turn the resonance knob to the max, it will never self-oscillate, but it will make those characteristic squelchy sounds – which was eventually called acid when the 303 had its rebirth. It wasn't the sound I wanted, but I left the synth architecture intact to keep the possibility of sound design open to users.
>
> (Kikumoto in Brotman 2022)

As such, perhaps we might better understand the initial fate of the TB-303 less in terms of a mismatch between designer and musician and more between designer and marketing department. That said, the timbral movement that is so characteristic of Acid House and almost all contemporary uses of the instrument, is not especially well-supported by the instrument's industrial design. The six rotary controls at the top of the instrument that provide the only means of manipulating the instruments sound circuits and unlocking the liquid manipulation of the synthesis controls are awkwardly recessed into the casing making them rather difficult to twist and turn. The design of these controls works well for the TR-606 where they are used to set the levels of the different built-in drum sounds and present no opportunity for tonal modification. It might surprise us little that the recessed controls appear to better suit the functionality of the TR-606 than the TB-303 given that they were designed for the former and incorporated into the latter. However, the impact of their tactility is rarely commented upon and the combination of their already shallow profile set into the casing of the instrument makes them extremely hard to grasp. Placing one's hands on the six rotary controls of the TB-303, it is immediately evident that these were not designed for real time sound shaping. Or perhaps we might be reminded that real time sound shaping was not considered to be a function of the instrument when it was designed.

In fact, remembering the Moog Taurus pedals and the perspex cover protecting their control settings, the relative difficulty with which Roland's small, sunken silver dials are manipulated in real time might be read as an uncontroversial design. Given that the TB-303 Owner's Manual illustrates a number of 'preset' Bass tones but makes no mention of dynamic changes, we could read the recessing of the synthesis controls less as an unavoidable consequence of an inherited interface and more as a design decision actively intended to help avoid accidental adjustments. And let us not forget that Roland's marketing materials place both of Oscar Peterson's hands on the piano keyboard with the TB-303 and TR-606 ProForming away without the need for manual intervention. What makes the TB-303 seem peculiarly awkward today is that it's hardware interface is comparatively poorly suited to the ways in which the instrument is now understood and used.

In recognition of the centrality of sound manipulation, and responding to the inability to record the dynamic knob movements into the TB-303 sequencer and the absence of any backup or save capability outside the instrument, contemporary players have developed their own notation systems for archiving, sharing and recreating patterns. As Davis (2022) notes, this device-specific notation typically comprises a transcription of the per-step Pitch, Octave and Slide values stored in the sequencer combined with a visualisation of the various sound-shaping controls. Crucially, many of these visual notation systems do not simply set out static values for synthesizer parameters such as Cut Off Freq and Resonance, for instance, but often also notate how the knobs should be manipulated over time or within which ranges for a given effect. In this way, these community-created pattern sheets both reflect and seek to capture the ways in which the sequencer and synthesizer have become indivisible in the contemporary understanding of the TB-303.

In fact, if we want further evidence of the central role of real time synthesis manipulation in the emergent history of the TB-303, we need look no further than the instrument's case. After many decades of being repeatedly twisted and turned by the hands of musicians and performers, it is common to find that the silver paint surrounding the rotary controls has worn away revealing the bare plastic underneath. This is partly a result of the recessed design that sees only a small portion of the control caps protruding and that requires the performer 'reach inside' the instrument to grasp them, pressing their fingers against the case as they do. In fact, the halos have more to reveal and bear witness to the fact that the six controls are not equally used with the Cut Off Freq, Resonance and Mod Depth typically having deeper and wider wear marks than the other three controls. If we were still curious about the importance of the TB-303 filter and the centrality of its manipulation in contemporary electronic music genres, the instrument has a memory that leaves us in no doubt. While it might be the case that these telltale 'halos' are the result of friction and abrasion, it is rather more poetic to imagine they are the result of successive performers' acidic sweat generated in the heat and excitement of the live performance. In this way, several decades after its initial release, it is commonplace to find that the sound of Acid is literally etched into the case of the TB-303.

Devilish Modifications and Rebirths

Even though the synthetic sound of Acid could be coaxed from the TB-303 through the use of its sequencer and the tone controls on its front panel without any internal alterations, there have been a large number of modifications to the circuits that have further transformed the character and capabilities of the instrument. The most well-known is the 'Devil Fish' created by Robin Whittle. So comprehensive are the alterations here that the unit has to be shipped to Whittle in Australia for the work to be completed. Among the key changes brought by the Devil Fish are CV and Gate inputs that allow the sound engine to be triggered by an external keyboard or sequencer. This modification alone completely transforms the TB-303 as a musical instrument by disconnecting the internal sequencer and refashioning it as a synthesizer that can be played like any other. Much attention is also given to the filter which has new modulation inputs giving more flexible control as well as a wider tonal range including a compensation for the reduction in volume that comes with high levels of resonance and which is typically perceived as a drop in bass. The Devil Fish filter is modified to allow it to self-oscillate, again confounding one of the original design decisions and aligning the instrument with other synthesizers in terms of tonality and capability. Indeed, many of the controls have modified scaling that allows them to exceed their original bounds. Maximum Env Mod is now three times higher than the original and fully closes to zero unlike in the original design where it always exerted some influence.

To further process the sound, the level of the oscillator passing into the filter can be increased by a devilish 66.6 times its normal level which results in an extremely overdriven sound, and a Muffler circuit adds soft-clipping distortion after the filter to generate further dirt and grime. While the modifications and new controls can be bypassed to restore the original, unaltered TB-303 sound, if there were any doubt about the intention behind the Devil Fish, the manual soon clarifies the situation in celebrating the 'diverse and extreme' tone. 'If you are interested in solid, heavy bass sounds, idiosyncratic synthesis or weird-and-wonderful sounds you have never heard before, then I promise you that you will really enjoy the Devil Fish.' (Whittle 2019: 9).

It is clear that, beyond simply tweaking the circuits of the original, Whittle's aim is that the Devil Fish transform the TB-303 into a fundamentally new device and we can perhaps detect a degree of frustration in the commentary that notes, 'I think, most people see the Devil Fish as a TB-303, typecast in techno or acid music, rather than an instrument, musical character/critter etc. in its own right.' (Whittle 2019: 30).

While it may be the most well-known, the Devil Fish is far from the only commercial modification to the TB-303. Social Entropy, creators of the Quicksilver 606 we saw in the previous chapter, also manufacture a Quicksilver 303. Like its drum machine counterpart, this modification includes a full CPU and a new operating system that offers additional interfacing possibilities via MIDI and significantly alters the behaviour of the sequencer.

> One comment that is always made about the TB-303 is that the sequencer is difficult to understand and program. This complexity would lead to happy accidents and a different way of thinking about programming patterns. Most modern TB-303 clones focus on the analog synthesis components of the machine, and combine this with a more conventional sequencer engine. Our desire is to keep the 'spirit' of the original sequencer intact, but take it to a new level and prove that the TB-303 still has a few new tricks left to uncover.
>
> (Social Entropy 2012: 2)

It is interesting to note the slight tension in the overview which, as with Whittle's outline of the Devil Fish, is motivated by a desire to explore the potential of the instrument and even add to its capabilities and uses, while remaining reverential of the original workflow and sound. This becomes particularly evident in the design of the new sequencer which does not simply rewrite the operations in the style of a modern device. Rather, Roland's fundamental concepts remain, with Pitch and Time information still being entered separately but, in this new iteration, the inflexibility of only being able to perform certain actions in one or other mode are removed. Similarly, patterns can now be created and edited while the sequencer is still running thereby doing away with the need to stop and start and shift between Play and Write modes. As such, while it is a complete rewrite of the operating system with new features such as a per-step 'Bump' feature that makes some degree of filter automation possible by increasing the cutoff frequency by a fixed amount, the spirit and even some of the apparently anachronistic concepts of the original are retained.

That the sound of the TB-303 has remained desirable since the late 1980s, and with comparatively few instruments produced (in relatively flimsy plastic cases that are not built for longevity), it is hardly a shock to find a good number of clones, remakes and emulations. The sheer number of them might come as a surprise, however. In fact, the TB-303 must surely rank as one of, if not the most, emulated electronic instruments. From the 1990s, the influence of the TB-303 can be felt on dedicated bass synthesizer instruments such as the Doepfer MS-404 and Novation Bass Station even though neither explicitly referenced the Roland original in their design. Moreover, like a number of subsequent instruments such as the MAM MB33 and the various Eurorack-compatible modular components such as the (in)authentic 18dB/Oct filters we noted above, many of the earliest homages focused solely on the sound of the TB-303 and included no built-in sequencer.

More recently, and perhaps in recognition that, despite (or perhaps even because of) its complexity, the sequencer with all of its esoteric operation and idiosyncratic Accent and Slide features, is a key constituent of the instrument's character, recreations have been more holistic in their emulation. Roland's TB-03 is the company's current hardware offering and sits within the Boutique range. This 2017 release followed the earlier, and similarly all-digital, TB-3 (2014) which sought to capture the essence of the instrument with a new touchpad, a host of additional sounds and a rather garish green colour scheme. The Boutique instrument eschews these novelties and returns to the sleek silver aesthetic of the original and replicates the controls and sequencer operation of the original. That said, Roland toes an interesting line in balancing the authenticity of 'an inspiring instrument with an instantly recognisable sound' with one whose workflow means that 'many users will agree it has a number of drawbacks, particularly when being used in a live context.' (Roland n.d.) As such, in addition to the '1980's style interface with all the good and bad of what that entails', the TB-03 also offers 'an easier-to-program, Step Sequence mode' that will be considerably more familiar to users of more contemporary sequencers. Tellingly, the TB-03 also addresses the design of the six synthesis controls which, while still small, are no longer recessed and are altogether easier to manipulate in the manner that modern uses of the instrument require. Most recently, Roland have added a TB-303 emulation to their 'Cloud' library of software instruments where it sits alongside the SH-101 and other synthesizers and drum machines.

This was far from the first software emulation, however as in 1996, Swedish software developer Propellerhead released the first version of its ReBirth RB-338. As the name might imply, this self-contained music production suite was based around emulations of the two TB-303s and a TR-808 drum machine. Although no reference was made to Roland or the model numbers of their instruments, their appearance on screen and sound were unmistakable. As well as both responding to, and helping to drive, interest in Roland's instruments, ReBirth played an important role in the development of software synthesizers and was one of the earliest examples of what we now term 'virtual instruments'. In a period in which Roland were still following founder Kakahashi's instruction not to chase ghosts, it seemingly fell to other companies to satiate the desire to revisit and reimagine these instruments. Intriguingly, Propellerhead's account of the launch of ReBirth notes that many of the initial downloads of the software originated from Internet addresses from within Roland's offices. However, rather than issue any instruction to cease production over rights infringements, Roland subsequently required the inclusion of the following text in ReBirth's advertising and documentation.

> ReBirth RB-338 was inspired by the TR-808 and TB-303, originally created by Roland Corporation. Their unique sounds and visual images have been re-born through digital simulation by Propellerhead Software.
>
> (Propellerhead 2005)

Although Propellerhead present this as a tacit endorsement, the software was later withdrawn from sale in 2017 apparently as the result of an intellectual property infringement claim from Roland (Wilson 2017). Although it is no longer available, ReBirth was certainly pivotal in continuing to cement Roland's place in popular electronic music, especially when version 2 added the TR-909 to the mix. Following this lead, software emulations are legion and include D-16's Phoscyon, Audiorealism's ABL 3 and Arturia's Acid V. Where once it was extraordinarily scarce, such emulations ensure the TB-303 exists across almost every contemporary computing platform including Mac, PC, phones and tablets.

In the hardware realm, more recent instruments such as the Mode Machines x0xb0x and Cyclone Analogic TT-303 each explicitly reference the TB-303 while as part of their growing line of instruments, Behringer's TD-3 also offers an analogue reimagining of the original instrument with the TD-3-MO adding the circuitry of the Devil Fish modification, albeit without the blessing of its designer, Robin Whittle. For all the references to the original instrument in the names, appearance, synthesis and sequencer design of these instruments, however, none are as meticulous in their recreation than the DinSync Re-303. Looking almost indistinguishable from the original and proudly defining itself as a 'replica' rather than a clone or emulation, this kit makes use of a combination of original components where possible and modern equivalents where not. As such, it is perhaps the most accurate, and reverential, recreation of the TB-303 to date.

With this example notwithstanding, it is notable that most hardware and software recreations do not simply remake the standard TB-303 and it is important to note the degree to which these recreations contribute to the TB-303 being synonymous with Acid. Sequencer randomisation features seek to automate the composition strategies we have noted above while some remade sequencers even include a range of algorithms that simulate different types of Acid pattern. Roland's own TB-03, for instance, allows the randomisation of notes and timing which can be further transformed with a secondary randomisation routine that adds octave variations, Accents and Slides to the underlying sequence. This two-stage process offers a degree of iterative control that encourages the repeated generation of randomised notes which, when considered satisfactory, can be frozen to form the basis of additional mutations. Procedures such as these continue the discursive production of the TB-303 as a happy accident machine based around repeated one-bar Patterns and steer away from the deliberate programming methods self-evidently encoded into the documentation, marketing and, most importantly, the design of the instrument itself.

In a similar fashion, most TB-303 recreations add various third-party, aftermarket alterations and incorporate widely used external processing circuits such as guitar effects pedals. For example, the popularity of the influential overdriven sound used by Daft Punk, Josh Wink, the Prodigy and others that grew form the mid-late 1990s, means that modern TB-303 recreations almost always include a distortion or drive circuit with many, such as Roland's TB-03 offering multiple varieties (based on Boss guitar pedals) along with reverb and delay effects. Reflecting the way in which these processed sounds have come to define the character of the instrument in certain genres, it is revealing to note Roland's own cautionary advice when approaching their recreations. 'Bear in mind the original 303 on its own doesn't sound like you'd expect it to – many producers connect it to a distortion unit for that aggressive growl, or delay unit for the spacey effect' (Roland 2019a). Ultimately, this rather neatly summarises the history of the TB-303 as it is an instrument that continues to be reshaped, reworked and revered by generations of musicians. Reimagined perhaps, but never replaced. A successful failure indeed.

6 SH-101 Synthesizer

Figure 6.1 Roland SH-101 Synthesizer (image credit: James Newman).

DOI: 10.4324/9781003374510-6

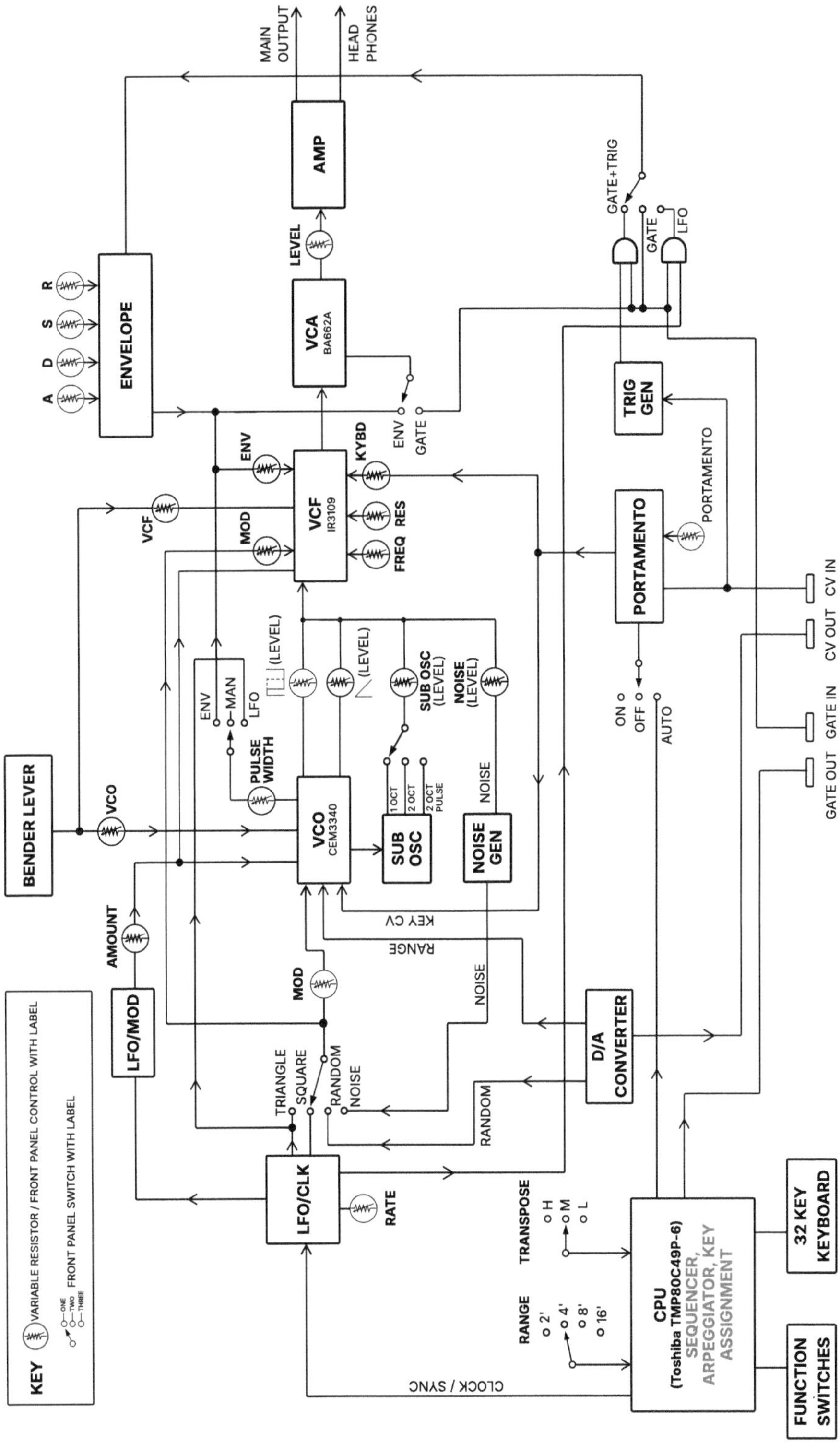

Figure 6.2 Roland SH-101 Simplified Block Diagram (adapted from Roland 1982e SH-101 Service Notes).

SH-101 Synthesizer

Released
1982 (discontinued 1986)

Sound source
Curtis Electro Music CEM3340 Integrated Circuit
Simultaneous Saw, Pulse and Sub Oscillator and Noise waveforms with independent Level control
Sub Oscillator can be set 1 octave below (Square wave) or 2 octaves below (Square or thin Pulse)

Filter
Roland IR3109 Integrated Circuit
24 dB/Oct Low Pass with resonance
Self-oscillating filter Tracking

Modulation
LFO with Triangle, Square, Random and Noise waveforms
Sine Wave is available for pitch modulation (vibrato) via Bender lever
ADSR envelope
Modulation destinations include Amplitude, Filter Cutoff and Pulse Width Modulation

Keyboard
32 keys, F-Scale

Sequencer
Load and Play up to 100 steps

Arpeggiator
Up, Down and Up & Down directions, with Hold control

Connectivity
Line Level and Headphone audio outputs (monaural)
CV and Gate Inputs and Outputs
External Clock input for synchronising LFO, Arpeggiator and Sequencer

Power
9V (centre negative) power supply or 6 C-type batteries

Dimensions
570 (W) x 311 (D) x 80 (H) mm
22-7/16 (W) x 12-1/4 (D) x 3-1/8 (H) inches

Weight
4.1kg / 9lb (without batteries)

Options
MG-1 Modulation Grip Set
Available in three colours: Silver, Red and Blue (with matching MG-1 modulation grips)

Suggested listening
Portishead 'Mystärens' (Go! Beat, 1994)
Boards of Canada Roygbiv (Warp, 1998)
A Guy Called Gerald 'Voodoo Ray' (Rham!, 1988)

Flops and Bombs

The accepted wisdom of the SH-101's early life is that it was an unmitigated failure. Modern reviews such as that offered in Stokes (2022a) are illustrative of this dominant narrative which tell us that, 'Released in 1982 and discontinued in 1986, the SH-101 was something of a commercial failure in its own era'. Despite no definitive sales or manufacturing figures, some commentaries go even further by positioning the SH-101 as an unequivocal failure at launch with Skytracks (2019) suggesting that 'It was a total flop' and Anniss (2017) simply concluding that, 'It bombed.'

On the face of it, however, it should have been a sure-fire winner. After all, at its heart is the Curtis CEM3340 oscillator chip that had powered the hugely influential and still popular Sequential Circuits Prophet-5 along with a filter based around Roland's own IR3109 which is the essentially same chip that gave the Juno-6 and Jupiter-8 their luscious sound. And these illustrious components were all housed in a snazzy-looking keyboard that could be battery-powered for portability and which even came with a leather guitar-style strap and modulation grip to transform it into a spotlight-grabbing 'keytar'. Where previous Roland synthesizers like the Junos had wooden end cheeks just like the previous generation of 1970s Moogs and Prophets, the SH-101 looked shockingly modern even compared with other instruments in the SH range. Made entirely from plastic, the SH-101 was eventually available in three colours with metallic blue and red models joining the original silver/grey instrument. The keytar stylings, strap and modulation grip (also available in matching silver, blue and red) and some of the most quintessentially 1980s lifestyle advertising all gave rise an instrument that, in a 2014 retrospective was declared by *Computer Music* magazine as 'tailor-made for the camera-conscious musician of the 80s.'

In a sense, the SH-101 fell slightly between two stools. It was certainly a professional-grade synthesizer with the kinds of features, sound quality and playability that would be expected of such an instrument. However, it was clear that Roland also intended the instrument to provide an accessible and comparatively affordable route into synthesis for new players. With interests piqued by the waves of international synth pop and electronic music, the SH-101 can be seen to follow in the footsteps of other pioneering 'entry level' instruments like the Casio VL-1. Perhaps most immediately indicative of the SH-101's beginner-friendly design is the provision of an arpeggiator to automate the performance of complex patterns which could also be latched on thereby freeing up the player's hands to manipulate the sound in real time. The SH-101 also included a sequencer that could be programmed with up to 100 notes. Like the arpeggiator, this could be synchronised with another instrument such as a drum machine enabling it to fit seamlessly into a larger set up. While a sequencer was already a comparatively unusual addition to a monophonic synth (though it was a key part of the Casio VL-1), the SH-101's method of inputting notes is considerably easier to use than Roland's dedicated sequencer units or the TB-303 and remains influential and widely emulated even today.

As we shall see, although the idea of the SH-101 being an immediate and resounding flop is somewhat overstated, it is certainly the case that its popularity would begin to significantly climb some years after its release where it would be 'rediscovered' alongside instruments such as the TB-303 in what Jenkins (2020) and others have called the 'analogue revival'. In fact, revival possibly undersells the degree of interest in the SH-101. Not only are the original instruments highly desirable, especially in their more limited special edition colours, but there are also myriad emulations and clones available in software and as analogue and digital hardware instruments. In addition to recreating the distinctive sound of the instrument with the unmistakable growl that comes from the combination of its oscillators and sub-oscillator and the 'rubbery' filter and snappy envelopes, many emulations and remakes also incorporate some of the numerous third-party modifications and adaptations that companies and individuals have developed over the years to further explore the sound.

Armed with the knowledge that the SH-101 would go on to become an iconic instrument as coveted and revered as almost any synthesizer in Roland's history and, for some players at least, one of the best and best-loved synthesizers ever created (Sunshine Jones 2017), the idea of a misunderstood, unliked and unappreciated in its time and rediscovered years later as a 'hidden gem' makes for a compelling origin story. However, like so many of the instruments under examination throughout this book, the history of the SH-101 is actually rather more complex and requires we learn a little more about the context in which it was developed and released.

The Right Synth at the Wrong Time?

The synthesizer landscape at the beginning of the 1980s was fast-moving and a place where instruments just a few years old could look hopelessly old-fashioned and out of date extremely quickly. Indeed, by 1985, the broad market survey in David Crombie's *The Synthesizer & Electronic Keyboard Handbook* concluded that the modular synthesizers which had dominated the landscape for many years were now relics of a bygone era. With the 1979 Roland System 100m as the example, Crombie (1985: 61) notes, 'These machines are neither very common nor widely useful. Modular synthesizers are essentially studio tools and since studios offer multitracking, a polyphonic modular instrument is seldom required.' With a new generation of synthesizers such as the Sequential Circuits Prophet-5 (1977), Oberheim OBX (1979), Korg Polysix (1981) and Roland's own Jupiter-8 and Juno-60 offering not only the ability to play multiple notes at once and harnessing digital technologies to offer patch storage and presets, the SH-101 was all analogue, monophonic and had no patch memories. On paper at least, it might look like yesterday's instrument and a flop in the making.

Yet, it if we survey previews, reviews and commentary from the early 1980s, we reveal a rather different picture. Contemporary sources reveal that the SH-101 was actually very favourably, even enthusiastically, received. Writing in *One, Two Testing* magazine in December 1983, Paul Colbert offers a gushing review.

> But there's scarce doubt that in terms of value, facilities and fashionable desirability, the SH101 dusts the sideboard with every other match-priced mono on the market. ... Though Roland are bound to be working on a follow up (when are they not?), they've already arrived at a classic.
>
> (Colbert 1983a: 68–69)

Even in 1985, there is little evidence of any dilution of enthusiasm for the instrument as Julian Colbeck offers similarly effusive praise in the first of the influential *Keyfax* books. Awarding the SH-101 the highest grading of 'QQQVVV' for Quality and Value for Money, Colbeck (1985: 29) declares that, 'If you want good, strong sounds and want to *use* those sounds in a variety of ways, then the SH-101 will do you proud. ... The SH-101 is probably the easiest, and most adaptable, mono around.'

Alongside these glowing reviews, it is important to note just how present the SH-101 is in popular publications of the time. Not only does the synthesizer receive its glowing review in the pages of 'Keyfax', but it is also one of only four instruments to feature on the cover. Moreover, in setting itself the ambitious task of guiding readers through the 'new-generation of keyboard instruments', Crombie's *The Synthesizer and Electronic Keyboard Handbook* dedicates a double-page spread to the SH-101 as the exemplar of monophonic synthesizers. Replete with annotations and commentary on the kinds of features and architecture one might expect to find on such an instrument, the SH-101 stands here not merely as an example, but as the archetypal monosynth. If nothing else, its sheer presence in publications of the time makes it hard to ignore.

However, it is a feature in the September 1983 issue of *Electronic Soundmaker and Computer Music* that is most intriguing. Exploring the available range of 'lead synthesizers' (instruments used to perform melodic solos, a task for which monophonic synthesizers are ideally suited) the article opens its ebullient consideration of the SH-101 with a tantalising sentence. 'The market leader, and may I say what a lovely little machine this is.' (Mowat 1983: 30). Given the more recent narrative that has positioned the instrument as a 'flop' that 'bombed', talk of the instrument being the market leader comes as something of a surprise. Yet, some more digging in the archives reveals further support. In the January 1983 issue of *Electronics & Music Maker*, a feature in the semi-regular 'Talking Shop' series showcasing UK-based electronic music technology retailers, discusses the opening of Capital Keyboard Centre, a new store in the centre of London. Following a rundown of the facilities and current stock which includes 'the best-selling Roland Juno 6, Juno 60 and SH-101', the article concludes by noting that, 'Business seems good, with twenty SH-101's selling in two days.' (Talking Shop 1983: 16).

It goes without saying that the fortunes of one store cannot speak to the fate of the SH-101 as a whole but, taken in combination with other factors such as the glowing reviews and sheer visibility of the instrument in popular publications of the time, a significantly more complex picture of a well-regarded and perhaps even successful synthesizer begins to emerge. In many ways, the revelation of positive responses to the SH-101 invites us to consider some altogether more interesting questions that are overlooked by the commonplace assertion of the synthesizer's status as an initially unloved failure. Given the rise of the polysynth that leads Colbeck to reasonably postulate in the opening of the Keyfax review that this is 'The latest, and probably last, of Roland's monophonic synths' (1985: 29), what was it that so appealed about this diminutive instrument?

The sound of the SH-101 is certainly a key talking point. 'For a single oscillator keyboard it has a vast sound, and bass lines work a treat.' (Colbert 1983a: 68). However, in addition, reviewers and commentators conspicuously and reliably draw attention to the SH-101's price. This is hardly surprising as Roland had priced the instrument comparatively aggressively at ¥60,000/$US495/£249. With polyphonic synthesizers like the Juno-6/60 retailing for nearer four or five times the price, the SH-101 was the cheapest keyboard in the Roland lineup by some margin and, for that reason alone, was a potentially compelling instrument. It would be tempting to think that this might be the major reason for such a positive reception but, in fact, this was not the first time that Roland had released a comparatively inexpensive synthesizer. While the general trend throughout the life of the SH series had been one of expansion and more complex sound production and manipulation features, Roland had already experimented with leaner implementations.

Released two years before the SH-101, the SH-09 was the company's first attempt to leverage its expertise and technology to expand its range beyond the professional musician and studio market. Roland's UK marketing left no doubt about the target audience for the instrument with print advertisements reading, 'When you buy your first synth, make sure you can create and expand. Roland consider this essential to all design concepts – the ability to interface. … Budget price synth with expandability.' (Roland (UK) Ltd 1981: 2)

Not the snappiest advertising copy, without doubt, but an unambiguous statement, nonetheless. With its stripped back set of controls and reduced functionality that offered, for instance, just a single oscillator compared with two found in the larger (now seemingly more professional) SH models, the SH-09 was an entry point into the world of synthesizers. However, in amongst that somewhat clunky marketing prose lurks a potentially conflicting set of messages that might actually serve to dissuade entrants in the new world of music technology. While the SH-09's 'expandability' could be read as indicator of its latent potential and futureproofing, it could be seen to speak to its incompleteness and the need for it to be built upon. Indeed, in the image that accompanies the advertising copy, the SH-09 is pictured with the matching CSQ-100 sequencer (sold separately, of course).

Roland's invitation to 'Safeguard your investment – compare our features and value for money' makes their preferred reading of the advertisement clear but it is rather difficult to confront a list of features that includes references to 'Unique Delayed Modulator', 'Hold-override settings for repetitive percussion effects' and 'Envelope Inversion' without feeling that at least a modicum of a priori synthesizer knowledge, or ownership, is assumed. Similarly, intermingled with the advanced synthesis concepts and jargon, there are myriad references to the ease with which the SH-09 interfaces with the TR-808 Rhythm Composer, the more advanced and expensive CSQ-600 sequencer, and the considerably more expensive SH-1, SH-2, SH-7 synthesizers, or even another SH-09 (for those already planning that second synthesizer purchase before reaching the end of the advert). It is possible that this could be seen as a safe first investment in synthesizers given the reassurance that Roland's entire range including its high end, flagship equipment is fully interoperable. Alternatively, it might signal that this initial purchase could actually be a gateway synth whose true potential is to provide opportunities to spend – or rather invest – yet more money.

What's Old is New Again

As part of Roland's family of monophonic synthesizers that included the more costly SH-1 (1978), SH-2 (1979) and the duophonic SH-7 (1978), the SH-09 was a cut down, comparatively low cost instrument. Where those instruments had two oscillators that could be independently tuned, for instance, the SH-09 reduced this to one. Similarly, where previous SH synthesizers had multiple LFOs, envelopes, different waveforms and flavours of noise, the SH-09 was stripped back to the barest essentials.

On the face of it, the functionality of the SH-101 and SH-09 appear remarkably similar. Both are powered by a single oscillator capable of producing multiple waveforms including Sawtooth, Pulse and Noise. The duty cycle of the pulse waveform can be set manually or modulated via the LFO or envelope. Alongside the main oscillator, there is a Sub Oscillator pitched either one or two octaves below the root note. This offers a choice or Square of thinner Pulse waveform. With faders to adjust the relative mix of the oscillator and Sub Oscillator waveforms, the output is fed into a 24dB/Oct (4-Pole) Low Pass Filter with manually adjustable Cutoff and Resonance to add a pronounced peak to the sound and, at the extreme, a sine wave generated as the filter goes into self-oscillation. Key

Tracking allows the self-oscillating filter to be 'played' thereby adding an extra flavour to the existing oscillators

As the simplified block diagram in Figure 6.2 shows, modulation is provided by a single LFO with multiple waveforms and an extremely snappy ADSR Envelope that can be set to affect combinations of oscillator pitch (for vibrato effects), the Pulse Width of the Pulse waveform (for PWM effects), the Voltage Controlled Amplifier and Filter to further contour and add movement to the sound over time. Both the SH-09 and SH-101 include a 32-note keyboard (running from F-C) the range of which can be extended to at least five octaves via the oscillator frequency knob. The SH-101 does omit a few features from its predecessor including the ability to process an external audio source but, importantly, gains some new ones including the sequencer, that help mark it out as an altogether more modern proposition.

Nonetheless, by simply using the specification sheets and block diagram in Figure 6.2 as our guide, it appears far easier to spot the similarities than the differences between the two instruments. However, while the core building blocks of the synthesizers are designed to offer similar functionality, the analogue circuits at the hearts of the SH-101 and SH-09 are remarkably different. Where the SH-09 oscillator used Roland's JFET Op Amp design that had featured in the System 100m modular and Jupiter-8, the SH-101 switched to the third-party CEM3340 integrated circuit. This self-contained chip was designed by influential US designer Doug Curtis and is found in a number of premium synthesizers created by Sequential Circuits, Oberheim, Moog and others. Interestingly, while the CEM3340 generates a Triangle wave, it is unused on the SH-101 which taps only the Sawtooth and Pulse waveforms as its default sound sources. The SH-101 filter eschews the OTA (Operational Transconductance Amplifier) designs of earlier Roland SH instruments and switches to a version of Roland's proprietary IR3109 integrated circuit that bestows so much of the classic early-80s 'Roland' character to the flagship Jupiter and Juno series instruments. As Craig Sue and Steve Jones' 'Secret Life of Synthesisers' blog concludes, 'With the oscillator from the Prophet 5 and the filter from the Jupiter 8 it's no wonder that the SH-101 has such a big sound.' (2017).

Of course, it is not necessary to delve into their circuits to reveal the differences between the SH synthesizers and even the most cursory glance at the SH-101 marks it out as unique. The SH-09 looks like it belongs to the SH family. Its metal case is entirely black with contrasting white text delineating the various sections of the instrument and clearly labelling each of the individual faders and switches. Within the sea of 18 black faders and the black plastic pitch bender that almost recede into the panel, there are knobs for controlling the oscillator 'Range' and 'Waveform' selection and the overall 'Volume' of the synthesizer. While there are just three of these controls, their almost disproportionately large size and shiny silver skirts lend them an extraordinary prominence. However, what really leap from the front panel are the words 'Synthesizer 09' which are set in a most distinctive font. Designed by Bob Newman in 1970, 'Data 70' is a typeface that almost perfectly captures the futuristic science fiction aesthetic of the decade and was used extensively on film posters, book and record covers. However, as Nørgaard (2009: 156) notes, understanding the typeface's original design intention and reference points provides us with a deeper understanding of its semiotic impact. 'A significant and very salient feature of Data 70 is its boldness at unexpected places, which was originally … motivated by readability for machines rather than for human beings.'

As such, by employing a typeface on the earlier SH instruments that takes its inspiration from designs aimed at assisting in the mechanised process of optical character recognition, Roland place these instruments within a complex network of meanings that negotiate the relationship between humans and machines. While its subsequent use and overuse in the reverie of a retro chic aesthetic might have destabilised Data 70 for modern eyes, there is no doubt that, in the brave new decade of the 1980s, this typeface's well-understood brand of futurism firmly anchored it in the past. It perhaps comes as little surprise to find that it does not feature on the SH-101's front panel. Rather, this new instrument of the 1980s has its name rendered in bold characters comprised of discrete individual points that mimic the pixels of a video display or the output of a dot matrix printer thereby locating this synthesizer not merely within an age of automation and mechanisation but the world of the personal computer. Even before it made a single sound, it was clear that the SH-101 had been designed to look like a modern synthesizer.

And to underscore this, the SH-101's case is strikingly different from its forebears. The basic layout of the panel remains largely similar to the SH-09 with the signal flow moving from left to right in a manner familiar not only to Roland synthesizer users but to players and programmers working with instruments dating back as far as the Minimoog. Gone, however, are the previous generation of SH's metal cases, black faders and large silver-skirted rotary controls. Instead, the SH-101's front panel is made entirely of plastic. Dark grey faders and smooth circular knobs each sport a coloured strip whose hue corresponds to the function they perform, while a new row of rectangular silver buttons bear intriguing legends such as 'Load', 'Hold' and 'Key Transpose'. Stylistically, the SH-101 could scarcely look more different than its predecessors. Where those instruments built a familial aesthetic through their controls, material, construction, layout and typography that connected them back as far as Roland's modular heritage, here was an instrument that broke the mould entirely. Where those 1970s instruments had communicated a reassuringly weighty professionalism with their sober, tasteful aesthetic, here was something shockingly new and perhaps even brash.

To be clear, this was in its standard configuration because the SH-101 could also be expanded with the optional MGS-1 kit that added a 'modulation grip'. By way of two screws and a proprietary wired connection, this peripheral attached to the left hand side of the instrument allowing for real time performance control which might seem unhelpfully awkward for an instrument sitting on a table or keyboard stand but which makes much more sense when combined with the Roland-branded leather strap that was also included and which allowed the SH-101 to be worn like a guitar. The MGS-1 kit transformed the SH-101 from a keyboard ready for mounting on a stand into a stage-ready keytar to be slung around the neck. Or, as Roland more prosaically put it in their December 1982 product brochure, allowing you 'To play the SH-101 while standing up'.

It is interesting to note that, for such a strikingly different looking synthesizer, these initial promotional materials are really rather staid, if a little incongruous. The Japanese brochure first published in 1982, includes a moodily-lit image of musician Mickie Yoshino (credited in the brochure as Micky Yoshino) of the band Godiego. In the image that occupies most of the four-page brochure's cover, Yoshino is pictured casually leaning against the bonnet of a chrome-clad car, staring off into the middle distance with an SH-101, complete with modulation grip, hanging around his neck. Just as incongruously, the cover image of the late 1982 US brochure places the SH-101 on a stylishly modern angular armchair. On closer inspection, however, sitting on the wooden floor in the background of the shot, almost but not quite out of focus, are three equally stylish polished metal beakers and what looks to be a bottle of Jack Daniels whiskey. Scanning back to the foreground, we find that a fourth beaker sits on the seat of the armchair next to what, on subsequent readings of the image that assess the full scene, increasingly appears to be an anthropomorphised synthesizer casually reclining in the chair. Thankfully, inside things are far more intelligible with a remarkably thorough breakdown of every one of the instrument's many controls and functions. In fact, the level of detail offered here both in terms of the SH-101's features and functionality and subtractive synthesis more broadly actually exceeds many of today's printed manuals (including Roland's own 2018 digital recreation of the SH-101 which we explore later in this chapter). This broader context and explanation of some of the foundational principles of synthesis also speaks to the potential range of audiences for the SH-101 and the targeting of beginners taking their first steps.

The Complete Package

Certainly, one of the things that made the SH-101 remarkable was the quantity of new features and functionality it packed into its plastic casing and the ways these breathed fresh life into what could still be seen as the essence of a 1970s monosynth. As Mowat's review notes,

> I have reserved till the very last the features on the SH-101 that undoubtedly make it sell. The sequencer will take up to 100 events (notes and rests) and is transposable as it plays ... The sequencer is remarkably easy to use; it plays in clock time, which means that it will correct your bad timing into a robotic pattern, though you can program in slurs and rests to approximate more or less any tune you have in mind. You can also edit the pattern; if you make a mistake, just stop the sequence before the bad note or passage and load in your corrections. A joy to use. As for the arpeggiator, all you can say is that it goes up. And down. And up. And down.
>
> (Mowat 1983: 30)

And so in the SH-101 we find a huge leap forward in performance features over the previous generations of SH synthesizers. The innocuous row of silver buttons located just above the keyboard unlock a synchronisable arpeggiator and the promise to make basslines or melodic solos that might be too tricky to play in time by hand, especially for the beginner, a breeze. Where the SH-09 was marketed as being compatible and able to interface with Roland's range of external sequencing hardware, the SH-101 had a sequencer built into the unit meaning that even more complex musical parts can be programmed in advance (all via an enviably simple and accessible interface). Ultimately, it is this mix of a solid core sound that is inherited, borrowed and adapted from its predecessors as well as from other non-Roland instruments, combined with a raft of new features and functions that so impresses contemporary reviewers. The end of year 'Synthesiser Update' feature in *Electronics & Music Maker* (1982: 26) is illustrative in remarking that, '... with the MSG-1 the whole package becomes capable of enormous versatility and portability.' This sentiment is echoed in Colbeck's (1985: 30) assertion that, 'no other mono offers such a wide range of applications.'

If the SH-09 had been positioned as a component in a larger ecosystem of Roland music technology, the SH-101 was presented more as a complete instrument that already contained many of those previously additional elements, or 'modern goodies' as Mowat has it. More than this, it was presented as inviting, playful and fun with an aesthetic that was clearly in tune with the times. Crucially, it was also in tune with the significant developments that had impacted the synthesizer and keyboard market. From the very end of the 1970s, companies such as Casio and Yamaha had developed and released ranges of inexpensive keyboard-based instruments that leveraged both their existing and emerging technologies of sound production and which sought to reach new audiences for electronic music making. Casio's VL-1 (1979), commonly known as the VL-Tone had been a pioneering, if miniaturised instrument, that has been dubbed the first all-digital synthesizer (Igoudin 1997: 26). Retailing at $69.95/£39.99, the VL-1 was part pocket calculator, part monophonic synthesizer and part toy-like curio. Sharing a cultural space already established and occupied by educational and entertainment products like Texas Instruments' 'Speak & Spell' (1978) and 'Little Professor' (1976), the VL-1 was a playable musical instrument/plaything that offered a variety of different preset tones that could be played in real time with its tiny button keyboard. A series of pre-programmed and unalterable rhythms could provide a backing track while an onboard sequencer allowed phrases to be recorded, stored and played back. Beloved of DIY artists seeking to,

> ... go against the overproduced and smooth sound of big stadium rock. The Casio VL-1 ... was employed by numerous Neue Deutsche Welle (NDW, Germany's response to new wave) bands in the early 1980s. For example, the band Trio prominently used the Casio VL-1 for their deadpan 'Da Da Da' hit in 1982.
>
> (Skov 2023: 193–194)

In addition to this underground success, Casio's instrument also found mass appeal with Ellis (1981: 61) claiming that, 'Over 100,000 VL-Tones are being sold per month in Japan alone and in the UK, reports from Turnkey Ltd. and Tempus Ltd. show outstanding sales, in particular to younger musicians.' The success of Casio's instrument did not go unnoticed and the VL-1 paved the way for a host of electronic keyboards from a wide range of manufacturers each with their own libraries of sounds and increasingly complex automatic accompaniment features. To some extent, these home keyboards were closely connected to the ranges of electronic organs that boasted similar sounds and features. However, of particular relevance to our discussion, is the Yamaha CS01 (1982). This included no one-finger-backing tracks or drum patterns but instead repackaged the company's analogue synthesizer technology into an approximately 19-inch-long instrument with mini keys. This single oscillator monophonic synthesizer with its silver-grey case certainly shares some resemblance to the SH-101 although as Colbert (1983a: 73) notes, 'The modulation wheels are mounted on the top left hand corner of the front panel – odd you might think, but the CS01 is also designed to be played standing up, hence the strap buttons at each end.'

It may not have had a modulation grip or come in a variety of colours, but make no mistake, this was another introductory synthesizer and it was another keytar, albeit a rather smaller one. Incidentally, it is interesting to see the almost overcompensatory assertions made in relation to the keyboard's size: 'in no way is the CS01 a toy.' notes *One Two Testing* ('Yamaha CS01' 1982: 23) while Yamaha's own Dave Bristow (1982: 15) reassures readers that the CS01 is 'certainly not a "toy" keyboard.' Nonetheless, as the presence of the SH-101 and CS-01 attest, it is clear the home keyboard market was a key site of interest for synthesizer manufacturers and provided fertile ground for innovation and development. As Jenkins observes,

> The home keyboard market overlaps in unusual ways with the world of professional or semi-professional keyboards. Many bands starting off can't afford the better-known polysynths such as Junos or Korgs, and the home keyboards can offer a solution even if it's only a temporary one; alternatively the rhythm and accompaniment sections make the Casios, Yamahas, JVCs and so on ideal for learning and entertaining in the home.
>
> (Jenkins 1984: 33)

By considering this less often recounted context, we arrive at a deeper understanding of the SH-101's early life, Roland's intentions and the ways in which the instrument was, initially at least, marketed and received. If the messaging of the SH-09 was a little confusing and perhaps even daunting as it straddled the worlds of the beginner and professional musician, Roland's positioning of the SH-101 was rather more focused. The SH-101 was not a component in a larger system. Nor was it uncomplicatedly an improved synthesizer for a new decade as, while it added some important functionality, it lost some of the features of the already cut down SH-09. Rather the SH-101 was presented as a complete instrument that had been engineered to be comprehensive yet accessible and designed to make inroads into the burgeoning home keyboard market. Like many home keyboards and unlike most professional synthesizers including Roland's earlier SH models, the SH-101 could even be powered by batteries making it a truly portable and personal instrument that, with the addition of a pair of headphones, was completely self-contained and mobile. Considered in this light, the June 1983 issue of *Keyboard* magazine makes perfect sense. Sporting the headline 'Here Come The Portables!', the front cover image shows a flight case with more than a dozen keyboards spilling out of it. Alongside some of Casio's extensive collection of mini key home keyboard instruments sit the VL-1, the CS-01 and the SH-101.

In fact, if we explore the SH-09 and SH-101 Owner's Manuals, we find yet more evidence of the shifting nature of the instruments and a noticeably different inscribed audience. Following an even more detailed guided tour through the operation of the instruments and the principles of sound synthesis, and in recognition that neither synthesizer offers preset patch memories, both manuals present a series of patch sheets with example sounds that can be programmed by matching the physical faders and knobs to the positions shown in the diagrams. The SH-09's example patches cover a range of ground and include tones that seek to imitate real world instruments like electric pianos, flutes and trumpets along with some more overtly synthetic and electronic sounds and effects. The SH-101's patches, however, cover all these bases but veer far further into novelty territory and so, along with xylophones that showcase the creation of high-pitched metallic tones or gongs that give insight into the uses of noise as a modulator and filter self-oscillation, there are rain sound effects, barking and chirping. These tones firmly locate the SH-101 in the world of home keyboards whose manufacturers often vied with one another to include playful, unusual sounds as well as other features such as barcode readers (Casio MT-70, 1982) and printers (Yamaha MP1, 1983). Perhaps most exemplary of the sonic connection between Roland's synthesizer and this new breed of keyboard, however, is the SH-101's 'Funny Cat' patch whose sound, and name, would surely still fit effortlessly into the final preset bank of any home keyboard. In fact, although it screams of novelty, the name 'Funny Cat' has some history within Roland as it was the name of a relatively obscure and wholly unrelated distortion and autowah guitar pedal that the company released in 1973. The SH-101 sound, however, is most definitely a miaow.

Any Colour You Like

In keeping with the idea of a flexible, feature-rich instrument that was perfect for performance, Roland ran a full page advertisement in UK magazines that featured a close up of a young man, with grey SH-101 slung over his shoulder, modulation grip in hand, in front of a Roland Cube Keyboard amplifier. Presumably shredding a solo, the advert is curious in that the action is set against a plain blue background that speaks neither to a performance in front of an audience, rehearsal session or domestic jam for fun. Perhaps, the neutrality was intended to allow the viewer and potential purchaser to project their personal use case into the image in order to exploit the instrument's 'Freedom For Expression' as the advert's tagline had it.

By 1984, however, there was a dramatic change in the promotion and marketing of the SH-101. While the instrument had debuted in a light grey colour scheme, two 'limited edition' models were added to the roster. These presented players with a choice of bright metallic red and lustrous metallic blue. Each was also available with a colour-matched MGS-1 modulation grip to compete the package. Incidentally, the original grey colour was now officially referred to as 'silver' adding a touch of sophistication and prestige. Importantly, all three versions of the instrument are functionally identical with the only difference being the hue of the plastic cases. Yet, these nods towards the potential desirability of different colourways and the implicit connections to cycles of fashion are far from trivial and they are themes developed in Roland's reframed SH-101 print advertising strategy.

Sporting the strapline 'SH-101 Takes You Where You Want To Go', three full-colour print advertisements draw attention to the instrument's versatility, portability and style. Highlights from the copy include,

> Battery powered, sleek lines, compact design, lightweight (less than most guitars). Strap it on and it becomes part of you.
>
> With the Source Mixer to blend Sawtooth and Square waves (with PWM), full Filtering, Envelope and Interface capabilities, built-in Sequencer, Arpeggiator, you've got more power strapped onto your body than you've ever had in any kind of portable keyboard.

Perhaps most revealing, however, is the paragraph that notes that the instrument is,

> Available in Silver, Red or Blue it looks great on stage, and so do you when you're playing it. Plug in a set of phones and let the SH-101 take you where you want to go.
>
> (Roland Corporation US 1984)

The advertisement, that ran in music magazines like *Keyboard*, had three variations. One was predominantly silver, one was blue and the other red. Each contained identical text which was complemented with a unique image that sought to emphasise the freedom of the SH-101's 'no strings attached' performance capabilities. The silver version of the ad shows a player dressed in silver playing the silver SH-101 and listening through headphones while riding a skateboard. The blue version sees a player clad in tight blue Lycra also listening to their performance on headphones as they solo away on the blue SH-101 while downhill skiing. The final ad sees the rider of a stylish red moped listening on headphones to the performance of a player also dressed entirely in red who is shredding on a red SH-101 while awkwardly straddling the pillion seat as if simultaneously giving a private musical performance and dance routine. The advertisements are extraordinary works of art for sure and their message is clear in positioning the SH-101 as an effortlessly natural part of the accoutrements of the young and stylish whose musical instruments are as dynamic as their lifestyle and whose modulation grips even match the colour of their ski pants.

What brought about this seismic shift in the branding and framing of the SH-101 as an accessory of the young generation? Certainly, the home keyboard market was a competitive one and any edge could potentially yield an advantage. However, another force was about to shake the synthesizer market to its foundations and, for a while at least, render analogue synthesizers yesterday's news. It is almost impossible to underestimate the impact of Yamaha's DX7 synthesizer. Released in 1983 with a hefty price tag of £1495/$1995, it was an all-digital instrument that used a new form of synthesis to generate its sounds. Initially developed by John Chowning at Stanford University, Frequency Modulation (or FM as it was colloquially known) was licensed to Yamaha having been rejected by US musical instrument manufacturers Hammond, Lowry and Wurlitzer (Lehrman 2005: 5). Space here does not permit a full discussion of the development or history of FM synthesis, the DX7 or the various instruments that preceded it (see Johnstone 1994; Kovarsky 2020 for more detail). However, as Colbert (1983b: 10) noted, 'it IS one of the most important keyboard advances so far this decade.' This view was shared by myriad commentators who declared it 'destined to be one of the keyboard classics' (Jenkins 1983: 46) and 'the most exciting new development of the decade'. However, being recognised as 'one of the most important breakthroughs in the industry' (Crombie 1985: 64) is one thing, but translating innovation into commercial success is another. Yamaha, however, managed this to a degree that even they had not anticipated. According to Dave Bristow, co-designer of the DX7's (in)famous factory sounds that, for some, become the 'sound of the 80s' (Lavengood 2009), Yamaha hoped that the new instrument might sell 20,000 units. This was soon exceeded and the DX7 went on to become the first synthesizer to sell more than 100,000. Indeed, it remains one of the best-selling synthesizers of all time (Vail 2014: 59) and is often considered 'one of the most important advances in the history of modern popular music.' (B.R. 2020).

In addition to its startling new digital sound, the DX7 also ushered in a new industrial design aesthetic.

> In stark contrast to its wide-ranging capabilities, the DX7 looked surprisingly conservative. At a time when many American instruments sported multi–coloured control panels, graphic affectations such as blue pin-stripes, and large blocks of wood on either end, Yamaha had given the DX7 a slim, sober, and professional appearance. This proved to be a master stroke. In 1983, the world was poised on the brink of the digital revolution … large, knobby analogue synthesizers were out, and sleek digital synthesizers were very much in.
>
> (Reid 2001b)

In fact, aware of the potential influence that digital synthesizers could have, Roland president Ikutaro Kakehashi had known about the development of FM and had even visited Stanford University to discuss the technology and its potential exploitation with Chowning. However, with Yamaha already having agreed a license and Roland not possessing their own viable digital technology, they were left with their existing analogue mono and poly synths. And so, to some extent, the launch of the lower cost Junos and SH-101 in the first years of the 1980s could be seen as pre-emptive, or even defensive, measures. Viewed in this light, Roland's 1983 double-page product family advert reads rather differently. 'Roland presents its product line for 1987. The nice thing is, it's available today.' (Roland Corporation US 1983).

With a full-page image picturing the Jupiter-8 and 6, Junos 6 and 60, JX-3P, MC-202, SH-101 and even the venerable System 100m lurking at the back, a wall of text doubles down on the 'We Design the Future' corporate mantra as it opines the qualities and forward-looking nature of the Roland's technologies and protocols. 'In fact, we put features on Roland products that you won't even use until next year, and then you'll wonder how you ever got along without them.' (Roland Corporation US 1983).

The great irony, of course, is that, despite what the advert claimed, the SH-101 would be discontinued in 1986 and Roland's flagship synthesizer in 1987 would be its first all-digital instrument, the D-50. In technology, connectivity and sound, the D-50 Linear Arithmetic Synthesizer, bore almost nothing in common with the analogue instruments in Roland's imagined 1987 lineup. By this time, the marketplace had completely transformed. Casio had released its CZ range of digital synths, beginning with the £395 CZ-101 in late 1984. Like the DX7, Casio's 101 offered polyphony, had presets and MIDI compatibility and also sported strap pins so it could be used as a keytar. As Casio aggressively priced the CZ-101 at around a quarter of the DX7's asking price and 'Yamaha flooded the market with a plethora of low-cost FM synths' (Reid 2001a), the writing was on the wall for the SH-101 and analogue synthesizers in general. Set against the sobriety of a growing range of digital instruments with sombre aesthetics and minimal interfaces stripped of the knobs, faders and rotary controls of analogue synthesis, the increasingly flamboyant, colourful appearance of the SH-101 was an attempt to stand out. Similarly, Roland attempted to weather the storm by extolling the virtues of analogue synthesis in claims that, 'a lead line synth like the SH-101 will not and cannot be replaced by a digital instrument because an analogue sound source is much more punchy.' (Roland Newslink 1984a: 31). Unfortunately for Roland, this was not true. The sounds of digital synthesizers like the DX and CZ, often described with adjectives such as bright, glassy, crystalline and sometimes even 'clinical' (Jenkins 2020: 87) were absolutely in vogue. Crucially, this new sonic palette was simply unobtainable with analogue instruments and even the lustre of a limited-edition metallic paintjob was not enough to stop the SH-101's star fading as the decade went on.

Gone but Not Forgotten

Although the SH-101 had enjoyed some degree of early success, by the time techno DJ Vince Watson (2017) came to purchase one in 1988 from a music store in Glasgow, it was available for just £70 (along with the TB-303 for £50) (Roland UK 2017). In the hands of a new generation of musicians, DJs and producers, it would not be long before the SH-101 became a staple of the emerging electronic music scenes. As Anniss (2017) puts it, 'the SH-101 took on a life of its own and became one of dance music's most popular instruments – a synth that arguably did more to fuel the early '90s house and techno explosion than any other.'

At least part of the appeal of the SH-101 in these new dance music-oriented contexts arose from the instrument's flexibility and accessibility which, ironically, were precisely the same qualities that initial reviewers had been so enamoured of and which had been at the heart of Roland's marketing and promotional framing just a few years earlier. With such a sparse but clearly laid out control panel, the SH-101 invites and rewards playful experimentation and exploration. Unlike some instruments that offer a limited range of 'sweet spots' between which might reside things altogether more unwelcome or perhaps even total silence, the SH-101's controls are straightforwardly calibrated in such a way that, outside of pulling down all the individual oscillator faders or overall Volume to 0, the player is usually greeted with an immediately and eminently usable sound. On those occasions when a sonic change is called for, the comparatively limited number of controls and the immediacy of their feedback makes edits straightforward. The SH-101 is a quintessentially real time, performance synthesizer that excels in providing an accessible, intelligible immediacy of hands-on control. This is in stark contrast to stripped back text-based LCDs with increment/decrement/confirm buttons that provides access to the vast array of parameters with complex interrelationships that lurk deep in within the labyrinthine menus of many digital synthesizers.

Even the sequencer in the SH-101 is implemented in a way that generates none of the head-scratching that frequently accompanies some of Roland's other contemporaneous instruments like the TB-303 and MC-202. Importantly, the SH-101 workflow does not presuppose that any composition be worked out in advance and transcribed into the sequencer. There is nothing here that could resemble programming and the sequencer builds on the liveness and immediacy of the instrument. There is no convoluted separation of pitch and timing as with the TB-303, no cryptic symbols to memorise because there is no display. In stark contrast to those earlier instruments, the SH-101 sequencer is simplicity itself. Pressing the 'Load' button puts the instrument into Record mode whereupon any notes played on the keyboard will be memorised. Up to 100 notes can be stored with ties and rests (to extend successive notes or enter silences respectively) entered with the appropriately-labelled silver buttons on the front panel. There are no edit modes as there is no editing. Any mistakes are remedied by re-recording the existing note information. When the sequence is completed or 100 notes are entered, simply pressing 'Play' will start it running. The sequence automatically loops without any need for further settings and the speed of playback is controlled either by the LFO Rate fader or by sending the SH-101 a clock signal from another source such as a drum machine

Alongside the Sequencer, there is an Arpeggiator. This is also notable for its simplicity and limitation. Playback modes are limited to Up, Down or Up and Down with none of the Random modes we might now expect nor the ability to extend an arpeggio over multiple octaves. The comparative shortness of the SH-101's keyboard and the Arpeggiator's inability to extend the range of notes beyond those which are physically held by the player means that the SH-101 actively encourages the creation of the short repeating patterns that have become characteristic of electronic dance music.

Moreover, the Sequencer's recording workflow makes it trivially easy to enter odd numbers of notes which, when synchronised with a drum machine playing even the simplest four-on-the-floor kick pattern, lead to polymetric looping whose complexity belies the ease of their creation. The hands-off operation that is facilitated by the Hold button allows these repeating riffs to chug away while the synthesizer parameters are modified in real time using the immediate and tactile interface of faders and rotary controls. Given that there is no way of recording synthesizer parameter changes such as Filter Cutoff or Resonance into the SH-101 Sequencer or any way of controlling those parameters via external signals sent to the instrument, any sonic tweaks must be performed in real time. In this way, the SH-101 remains a 'what-you-see-is-what-you-hear instrument'. Even when under the control of an entirely automated Sequencer or Arpeggiator or at the heart of an interconnected setup of drum machines and other devices all operating in perfect synchronicity, the SH-101 is a fundamentally (a)live instrument.

Plays Nicely (with Others)

In terms of synchronisation and interconnectivity, that the SH-101 debuted in 1982 leads to some interesting consequences. Predating the soon-to-be ubiquitous MIDI connectivity standard that allowed synthesizers, drum machines and sequencers from any manufacturer to digitally converse with one another, the SH-101 was equipped only with analogue CV/Gate connections to externally control the pitch and duration of its sounds and an analogue Clock input to provide a timing signal to its Sequencer and Arpeggiator. While this made it hard to interface the SH-101 with the new crop of MIDI-equipped synthesizers that appeared and dominated from 1983 onwards, it made it a perfect match for the less in vogue and more affordable instruments of the late 1970s and early 80s. It almost goes without saying that Roland's pre-MIDI instruments all played together particularly nicely creating an ecosystem of 'obsolete' yet newly essential instruments.

The production of A Guy Called Gerald's (aka Gerald Simpson) 1988 Acid House track *Voodoo Ray* is a case in point. Along with the TB-303 that provides the characteristically 'squelchy' 'Acid' sequence, the melodic and rhythmic heart of the track is built around an SH-101 riff. As Doyle (2015) notes, 'Working with this pre–MIDI setup, Simpson used two of the three trigger outputs from the 808 to link to the clock inputs of the SH101s, while the drum machine's DIN sync out connected to the 303.'

Similarly, and more recently, electronic music producer John Clark (aka Debukas) notes that, 'The 101 is a mainstay of the studio and is usually used in sequencer mode with notes being triggered from the 909 or 707.' (Clark 2013). This ease of use and accessibility is echoed by techno producer Elias Landberg of Skudge who notes,

> The synth is really easy to use. It's very hands-on and has a funny built in sequencer. It's one of the nicest synths for making basslines ... Most of the early Skudge stuff has a sub bass made with the SH-101.
>
> (Skudge 2016)

This last point, of course, reminds us that interoperability and immediacy would be worth little if the SH-101 did not sound good. Fortunately, the instrument's combination of Oscillator and Sub-Oscillator make it ideally suited to bass. From its initial release, the instrument was recognised for the breadth and depth of its sound.

> In the past I've never been entirely convinced by the Roland sound. It's always seemed thin and piano-like. The SH101 was the machine that finally converted me. For a single oscillator keyboard it has a vast sound, and bass lines work a treat.
>
> (Colbert 1983a: 68)

This view of the SH-101 has only strengthened over time with live techno artist Matador (2016) asserting that, 'The sub-oscillator is one of the fattest, purest tones available to man!' In fact, while the Sub Oscillator does have a notable sonic purity when heard in isolation, once it is mixed with the outputs of the main oscillator, it subtly detuned, beating pitch adds an undeniable heft. Despite being a single oscillator synthesizer with a comparatively simple range of waveforms, the Source Mixer unlocks an extraordinary variety of tones. By carefully balancing the amount of Pulse and Saw, new shapes can be derived, while fading in the Sub Oscillator can bring anything from gentle supporting weight to a fierce growl. Speaking in *Attack* magazine (2016), Michael Penman concurs, 'The 101 is a marvellous synth for basslines – you get that instant bouncy feel that adds some real funk and groove to a track.'

The reference to the SH-101's 'bounce' reminds us that it is not just the deepness of the bass that characterises the machine. Over the years, many adjectives have been used to describe the sound of the SH-101 with 'rubbery' or 'plastic' being the most common (Moraga 2020; Behringer 2019). This is most usually attributed to the instrument's filter which uses a version of the revered Roland IR3109 chipset also found in the Jupiters 6 and 8 and Junos 6, 60 and 106 among others. Certainly, the SH-101 filter gives the synthesizer a palpably different, more restrained and refined character than instruments like the almost outlandishly screaming TB-303. In truth, the 'rubbery' tonality is not simply due to the filter and it is notable that the other instruments that use IR3109 chip are not described in the same way. Rather, the distinctive 'bouncy' or 'plastic' sound is a product of the interaction between the filter and the SH-101 envelope. Based around a dual Op Amp design that differs from previous Roland synthesizers, the envelope's 1.5ms Attack and 2ms Decay and Release are remarkably short and, by modulating the filter and VCA (based on the BA662 OTA chip), can create extraordinarily snappy and 'rubbery' contours.

It should be noted that, while the IR3109 circuit is much celebrated, its implementation in the SH-101 differs from the Jupiter-8 and Junos. Chief among the differences is the design of the filter resonance. The Jupiter-8 included a novel 'Q-compensation' circuit which fed the input signal back into the chip in order to make up for the natural drop in gain that accompanies high levels of resonance. This feature was absent in rival polysynths like the Prophet-5 and helped define the 'Roland' sound by maintaining the integrity and presence of the oscillators even as the filter approached self-oscillation. Notably, while the Jupiter-8 filter was designed not to go into self-oscillation, both the Junos and SH-101 were. However, unlike its polyphonic forebears, the IR3109 in the SH-101 does not include the Q-compensation circuitry which creates the audible effect of a loss of low frequencies as the resonance is increased.

Not that the SH-101 had any difficulty generating bass frequencies with its Sub Oscillator. However, more than this, with the Resonance set to maximum, the SH-101 Filter self-oscillates in a way that the TB-303 (and Jupiter-8) was expressly designed not to. The resulting sine wave generated by the Filter circuit not only gives an extra tonal flavour to add to the Sawtooth, Pulse and Noise generated by the Oscillator, but can also be independently tuned by way of the Filter Freq control to provide a counterpoint interval for pseudo-chordal performance. This technique was utilised on *Voodoo Ray* to create a higher pitched overtone that perfectly tracks the movement of the deep bass note that drives the track forward. Alternatively, lowering the cutoff of the self-resonating filter produces a rich, deep, pure sine wave that creates the truly room shaking sub bass that provided the root to countless drum and bass tracks throughout the 1990s.

However, as Moraga (2020) notes, the SH-101's synthesis capabilities make it far more than just a bass synth and Preve's (2023) list of SH-101 users including Boards of Canada, Squarepusher, Aphex Twin, and Devo reminds us of the 'vast stylistic range for a single synth'. As A Guy Called Gerald notes, 'The SH-101 had a lot more range than my TB-303, so I could use it to record high, bell-like sounds, really low, deep tones and everything in between.' (Anniss 2017). The sinuous, mid-range filter sweeps of Aphex Twin's 'Polynomial-C' (1995) similarly showcase the instrument's versatility which see it sit across a wide range of musical genres. If the SH-101 has become associated with one group in particular, it would likely be Boards of Canada, however. Of course, the bassline that drives 'Roygbiv' (1996) draws on the instrument's much-vaunted low frequency capabilities, but the SH-101 is just as comfortable contributing melodic lines to the tracks such as 'Kaini Industries' and 'Bocuma'. Indeed, while the hauntological retrofuturism of their wilful embracing of tape noise, wow and flutter and the artefacts of analogue reproduction and degeneration might be the most widely celebrated aspects of their work (see Sexton 2012: 578–579; Pattison

2013), the SH-101 is perhaps the most recognisable instrument found within Boards of Canada's influential repertoire. Heard throughout the duo's official commercial releases as well as in the apocrypha of their early cassettes and bootlegs that continue to circulate online, stripped-back tracks like 'Alpha Rainbow' are virtual paeans to the SH-101.

Nonetheless, while it has a distinctive sonic fingerprint, the SH-101's range means that it is not as singularly identifiable as a TB-303 acid line or even a booming TR-808 Bass Drum, for instance. Yet, far from being a negative, it is the flexibility of the SH-101 that, for many, is the instrument's key quality. Bristol-based DJ October offer a succinct and celebratory commentary,

> The SH-101 has always been more of a chameleon. A lot of the time, you won't be able to tell that a 101 has been used. Out of all the machines Roland made in the '80s, it's the least definable by its sound, but that versatility was its strength. It deserves wider recognition.
>
> (Anniss 2017)

Yesterday and Today

One of the great ironies of the eminently portable, performance-oriented SH-101 and its promise to 'take you where you want to go' and keytar stylings is that the instrument has now become firmly established as a studio fixture. In addition to the desirability of its sound, this must be at least partly attributable to the delicacy of decades-old analogue electronics, the fragility of plastic cases and the considerable rise in the monetary value of the instrument. Beginning its life as one of the most inexpensive products in Roland's lineup and passing through a period in the doldrums of the late 1980s in which they changed hands for less than half the initial retail price, the SH-101 has become an instrument that routinely commands a four-figure sum on the used market today. It is not an instrument to be played on the back of a moped.

The SH-101's move into the studio is far from the only transformation that has been witnessed over the decades, however, and while some continue to celebrate the simplicity of the instrument in its original form, others have sought to expand its potential via modifying the hardware. Without doubt, the most well-known reworking is the 'Nova-Mod'. Actually not a single modification but a growing collection of adaptations, the Nova-Mod was originally devised by Phil Pilgrim in 1995 but versions of it have been offered as installed products and services or documented as DIY projects by companies such as Din Sync and Circuitbenders (who also offer a modification to compensate for the instrument's 'inexcusably inadequate external power filtering'). The Nova Mod adds a variety of new synthesis functionality to the SH-101 including the ability to overdrive the filter in order to create additional grit and distortion; extending the range of the LFO to allow it to operate at very high speeds which, among other things, allows parameters such as oscillator pitch and filter cutoff to be modulated at audio rate. The effect of this is the creation of 'sidebands' of additional frequencies (Hass 2020) or 'all kinds of clanging cross modulation and bizarre alien splatters and sweeps' as the Circuitbenders guide has it.

Some modifications simply alter the scaling and calibration of existing internal components or the inclusion of new resistors and capacitors to change the behaviour of the circuits but most are rather more invasive and require the addition of new physical controls, hardware inputs or outputs to the case itself. However, for those happy to take a drill to the silver, red or blue plastic, the rewards are an external input that allows any audio to be fed into the synthesiser's synthesizer's signal path and processed through the rubbery filter, and voltage control of the filter cutoff so that it can be modulated, and therefore automated, by an external device such as a sequencer.

Taking the SH-101 even further, German company Tubbutec's 'SH-1oh1' modification dramatically extends the functionality of the instrument and retrofits MIDI input and output to bring compatibility with a wider range of external devices and allows those devices to dynamically control the filter cutoff and envelope. Where the Nova Mod adjusted existing elements of the original circuit, the Tubbutec mod greatly expands it. The SH-1oh1 adds a second envelope and LFO which are dedicated to the Filter rather than being shared with the VCA. More than this, however, the modification adds wholly new functionality to the sequencer including the ability to store 32 sequences each with 120 steps (compared with a single sequence of 100 steps in the original). A 'Song' mode allows multiple sequences to be chained together to create even more complex performances. It will be clear that the range of extensions here cannot be achieved through creative soldering or recalibration of the SH-101's analogue circuits. Rather, at the heart of the SH-1oh1 upgrade is an entirely new CPU and updated software. Essentially, the SH-1oh1 gives the instrument a digital brain transplant. Even in the face of such significant transformation, the reverence that is still palpably felt for the original instrument and its sound in Tubbutec's marketing reveals much about the status of the instrument today. 'Great care has been taken to avoid altering the original behavior of the synth while only adding features on top.' (Tubbutec SH-10h1 User Manual v1.1 nd: 4).

If we needed further evidence of the ongoing status of the SH-101, we surely find it in the host of new instruments that have been indirectly influenced or that have squarely imitated the original. Most obvious are Roland's own 'reissue' in the form of the miniaturised Boutique SH-01a hardware device released in 2018 and Behringer's 2019 MS-101 (subsequently renamed the MS-1). The former is an entirely digital emulation using Roland's ACB (Advanced Circuit Behaviour) modelling technology that seeks to replicate the sound of the original instrument by recreating the behaviour of its individual analogue components, including the inherent variations and instabilities. The slightly awkward SH-01a product name is broadly consistent with the other models in the Boutique range but the 'a' reminds us that Roland has long used the cachet of the SH name and x0x numbers with, in this case the 2010 'Gaia' digital/virtual analogue synthesizer having already snapped up the SH-01 soubriquet.

Using the same hardware form factor as other instruments in the Boutique series, the SH-01a mimics the front panel controls of the original but excises its keyboard altogether. As such, the new instrument dramatically alters the tactility of the original instrument. The lack of keyboard is the most obvious change yet the interface controls, while stylistically similar in their design even down to the multicoloured stripes on the fader and knob caps, are miniaturised almost to the point of being unusable. Dialling in precise values is certainly more challenging and it is comparatively difficult to find 'sweet spots' in the Filter or the right combination of waveform levels in the Source Mixer that are so key to creating a wider range of tones than appear possible with such an apparently simple oscillator. More pertinently, the SH-01a removes the all-in-one performance orientation of the original instrument. The Boutique instrument requires the use of another device to trigger its sounds whether that be an external keyboard controller or a computer-based Digital Audio Workstation. As such, what the SH-01a recreates is the studio instrument the SH-101 has become rather than the live, performance instrument that it was.

Like other instruments in Roland's Boutique range, however, while the SH-01a removes some features, it adds others. As might be expected, the SH-01a brings a plethora of new features such as multiple sequences, patch memories, MIDI inputs and outputs and USB connectivity. However, where there had been criticism of the earlier Boutique synthesizers for the reduced polyphony of their digital recreations (the JP-08 (Jupiter-8) and JU-06 (Juno-106) were given just four of their original eight and six voices, for example), the SH-01a expands on the original by offering four notes rather than just one. This might be explained by Reid's (2018) suggestion that, 'their similar voicings and their use of the same filter meant that many people thought of the Junos as polyphonic SH-101s'. Regardless, with the SH-01a, Roland transforms what has become one of its most iconic monosynths into the MIDI enabled, studio oriented, polyphonic synthesizer that might have made an excellent market proposition in 1982.

Behringer's MS-1/101 takes a rather different approach to the SH-101's legacy by recreating its analogue circuits and keyboard which are packaged in a plastic case that, although not a 1-for-1 clone definitely captures the spirit of the SH-101. This even extends to including the modulation grip and strap fittings as standard and, initially at least, making the instrument available in red and blue variants (although intriguingly Behringer opted for black as its standard model rather than the grey/silver of the original). In fact, Behringer's MS keyboard does make some alterations to the digital side of the instrument by adding MIDI and USB interfaces as well as incorporating a very different, and more operationally complex sequencer (which bears more than a passing resemblance to the one in Moog's Mother-32 in its functions and interface design). In addition, the MS-1/101 ships with a number of the analogue Nova Mod enhancements already included in its circuits. As such, the LFO can operate at much higher rates than the standard Roland unit, the Source Mixer includes the Triangle wave oscillator not present in the original and provision for an external audio input, while audio rate filter modulation is offered with a number of different waveshapes. It is interesting to note that, perhaps as a result of the comparatively low price for which its analogue circuits are made available, the Behringer MS-1/101 has already become something of a platform for experimentation and modding itself allowing the core of the SH-101 to be yet further explored. Thus, while the Nova Mods are seamlessly integrated into the case of the MS-1/101, many units are subsequently drilled into and have new controls protruding from them as modders and circuitbenders bring new sonic and performance affordances to the surface of their instruments.

While instruments like the SH01a and MS-1/101 along with the Superlative SB01 and DinSync RE-101, clearly reference the SH-101 and either expand or replicate it circuits and features, countless other instruments bear the influence of the accessibility, synthesis model or sequencer design. In the world of contemporary Eurorack modular synthesizers, for instance, Intellijel's Atlantis explicitly looks to the SH-101 for inspiration in its architecture and sound. It even references the Roland original in its use of linear sliders rather than the rotary controls which are far more commonplace in Eurorack instruments where front-panel space is at a premium. Mutable Instruments' Yarns module similarly draws on the SH-101 which might seem unusual given that it is a module dedicated to converting digital MIDI messages into CV and Gate signals. Certainly, this would be useful in allowing MIDI-equipped devices to interface with the SH-101, yet the influence of Roland's design is actually found in Yarns' built in sequencer which, according to the manual '…is inspired by the SH-101, in that the primary way of inputting sequences is to use a keyboard and record the notes step by step.' (Mutable Instruments 2013). The immediacy of the SH-101 sequencer has become almost as widely revered as Roland's TR-REC drum programming method. Indeed, Roland's MC-101 and MC-707 'Grooveboxes' both proudly boast SH-101 style sequencer modes and maintain the legacy and discourse of the original instrument's accessibility and approachability.

In addition to hardware instruments and Eurorack modules, the popularity of the SH-101 has seen software emulations abound with plugins such as TAL-Bassline-101 and Softube Model 82 more recently joined by Roland's 'Cloud' emulation. Each of these emulations make claims to the merits of their scrupulous digital modelling of analogue circuits and the authenticity of their recreation of the sound of the original. Simultaneously, though, they expand on the original instrument and its palette with TAL and Softube replicating a number of the analogue modifications collected in the NOVA collection, for instance. Regardless of their specific extensions, each plugin reveres but also resituates the SH-101within the thoroughly modern context. Tight integration with the Digital Audio Workstation brings every parameter under the direct control of editing tools and automation curves, further transforming the SH-101 into a studio tool to be laboured over rather than picked up and played. The popularity of the 'rubbery', 'bouncy' bass sound of the SH-101 has also seen such tones become a staple of digital synthesizer presets, waveform lists and sample libraries. Even the Nord Piano library includes a selection of characteristically deep but twangy SH-101 bass samples for those wishing to play the sound on an 88-note triple sensor hammer action keybed. The presence of these sounds across sound libraries and preset banks certainly cements the SH-101 name and number in the lexicon of digital music. Yet, they also reduce an instrument elsewhere lauded for its flexibility to a caricature. It is also important to note that, as it is integrated into new workflows, the SH-101 can be seen to be disintegrated. Separating its sounds from its sequencer, and even the instrument from its keyboard, is at odds with the original design and the drive to create the SH-101 as an all-in-one.

Ultimately, while it has become a fixture of many studios, the design of the SH-101, its portability and the modulation grip and strap, all remind us of instrument's genesis as well as its association with the somewhat preposterous poses of the keytar. Indeed, throughout Roland's Japanese marketing of the instrument, from the 'Micky Yoshino' product brochure through to its inclusion in broader Roland 'keyboard range' catalogues, the SH-101 is unequivocally presented for performance and is listed under the heading of 'Solo Synthesizer'. Perhaps it would be easy to dismiss the keytar as nothing more than a novelty item or even a quintessential example of the excesses of the1980s. However, in a 1985 editorialised interview with *International Musician and Recording World* in which he reflects on the company's newly released AXIS keytar and its progenitor, the SH-101, Roland President and founder Ikutaro Kakehashi offers some further insight.

> Paradoxically Mr Kakehashi is both the man who put the keyboard player in the guitarist's role with the now-imitated SH-101 and Axis, and the man who put the guitarist in the keyboard player's role with the Guitar Synthesizer and its powerful, layered harmonies. His reason was that he intended to offer both of them greater freedom, the keyboard player in terms of movement, the guitar player in terms of sound. Of the Axis itself he says simply 'Videos need action' – capturing in a phrase the impact of video recording on the way bands present themselves, and summing up Roland's response to it.
>
> (Kakehashi 1985: 79)

Video may have killed the radio star, but it also created a new opportunity for the portable, wearable synthesizer. This was an instrument that not only helped the player perform but that was created so they could be seen performing.

7 MC-202 MicroComposer

Figure 7.1 Roland MC-202 MicroComposer (image credit: James Newman).

DOI: 10.4324/9781003374510-7

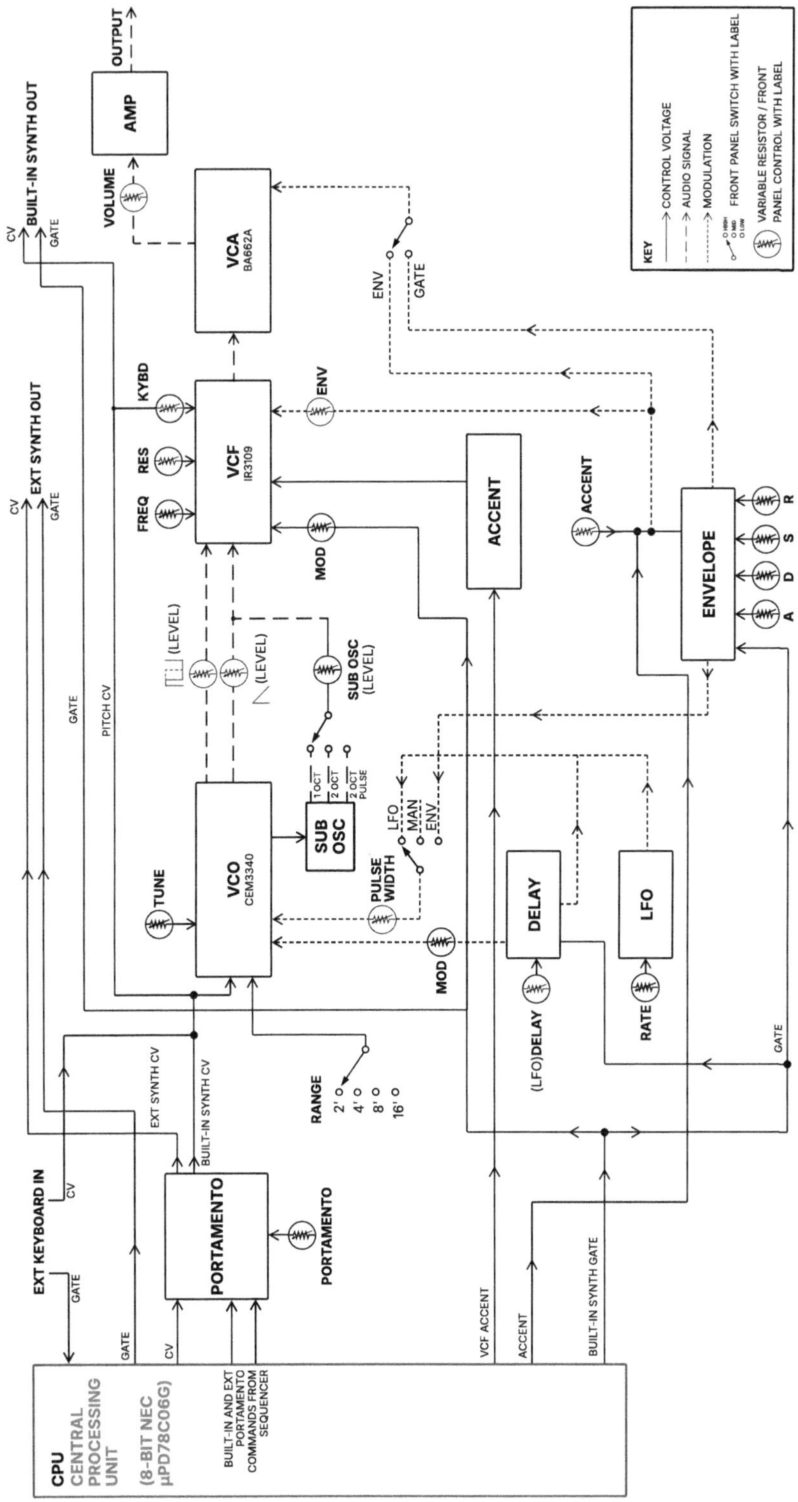

Figure 7.2 Roland MC-202 Simplified Block Diagram (adapted from Roland 1983a MC-202 Service Notes).

MC-202 MicroComposer

Released
1983 (discontinued 1985)

Sound source
Curtis Electro Music CEM3340 Integrated Circuit
Simultaneous Saw, Pulse and Sub Oscillator waveforms with independent Level control
Sub Oscillator can be set 1 octave below (Square wave) or 2 octaves below (Square or thin Pulse)

Filter
Roland IR3109 Integrated Circuit
24 dB/Oct Low Pass with resonance
Self-oscillating filter Tracking

Modulation
LFO (Triangle waveform) with Delay
ADSR envelope
Modulation destinations include Amplitude, Filter Cutoff and Pulse Width Modulation

Keyboard
32 rubber keys, F-Scale

Sequencer
2 Tracks (one for internal synthesizer, one for external CV/Gate instrument)
Maximum 2600 steps

Connectivity
Line Level and Headphone audio outputs (monaural)
Built-in Synth CV/Gate Output (Sequencer Track 1)
External CV/Gate Output (Sequencer Track 2)
External Keyboard CV/Gate Input
Tape Memory Save/Load
Tape Sync In and Out
DIN Sync
1 x Input
2 x Output

Power
9-12V (centre negative) power supply

Dimensions
343 (W) x 55 (H) x 204 (D) mm
13-1/2 (W) x 2-3/16 (H) x 8 (D) inches

Weight
1.35kg / 3lb (without batteries)

Notes
The MC-202 sequencer memory is volatile meaning that all data is lost if the instrument is powered off without backing up to cassette. Data Cassette with demonstration sequences including: Invention Vol.1 (J.S. Bach); Invention Vol.2 (J.S. Bach); Dream (Debussy); Prelude (J.S. Bach); Snow Is Dancing (Debussy).

Suggested listening
Alessandro Cortini 'Sonno' (Hospital Records, 2014)

The Sum of its Parts

To modern eyes, the MC-202 makes perfect sense. It brings together a flexible, fully analogue monophonic synthesizer with a sequencer that offers two completely independent channels of CV and Gate outputs for controlling instruments. Its comprehensive synchronisation capabilities mean that the MC-202 can operate as a centrepiece that keeps other drum machines and sequencers tightly in step or it can lock onto the timing signals of other external equipment to seamlessly integrate into an existing setup. Factor in that the MC-202's built-in synthesizer is not a cut down afterthought but is essentially an SH-101 with just a couple of omissions, at least one major new feature and the characteristic 'thundering' (Johnson 1994) sound, and the package starts to seem even more appealing. That all of this functionality is contained within a compact case that still manages to include a full complement of hands-on controls to manipulate the built in synthesizer in real time is nothing short of remarkable. That this much functionality and analogue circuitry can be battery powered for performance on the move is almost unbelievable and a feature much lamented in considerably more modern instruments. Even the MC-202's miniature button keyboard set out across the bottom of the unit would not look wildly out of place on a modern Elektron device such as an Analog Four or Octatrack and is remarkably similar to a Moog Mother 32. Indeed, with two and a half octaves of buttons the MC-202 offers a wider range than those instruments (and plenty of others before and since) and perfectly replicates the keyboard length, if not the playability, of the SH-101 on which it is partly based.

It is small wonder, then, that the MC-202 is sometimes seen as one of the earliest examples of a 'Groovebox' (a term first used by Roland but not until the mid-1990s in relation to the MC-202's all-digital, spiritual and numerical successor, the MC-303). In fact, in addition to predicting these self-contained sequencer/synthesizer instruments, the MC-202 also plays an important and far less well-known role in the contrasting development of the computer-based music studio as it becomes the basis for one of Roland's pioneering Compu Music instruments.

Yet, for all its apparent prescience and the host of benefits and features that might attract the modern performer, the MC-202 was not an easy sell in 1983. Reviewers seemed somewhat confused about the instrument's true purpose and, while introducing any new product runs the risk of being misunderstood, even Roland's own advertising and promotion struggled to succinctly communicate exactly what the MC-202 was, precisely which problems it solved, and who it was for. To compound the issue, the instrument, and its sequencer in particular, quickly gained a reputation for being capable but complex. In an otherwise positive review, Colbeck warns that the MC-202 'may not be the easiest instrument to figure out' (Colbeck 1985: 194). As a result, the MC-202 remained in production for just 17 months and, while Roland are silent on the matter, it is sometimes reported as being the company's worst selling keyboard/synthesizer instrument of the decade (Computer Music 2018), though without much supporting data to corroborate such a claim.

Since then, while it is the case that the MC-202 has not enjoyed anywhere near the same level of notoriety, adulation or veneration as the TB-303 or even the SH-101 to which it is so closely related, it has steadily become more highly prized. With an initial retail price of £365, by 1995 retrospectives such as that in August issue of *Sound on Sound* magazine claimed that, 'It wasn't long ago that you could pick up an MC202 for around £100 … Now, MC202 prices have caught up with those of their keyboard counterpart, and the MC202 can sell for £200 to £250.' ('Roland MC202 Microcomposer [Retro]', 1995). Thirty years on, second-hand prices have certainly risen with the instrument typically commanding a £1000–1250 price tag depending on condition but, even at the high end this is between a third and a quarter of the asking price for a TB-303 today.

At least part of the reason for this is that, despite the closeness of its relationship, the MC-202 is not as immediately rewarding. Moreover, as its core sequencing workflows are geared for the transcription of existing musical ideas, it does not encourage the improvisation and experimentation common in modern approaches to composition, nor does its design lend itself to solo performance. Nonetheless, the instrument has a loyal and dedicated following and of all the 1980s x0x instruments we will encounter in this book, it is the MC-202 that has been most significantly transformed, modded and enhanced since its initial release. Where each of the x0x instruments have been differently 'probed' and have taken on new cultural and musical meanings, the MC-202 has been technically and functionally altered to a point where it is almost unrecognisable. With the appropriate analogue and digital modifications and upgrades installed, the MC-202 not only acquires many of the features, workflows and interoperability of far more modern instruments, but in some ways, it comprehensively surpasses them.

Identity Crisis

For those familiar with the plethora of post-1990s standalone instruments that combine sound synthesis with sequencing into a desktop form factor that eschews a traditional musical keyboard in favour of buttons or pads, it will be hard to imagine just how unusual the MC-202 seemed in 1983. Its small, plastic casing cuts an unassuming figure. Its front panel is littered with the sliders and knobs that suggest it is a high-end synthesizer while the row of buttons laid out to mimic a musical keyboard but made of tiny squares of rubber suggest it is anything but. Paul Colbert's December 1983 review captures the contemporary reaction perfectly. 'To be honest, the 202 looks terrifying. It's not much more than 2in thick, measures 13½ in across by 8in wide, and is packed solid with sliders, switches, knobs and a miniature keyboard.' (Colbert 1983c: 10).

It is certainly an instrument of two parts with the top section of the front panel being an almost perfectly miniaturised recreation of the SH-101, complete with appropriately shrunk down versions of Roland's characteristic faders with their multicoloured lines the only punctuation among the silvery grey. In fact, if one removed this section it would make a passable modern Roland Boutique SH-101. Yet the bottom of the MC-202 looks nothing like any other Roland instrument and, instead, appears to have been taken directly from Casio's VL-1 instrument/toy/plaything. If ever there was a musical instrument equivalent of a 'cut and shut' car, this might be it. Indeed, when Roland launched the MC-202, it was clear that even their marketing departments were uncertain as to precisely how the instrument should be framed. The print advertisement that ran in the specialist UK music technology press throughout 1983, sees a pull quote declaring the instrument as 'one of the hottest products for 83'. With hindsight, this might seem like overconfidence but just a little further down the page the exuberance is somewhat undercut as the advert continues by noting that, 'Basically, the MicroComposer is in two parts. MicroComposer and Monophonic Synthesizer.'

Although explaining that one part of this new MicroComposer product is a MicroComposer is not as helpful as it might be, it is the tone that is most revealing here. The informal, almost conversational style appears reassuring yet does not make it wholly clear what the relationship between these two parts might be and how they combine to form a coherent instrument. In fact, as the copy continues, the MC-202's sound generation is treated in decidedly pedestrian terms. 'Playback facility through built-in synthesizer with almost the same features as the SH-101'. There is an almost matter-of-fact relegation of the MC-202's ability to create sound as a functional 'playback facility' that belies Roland's enthusiasm for these sounds elsewhere in their product range. Given that the bulk of the advertising copy is dedicated to listings of the instrument's technical features and connectivity including its memory capacity, tape sync facility, built-in cassette interface and liquid crystal display (LCD), one could easily begin to think that this was not an instrument of two equal parts. Roland's unusually long eight-page November 1983 product brochure for the instrument does a similar job in the opening overview. 'The MC-202 is a 2-CHANNEL DIGITAL SEQUENCER equipped with a one-channel monophonic synthesizer' (original emphasis). (1983b)

Of the 15 bullet points that outline the instrument's innovations and capabilities, only two even mention the built-in synthesizer (which is elsewhere in the brochure referred to as a 'synthesizer module' further disintegrating it). Even then, the ability to create sound internally is positioned more as a functional add-on to allow the verification of inputted data rather than forming a key feature of the instrument, per se. Roland's (1983b) brochure continues by noting that, 'The built-in monophonic synthesizer (1V/oct.) comparable to the SH-101 allows data to be entered while hearing the actual sound with headphones.'

It is very clear that, while we might encounter the MC-202 today as a synthesizer with a built-in sequencer or, more pointedly, as the heart of an SH-101 with a more complex and feature-rich sequencer than that keyboard instrument possessed, in 1983 the MC-202 was a sequencer with a synth module built in. Perhaps the clue is in the name as this is, after all, an MC-202 rather than an SH-202. This naming certainly reminds us that to fully understand the design ethos and history of the instrument, we need to rewind several years earlier than the SH-101 to the birth of the Roland MC MicroComposer range.

Music Processing

Released in 1977, Roland's MC-8 became one of the first microprocessor-controlled sequencers. Roland's machine was actually based on prototypes developed from 1971 by Canadian composer and engineer Ralph Dyck. These had come to the attention of Ikutaro Kakehashi who subsequently assembled a team to build on and expand the ideas. The resulting product could store thousands of notes and simultaneously control up to eight separate synthesizers with analogue Control Voltage and Gate signals. Perhaps unsurprisingly given the amount of physical hardware required, the package was imposing, comprising of two desktop boxes with solid metal cases, wooden end cheeks, countless sockets for connecting instruments, and an array of computer keyboard keys including a 0–9 numerical keypad. For those familiar with the previous generation of analogue sequencers, this was a significant development. Where those sequences had typically been able to store somewhere between eight and 16 steps, the MC-8 could memorise more than five thousand. If the shorter patterns of those earlier devices such as the Moog '960 Sequential Controller' and Roland's own '182 Sequencer' module for its System 100m modular had given rise to the hypnotic ostinatos of Tangerine Dream's Phaedra and Rubicon, for instance, the MC-8 promised altogether more complex song structures. The MC-8's complex routing capabilities meant that it could function as a true centrepiece in a studio as its 5000 stored notes could be distributed to eight separate instruments. If further differentiation from the past were needed, where previous generations of analogue sequencers were typically arranged around physical interfaces that used the same knobs and switches as the synthesizers they controlled, the MC-8 looked quite different and more like a computer terminal than a musical instrument.

In truth, while it was a musical instrument, the MC-8 was also a computer terminal and by placing programming at its centre and the distribution of data to remotely controlled instruments, it clearly presages the workflows and arrangement of the modern electronic music studio. As Chris Carter's (1997b) feature states, 'The instruction manual for the MC8 is a huge tome an inch thick, weighing over 2lb and full of terminology left over from 1960s computer programming.' Indeed, Roland's print advertisements that appeared at the end of 1977 in publications such as *Contemporary Keyboard* magazine were very keen to make the connection between their new sequencer and other aspects of creative work that were being transformed by modern microcomputers with print advertisements boldly stating, 'You've heard of word processing ... and data processing. But have you heard of 'music processing? ... Music can be programmed mathematically on the integral ten key keyboard.'

If 'music processing' and entering notes on a numerical keypad sounds daunting, let us not forget that this was still the early days of computing. With 1984's Apple Macintosh still many years away, there were no graphical user interfaces, windows or mice. Instead, cryptic codes, repetition and immeasurable patience were the order of the day. Roland's MC-8 advert that featured in magazines such as *International Musician and Recording World* in 1978 is particularly illustrative. Under the promisingly user-friendly heading, 'If you can hum a tune, the MicroComposer will play it for you', the copy documents each step in the process.

> The Roland Micro-Composer does for the musician what the pocket calculator has done for the schoolboy. It is very simple to operate and it allows the operator to think of a tune, push the appropriate buttons and the Composer will make a synthesizer (or a wide group of synths) play the tune. The writer needn't even touch a musical keyboard!
>
> Programming the Micro-Composer has deliberately been made easy. If you think of a tune that starts with (for example) C, E, G and the first three notes you punch in the code for C, E and G. The code is so simple you can understand it immediately and learn it by heart in a couple of days. To give you an idea how easy it is middle C is 24 (that's the code number you would push on the calculator-type panel) and because the notes are numbered to correspond with the notes on a keyboard, the next note, E, is 28 because that's four semi-tones up. ... If you want it to be a crotchet (one beat) you punch in the crotchet code which is 32 ... The only thing you've got to decide now is how loud you want the note to sound. Naturally there's a code for loudness (it starts at 100 and goes down to 30 for very soft). So by punching three numbers in, you've chosen a note, decided how long you want it to sound for and decided how loud it will be. You can programme each note in about one and a half seconds.
>
> (Roland 1978)

Bearing in mind that this is an abridged version of promotional text rather than the 'inch thick' operating manual, and noting the gendered language, it is clear that the process of working with the MC-8 was a laborious one. Remember, also, that the MC-8 had the potential to simultaneously control eight separate synthesizers each requiring the entry of their own data. It is interesting to note the techno-futurism here which simultaneously seeks to embrace the 'compu-age' while reassuring the 'operator' that 'music processing' is not so hard to understand. Moreover, and following Dyck's initial model, the description reveals some crucial facets of the design and intention of the MicroComposer. Most obviously, the act of using the sequencer is rendered as data entry with music becoming a series of abstract two-digit codes to be input via a computer keypad and single line display. Furthermore, one does not enter musical phrases per se as each 'note' is further deconstructed into pitch, timing and duration with each of these elements becoming a separate stream of data entered by the 'operator'. This method replicates Dyck's original design and the computer-centric workflow and conceptual model it reveals speaks to a second important quality of the MicroComposer.

Complexity notwithstanding, the unrivalled functionality of the MC-8 (and its successor, the MC-4, that offered almost identical capability but with half the number of outputs), ensured that they found considerable use, albeit limited to those individuals, groups and studios that could afford the £4000/$4,795/¥1,200,000 price tag. Groups such as Tangerine Dream and Landscape used the MicroComposer widely as did Kraftwerk and Martin Rushent, whose work on 'The Man-Machine' (1978) and The Human League's 'Dare' (1981), respectively, are among the most famous examples of the sequencer in use. Similarly, Yellow Magic Orchestra made extensive use of the sequencer both in the studio where it can be heard throughout their eponymous first album, as well as live on their 1980 tour while Suzanne Ciani was another high-profile user with both MC-4 and MC-8 being used on her influential 'Seven Waves' album (1982) that, 'blended my classical-romantic sense for melodies with the astounding possibilities offered by electronics.' (Ciani 2016).

Isao Tomita's fondness for Roland's sequencers even went so far as to humorously list the MC-4 and MC-8 as 'Assistant Conductors' in the liner notes of the 1978 'Bermuda Triangle' album (Beecher 1983). And so, while the workflow might not have been quite as straightforward as Roland's initial promotional material suggested, the flexibility of the sequencer is self-evident in the breadth of the discography above. However, and perhaps representative of the onerous nature of the data entry and the separation of the process into musical creation and transcription phases, it should be noted both YMO and Tomita worked with programmer Hideki Matsutake (who also used the MC-8 in his own Logic System project) as an assistant (or perhaps an assistant to their assistant in Tomita's case). Ultimately, however, as Carter (1997b) notes, while the MC-8 was influential, its price and complexity inevitably made it a niche product, 'Only 300 were manufactured; but, while the MC8 may have been a financial failure for Roland, it was an undoubted milestone in sequencer history and was the first of a line of distinguished and sometimes quirky MicroComposers.'

In fact, other reports suggest there may have been as few as 200 MC-8s (Carpenter 1995). Nonetheless, with only that machine and MC-4 that followed it, the familial line arrives very speedily at the MC-202 which, in terms of quirkiness at least, significantly moves things forwards and dramatically reduces the price of entry.

8+101=202?

In light of the importance and innovation of the two earlier MC models, Roland's focus on the sequencing aspects of the MC-202 certainly make a lot more sense. Of the two 'parts' of the instrument, this is the one that is demonstrably novel and innovative, so it is hardly surprising to find sequencing dominating the description of the device in Roland's futuristically-titled 'Product Line for 1987' (published in 1983) brochure: 'The MC-202 is a two channel polyphonic sequencer that employs the same advanced programming system found on Roland's MC-4 and MC-8 Micro Composers.' (Roland Corporation US 1983). Indeed, the provision of two independent sets of control outputs means that the MC-202 can function as a sequencer with the 'playback verifying' internal sound generator completely bypassed.

This specification further reminds us of the complicated place of the MC-202 in Roland's product ranges. Where the MC-8 offered eight independent channels of control voltage, the MC-4 halved this, while the MC-202 halved it again. As such, at least one of the '2's in '202' speaks to the number of sequencer channels and locates the instrument in the MC lineage. However, the xOx format simultaneously reminds us the instrument is part of the growing family of synthesizers and drum machines bearing that numerical soubriquet. In fact, it is very rare to see the MC-202 pictured in isolation in any of Roland's promotional literature. Group shots see it not only positioned alongside the SH-101, TB-303 and TR-606 as part of the family, but also connected to them as part of a tightly integrated music making, or perhaps music processing, system. In this way, while we might come to the MC-202 today as an integrated multifunctional instrument – the MicroComposer of the xOx range, Roland's contemporary advertising invited us to consider it as much as the MicroComposer for the range.

To emphasise this, it is telling that when powered up, the MC-202 has been designed to default to its Edit rather than Play mode. As well as suggesting that this is a sequencer first and foremost, one consequence of this is that any key presses, whether intentional or otherwise, are immediately recorded. Comparing this with a modern instrument, even a modern hardware sequencer with no built-in sound source, this is extremely unusual and one would expect to have to deliberately put the device into a recording mode if only to preserve the accidental overwriting of existing material. The MC-202, however, is always listening and always waiting to process music and, as we shall see, the worry of overwriting data is not one that need concern the MC-202 user.

Of course, it is not quite true to say that the MC-202 immediately begins recording 'notes' because, like the earlier, larger MicroComposers, what the instrument actually first records is a series of pitch values based on the inputs from the row of small rubber keys that comprise its musical keyboard. Just as with the previous MCs (and TB-303), timing and note duration data to govern when and for how long those pitches sound are entered in separate passes as separate data streams once the entry of pitch values has been completed. Unlike the earlier MCs, however, the MC-202 possesses no dedicated 0–9 numeric keypad. Instead, all data entry is performed with the combination of the row of musical keyboard buttons for setting pitches and an array of keys dedicated to specific editing functions. The musical keyboard buttons do, however, have secondary functions that enable the entry of note durations and transform them into 0–9 number keys to replicate some of the earlier MC functionality.

The MC-202 actually has a number of discrete Edit modes which must be switched between in order to enter pitches, timing and duration data. In each mode, the keyboard buttons perform different functions. Most obviously, in Edit Pitch mode pitches are entered in sequence by pressing the appropriate black or white key which, when used with the transpose buttons, gives almost five octaves of range. Alternatively, pitches may be entered via an external keyboard connected to the MC-202's analogue CV and Gate inputs. At this point the MC-202 can be switched into Play mode and the sequence of pitches replayed. However, all notes will be performed with the default Step Time of a quaver and with Gate Time or duration, set to 50%.

Switching to Edit Step mode allows the Step Time of notes to be altered. This is also achieved with the musical keyboard buttons whose function is automatically altered to match the 'secondary' text and image labels. The first nine white keys, for instance, have musical values from a semibreve/whole note through to sixteenth note/semiquaver (plus quarter, eighth and sixteenth note triplets). There is also provision for dotted notes via an additional key combination. Edit Gate mode allows the durations of notes to be set with the first five black keys which are also labelled '0 Rest', '1 Stacc', '2 Non Leg', '3 Legato', and '4 Tie'. In both Edit Step and Edit Gate modes, more precise settings can be made using the final ten white keys which double as the MC-202's numerical keypad.

It will be clear already that the MC-202 offers considerably more control in its sequencer than instruments like the SH-101 and even the TB-303. Indeed, while the enthusiasm for the coming of the computer age is undiluted, it remains fascinating to see how many discussions of digital sequencing focus on (re)creating a human feel and avoiding the overtly machine-like effects of strict quantisation. Celebrating the level of control, Colbert notes.

> Though step time generally plays the most important part in determining the feel and timing of a piece, it's the gate that most often introduces 'human' idiosyncrasies. If you've got a repetitive bass line, for example, it's the short gate time of the final note that makes it sound as if you've 'snatched' it off the keyboard before dashing back to the beginning of the riff.
>
> (Colbert 1983c: 11)

To deliver control with this degree of precision, although the MC-202 presents more dedicated controls than the MC-8 or MC-4, the programmed data are still relayed as the same series of two-digit numerical codes via an LCD. There is one important caveat, however, the smaller size of the MC-202 LCD means it is not possible to see all the inputted data pertaining to any given note simultaneously. Instead, to reveal all the pitch, timing and duration data for a given note, it is necessary to cycle between the three Pitch, Step and Gate modes by repeatedly pressing the Edit button to the left of the display. Similarly, while additional articulations such as Portamento and Accent (which can be set to accentuate the amplifier and/or filter for a more aggressive sound) have their own dedicated 'Accent' and 'Porta' buttons on the MC-202's front panel, these data can only be entered (and displayed) in specific Edit Modes. The use of Accent and Porta(mento) can lend the MC-202 a distinctive tone, however, and enables a dynamism in the sound that is not possible to recreate with the SH-101 sequencer alone despite the apparent similarities in their synthesis engines (Hub 2015).

However time-consuming data entry might be, even with the MC-202's new interface, one of the benefits of such a complex digital sequencer is that, once entered these data can be endlessly edited. Like its forebears, the MC-202 offers a full complement of commands for modifying, inserting and deleting individual steps or even whole measures. As such, sequences can be built up a piece at a time, previewed by switching out of Edit and into Play mode and then further honed back in the Edit screen. To use the MC-202 is to glimpse the future of electronic music–making workflows. The constant shift between programming, listening, editing and tweaking will be familiar to users of modern hardware instruments as well as software-based Digital Audio Workstations. It is important to note, however, that for all these apparent similarities, there is some important differences. Chief among them is that, like the MC-4 and MC-8 before, the MC-202 is a linear sequencer. That means that, even though individual notes and whole measures can be inserted, copied and deleted, there are no Patterns to be chained together or looped to create a lengthier song or 'Track' to borrow Roland's TR parlance. Rather, the MC-202 sequencer runs from start to finish playing every note in order according to its pitch, timing and duration data. In this way, the MC-202 is quite unlike sequencers such as that underpinning the TB-303 and even the entirety of Roland's TR-x0x drum machine lines which are wholly based around short, typically one to four bar patterns that can be individually looped and/or assembled into Chains and longer Songs.

The MC-202 does offer a Cycle mode which, upon reaching the last programmed step, seamlessly restarts the sequence from the beginning thereby allowing the musical performance to loop indefinitely. This might seem to present a way of working around the linear design and bringing some of the functionality of the pattern-based x0x instruments to the MicroComposer. However, because the MC-202 can only hold one sequence in memory and that sequence is always played in its entirety from start to finish regardless of whether it comprises just a single note or the full complement of over 2000, there is, in effect, only a single loop available to cycle.

Recording and Play Bach

Digging into the Owner's Manual, we find much evidence of the MC-202's design ethos, the workflow around which its operation is based and, crucially, the depth of the prior knowledge of Western music theory that is assumed. Most obviously, there is traditional musical notation throughout. Examples abound of staves of melodic lines that are explained and translated into the requisite codes and commands expected by the MicroComposer in a manner we have seen in the TB-303 guidance. As such, phrases of notated crotchets and minims sit alongside procedural flow charts detailing each step in the music processing workflow. It is fascinating to note that there is no explanation of the musical notation or terminology itself which is assumed to be self-evident. The entirety of documentation from the first pages takes for granted that the reader will be familiar with the conventions and will be able to interpret note pitches, rests, bar markers and time signatures from the reproduced staves and that these will have meaning when they are translated into the machine codes. This level of a priori knowledge and familiarity with the modalities of modern classical notation is clearly the baseline and it is upon this presumed foundation that the manual sets itself the task of explaining the conceptual transformation of a musical score into pitch, timing and gate data along with the operational flow of button presses required to enter those data into the system.

Interestingly, this level of detail and assumed knowledge is not only found in the Owner's Manual but actually extends back to the marketing and promotional literature for the instrument. The centre pages of Roland's 1983 MC-202 product brochure, that so struggled to commend the internal synthesizer, are perhaps unsurprisingly, dedicated to the sequencer aspect of the MicroComposer. What might be surprising, however, is the extraordinary level of detail that walks through every step of the data entry procedure. Again, guidance on which pitches to enter is given in musical notation with a number of further staves annotated to illustrate the relationship between the score as written and the data and codes required by the MC-202. As in the Owner's Manual, musical terms such as legato, slur, tie are given no further unpacking while note durations are shown both in visual notation and as their numerical value in pulse per quarter note.

What is even more fascinating is that, following this thorough guide through the MC-202's manual data entry, an entire page is dedicated to a reproduction of a musical score. Under the headline, 'Play Bach. Let the MC-202 perform Bach's Invention No 13' are printed 50 bars of musical notation ready to be entered into the instrument. Quite apart from leading us to question exactly what the intended use-case of this pre-purchase promotional brochure might have been, what we find here again speaks to the presumption of forms of musical literacy. In this hybridised score, all the pitches are written in notation on the five-line stave while, beneath, are numerical and visual instructions that translate the timing and duration information into the MicroComposer's codes and button presses.

The choice of a piece from the European classical repertoire might not seem like the obvious choice today but was far more understandable at the time of the MC-202's release. In 1968, Wendy Carlos' 'Switched On Bach' had gained extraordinary popularity and, in the process, had propelled synthesizers into the mainstream. The album is a collection of pieces composed by Johann Sebastian Bach performed on a Moog modular synthesizer. Not only did 'Switched On Bach' become an international best-seller, it spawned a number of follow-ups ('The Well-Tempered Synthesizer' [1969], 'Switched-On Bach II' [1973], 'Switched-On Brandenburgs' [1980] as well as the 1972 soundtrack to 'Stanley Kubrick's A Clockwork Orange'). The album also spawned a host of imitators and recordings inspired by the combination of popular classical composition and modern synthesizers such as such as Moog Machine's 'Switched-On Rock' (1969), Gershon Kingsley's 'Music to Moog By' (1969), and Hans Wurman's 'The Moog Strikes Bach' (1970). Speaking on the impetus for the album, Carlos herself recounts that the first exploratory recording was of a Bach Invention giving it an important place in the pantheon of influential synthesizer music (Wright 1999). Beyond Bach, some musicians broadened the repertoire yet further. Isao Tomita's Snowflakes Are Dancing (1974), for instance, turned its attentions to Debussy with subsequent classical albums interpreting Stravinsky's 'The Firebird', Modest Mussorgsky's 'Pictures at an Exhibition', and Gustav Holst's 'The Planets' all released throughout 1976 and the 1978 'Kosmos' exploring space and science fiction themes and including John Williams' 'Star Wars' soundtrack (1977). Decades later, electronic music producers such as William Orbit have continued this interpretative tradition with albums such as 'Pieces in a Modern Style' (1995).

The deep connection between classical music and synthesizers that had been cemented by these recordings throughout the late 1970s and 1980s also helps us better understand another part of the MC-202 package. Along with the instrument itself and its operating manuals, users received a cassette. Perhaps this contained recordings of the instrument in action? In fact, attempting to play to this cassette in a Hi-Fi, the listener would be greeted with nothing but a shrill piercing tone followed by bursts of digital noise. Perhaps an avant grade performance? In fact, this was Roland's 'MC-202 Data Cassette' and on it were not recordings of music per se but rather the audio encodings of the MC-202's data that could be loaded into the machine. Like a number of computer-based instruments at the time, including Roland's earlier MicroComposers as well as a host of home computers, the MC-202 included a 'Tape Interface' that allowed data to be saved and loaded to a standard compact cassette in much the same way as other software and computer games. Although Roland sold the MTR-100 'Digital Cassette Recorder' for the purpose, any cassette deck with a record and playback facilities would fit the bill. In the case of the MC-202, this meant that the entirety of the sequencer's memory could be stored and later re-loaded into the machine preserving one's compositions for ever (or for as long as the cassette lasted).

The MC-202 Data Cassette actually contains five demo tracks which can be individually loaded into the instrument. Two are J.S. Bach 'Inventions' with is accompanied by a further Prelude while the other two are Debussy compositions named on the cassette inlay as 'Dream' and 'Snow is Dancing'. As such, not only does the MC-202 Data Cassette speak to the specificities of 1980s data storage technologies but also continues to solidify the 1970s–80s relationship between synthesizers and interpretations of the Western classical repertoire. With the Bach 'Inventions' charting a link back to Carlos' influential work and Debussy making the connection to Tomita (who was also a notable user of the MicroComposers), the MC-202 is both embedded within and responsible for deepening a complex network of relationships between new technologies and reinterpretations and remediations of existing music. It is also notable that the pieces on the cassette comprise two-part compositions that take advantage of the MC-202's ability to control both its own internal synthesizer and another external sound source reminding us again of the MC-202's MicroComposer heritage and its ability to sit at the centre of a larger setup. With the Data Cassette tantalisingly including a '1' printed on its inlay and cover, there is some discussion among MC-202 enthusiasts online that points to the existence of a second cassette collection that contained information and data for excerpts of popular chart hit 'Popcorn' made famous by Hot Butter in 1972 as well as the 'Cantina Band' theme from Star Wars, among others. While the sci-fi track makes the connection to Tomita even more explicit, a rendition of 'Popcorn' firmly anchors the MicroComposer in the world of synth pop.

Use It Then Lose It

Whether there were two or just one, the Data Cassette also reminds us of another important feature of the MC-202 which it also shares with the earlier MicroComposers. Under the heading 'Cautions', the Owner's Manual notes that,

> The MC-202's memory is volatile (i.e. the data is not retained if the MC-202 is switched off. Therefore, to avoid accidental erasure of the data, you may be required to save the data ... before turning the MC-202 off.
>
> (Roland 1983c: 6)

That the instrument's memory is wiped as soon as it is powered down helps us understand why the MC-202 starts up in Edit mode and why there is no concern about inadvertently overwriting a carefully crafted sequence as it always powers up empty. Even though the MC-202 shares this behaviour with the earlier MicroComposers, it is still a little surprising to find such sanguine responses to this volatility with Colbert matter-of-factly noting that, 'Needless to say, the 202 forgets everything as soon as you switch off the power' (Colbert 1983c: 88). This resignation is particularly notable given the MC-202's other great innovation as, unlike the MC-4 and MC-8 that were unequivocally studio instruments, the MC-202 is a portable device.

Indeed, the May 1983 issue of *Electronic & Music Maker* in which the MC-202 is reviewed ('Microcomposing with Roland' 1983) includes an image on its contents page of two musicians on the beach. Accompanied by the caption 'Portable music with Fostex X15 and Roland MC-202!', both wear headphones with one singing into a microphone connected to the Fostex multitrack tape recorder while the other plays the small rubbery keys of the MC-202. Not the most likely scenario, maybe, but a clear indication of the way the instrument(s) were understood. Like the SH-101, TB-303 and TR-606, the MC-202 can be 'battery operated for go-anywhere convenience' as Roland put it in their 1983 description of the instrument. This feature of the MC-202 was not lost on reviewers either upon its release.

> Roland's MC-202 MicroComposer represents yet another step forward in a compact portable design that lets you create music anywhere. ... Plug in your stereo headphones and wherever you are you can compose your music!
>
> ('Microcomposing with Roland' 1983)

Similarly, retrospectives of the instrument such as Carpenter (1995), draw attention to the novel nature of the instrument's portability. 'The MC202 can be run off batteries and used with headphones, so it's ideal for working on your latest track whilst you're on the move. Does this make it one of the earliest "walkstations"?'

While forward looking it might seem with the benefit of hindsight, the MC-202 clearly draws from the aesthetics, functionality and popularity of the new breed of portable instruments such as the 1979 Casio VL-1. In particular, the MC-202 does actually include some alternative data input modes. Most pertinent here is the TAP mode which replaces the manual entry of timing and duration data. Having entered the string of pitches in the same manner outlined above, activating TAP mode begins a built-in metronome. As the manual encouragingly puts it, 'Start tapping after two measures of the metronome. Please try to tap in the right rhythm.' (Roland 1983a: 23). This method of rhythmic programming will be very familiar to users of the Casio VL-1 as will the two blue buttons which are remarkably similar in shape, colour and size on the two instruments.

However, while the VL-1 was also a portable device with volatile memory, it was more of a musical plaything that offered nowhere near the complexity of sequencing and none of the connectivity of the MC-202.

> Unlike other portable keyboards that offer melody recording and playback, the editing via the liquid crystal display and the direct communication with analogue and synth-controlled machines puts the MC-202 into a special category of its own for the more serious musician.
>
> ('Microcomposing with Roland' 1983)

As such, the level of investment that the MC-202 encourages with its complex sequencing capabilities means that the combination of battery power and portability with volatile memory does not feel like a particularly robust one. Indeed, the MC-202 manual offers plenty of cautionary advice.

> Specifically, when the MC-202 is operated by battery power, take good notice of the battery consumption. If the Battery indicator on the right side of the Display Window flashes showing the batteries are getting low, immediately replace them with a new set. Otherwise the MC-202 stops operating correctly in about 1 hour.
>
> (Roland 1983c: 6)

In case we read this believing that there is a window of opportunity while the MC-202 retains its sequence data while the exhausted batteries are switched out for a new set, the manual is very clear, 'Be sure to save the data on tape before exchanging the batteries.'

In other words, the moment the batteries are removed or depleted, the sequence is lost. The implications for the true portability of the instrument are significant and with such a precarious arrangement, simply packing a pair of headphones is a risky endeavour. With approximately eight hours of battery life expected from a new set of six C cell (LR14) batteries, the clock is ticking for the portable musician whose MC-202 needs to be back at base to be connected to a tape recorder to backup that carefully entered data. The other solution is to add a portable, battery powered cassette recorder to backup and restore data between battery changes. Perhaps this is the real reason the portable musicians pictured in *Electronics and Music Maker* magazine had also packed a multitrack tape recorder along with their MC-202 for their trip to the beach.

> Even while using the AC Adaptor, be sure to keep the battery securely connected, so that even if the AC Adaptor cord comes out during performance, the MC-202 will continue to operate. (If, however, the Adaptor gets disconnected from the AC outlet, the data will be erased.
>
> (Roland 1983c: 6)

Ultimately, the sheer number of caveats and ways of immediately losing data do suggest that the MC-202 is rather more at home indoors, securely attached to a mains electricity supply with the batteries acting as a backup in case the cable is accidentally disconnected. Indeed, the portability of the device seems almost confusing given the frequency with which it is described as a sequencer and the centrepiece of suite of connected instruments.

The Synthesizer 'Playback Facility'

The design decisions arising from the MC-202's portability have other consequences notably in the implementation and functionality of the synthesizer component. We have already noted that the MC-202 is clearly presented as a MicroComposer with a built-in sound generator, however, it remains the case that the sound generator bears a very close resemblance to the SH-101 which, released just a year earlier, remained a much-vaunted instrument. As a comparison of Figures 6.2 and 7.2 reveal, the MC-202 and SH-101 are almost identical in terms of their sound circuitry with both instruments employing the CEM3340 oscillator chip and Roland's own proprietary IR3109 filter circuit. There are, however, some important differences and, for all the similarities in their architecture, the MC-202 and SH-101 do not sound exactly the same.

It would be hard to claim the MicroComposer sounds thin, yet it does have an appreciably less weighty tone in its core oscillators especially when the waveforms are mixed. Given the importance of the Source Mixer in creating the fine balance of waveshapes, we might begin by thinking any difference is attributable to the MC-202's shorter faders making it harder to dial in the precise sweet spots. However, under the oscilloscope, the reason becomes clear. Where the SH-101 Sub Osc is phase locked to the end of the Sawtooth waveform cycle, in the MC-202 it is shifted and locked to a point part way through the Saw wave. In isolation, this is not noticeable, yet in combination with other waves, the mix is definitely lacking some of the SH-101's depth. It is not clear whether this was a choice on Roland's part but comparing the schematics of the two instruments, this change of behaviour may be attributable to the replacement of a transistor with an op amp.

Regardless of the reason, however, it is a quirk that further differentiates two synthesizers that are all too often described as repackaged versions of one another.

More deliberately, the MC-202 adds a delay to its LFO so that modulation does not occur instantaneously. In truth, this addition is likely a result of the absence of the SH-101 bender lever or optional MGS-1 grip that allowed real time performance in the application of modulation. Nonetheless, the LFO delay is a feature that, when used in combination with the sequencing of note durations, adds a distinctive flavour to the MC-202 as short notes go unembellished and longer ones sing out with vibrato for articulation. Elsewhere, absences are not so well accommodated and the MC-202 LFO offers only a Triangle waveform compared with Triangle, Square, Noise and Random on the SH-101. Similarly, the MC-202 removes the Noise waveform from the Source Mixer robbing the sound designer of some flexibility and making it harder to synthesise some of the drum and percussion tones to which the instrument's snappy envelopes are well-suited. As the Noise waveform is generated by the CEM3340 oscillator that is at the heart of both the SH-101 and MC-202, this decision is a curious one. Given that Noise is very often used to create drum and percussive sounds, perhaps we could argue that the MC-202's synchronisation capabilities allow it to sit at the centre of a larger system where a dedicated drum machine such as the TR-606 might handle such duties. It is rather more likely, however, that the desire for compactness and portability and reduced space on the front panel simply led to a reduction in the number of physical controls and a slight streamlining of the synthesizer's functionality.

Perhaps these alterations offer some explanation for Roland's apparent reticence to foreground the MC-202 sonic credentials and potential. However, taken as a whole, the marketing and promotional materials, the operation manual and even the contents of the Data Cassette(s), paint a narrative of an instrument primarily designed as a sequencer. More than this, by interrogating the annotated scores, a priori knowledge and complex methods of data entry that require the conceptualisation of notes as separate pitch, timing and duration components, the MC-202 is a sequencer that seems primarily designed for the transcription of existing music ideas and phrases.

The Groovebox

This combination of design and marketing factors means that, not only was the MC-202 not well-received (or perhaps understood) when it was initially released, but that it was also rather disregarded even during the renaissance of interest in analogue instruments during the 1990s with Carpenter (1995) noting that, 'Almost overlooked in the rush of people desperate to get their hands on a Roland TB303, the MC202 is just beginning to find favour with the all-analogue dance crowd.' Even *Computer Music* magazine's far more recent 'Blast from the Past' review points to the relative obscurity of the device. 'Even the most studied aficionado of analogue esoterica could be forgiven for overlooking Roland's MC-202.' (2018). Indeed, of all the 1980s x0x instruments, the MC-202 remains one of the few instruments that has not been celebrated and re-released in digital 'Boutique' format. From a functional perspective, with the removal of the keyboard, there is a surely strong argument that the Boutique SH-101 would have been more properly identified as an MC-202. However, the recognition of the SH-101 clearly far exceeds that of the MC-202.

That said, Roland did not let ether the MicroComposer name or the underlying concept of the instrument wither away with the MC-202. They did, however, use them in very different ways. The principle of the combined sequencer and sound source laid low for a little while but was dramatically rejuvenated in 1996 when Roland released the MC-303. Superficially, there are some similarities with the MC-202 in that the unit is structured around a short row of buttons laid out as a musical keyboard (far shorter than the MC-202's with just 16 notes) with realtime controls for the built-in synthesizer section occupying the top of the unit. That these controls are all rotary encoders rather than the MC-202's sliders, and that there are only eight of them, gives the unit an appearance that appears to fuse elements of MC-202 and TB-303 together. While the MC element of the instrument's name connected the machine back to its roots, the allusion to the TB-303 proved to be rather contentious. The MC-303 uses an entirely digital sound source with samples of Roland's TB-303, TR-808 and TR-909 instruments along with various other waveforms. In a marketplace that had long-been celebrating these out of production instruments and that had been clamouring for authentic analogue reissues, the large number of unalterable preset sounds were somewhat underwhelming.

Similarly, although the sequencer eschewed the MC-202's linearity for a pattern-based arrangement much more in keeping with structures and performance regimes of the dance music that had so popularised the machines which the MC-303 referenced, it offered more presets than locations for original compositions. Moreover, the preset patterns tended towards rather polite, mainstream flavours of Techno and Trance thereby framing the instrument as lacking the crucial edge and cachet that the original xOx instruments had been bestowed with. Interestingly, in these preset sounds and patterns we see and hear a very particular cultural and musical contextualisation of the MC-303 as an instrument. In principle, this is more than a little similar to the ways in which Bach's 'Inventions' and Debussy's 'Snowflakes' framed the MC-202 and spoke of the manufacturer's sense of what, and who, the instrument was for (just as the Oscar Peterson advert had, to some extent, with the ProForm TB-303 and TR-606 pairing). Later Roland Grooveboxes moved through the xOx numbering with the same lurching imprecision as Roland used in the 1980s with the MC-505 (1998) followed by the awkwardly named MC-307 (1999) and the MC-909 (2002) and MC-09 (2002) preceding the MC-808 (2006). Broadly speaking, the 1990s MCs each retained the same core design principles and stood as all-in-one instruments of varying complexity. It is notable, though, that in the coverage of its internal waveforms, preset sounds and patterns, each was squarely aimed at the electronic dance music producer reflecting the ways in which the 1980s MC, TB and TR ranges had become embedded in those musical genres and practices.

These were not Roland's only MC-named devices, however. A few years after the MC-202, the company regrouped around the concept of the standalone hardware sequencer and released the MC-500. Launched in late 1986, the MC-500 brought back the 0–9 numerical keypad of the MC-8 and MC-4 although, as the recording workflow of the new devices was primarily centred on capturing real time performances, this was used mostly for undertaking editing rather than data entry functions. Unlike the later Grooveboxes that also bore the MC name, the MC-500 and its later Mk2 revision, were sequencers through and through with no ability to generate sound. They did, though, add a floppy disk drive so that sequencer and songs could be saved and, while they were small, there was no suggestion that these were anything other than studio tools.

Crucially, the MC-500 had done away with the previous MC's analogue control voltage and DIN Sync connectivity which were replaced with MIDI sockets. The near ubiquitous MIDI standard had seen the development of a plethora of new instruments including a new trend for keyboard-less sound modules. Typically housed in 19 inch-wide rack mountable enclosures, these instruments were designed to be played remotely via a controller keyboard or sequencer. Such an arrangement is very familiar today where the computer-based Digital Audio Workstation (DAW) has largely taken the place of the hardware sequencer but still performs a similarly central function. At the time Roland released the MC-500 MicroComposers, a number of other companies had similar products on the market including Yamaha with their QX range, for instance. However, in reality while these companies and others like them which were producing similar products, there was a more major challenge.

Compu Music

> We no longer need to think of the computer as the future, it is here now! It is in nearly every facet of our lives to help us, teach us, and entertain us. The idea of using a computer for making music is not a new one. Scientists and artistic collaborators have been working on this problem for some time and with a variety of approaches. We are now ready to reap the benefits of their labors.
>
> (Roland 1983d: 1)

These are the opening lines of the Operation Manual for the Computer Music CMU-800. Initially released under their Amdek subsidiary and later rebadged as a Roland DG (Digital Group) product, this little-known instrument marks a considerable shift in Roland's strategy. A combination of hardware and software, the CMU was a polyphonic desktop synthesizer that connected to a home microcomputer running Roland's proprietary software to provide sequencing and control. Initially compatible with the Apple II, Sharp MZ-80K and NEC PC-8001 and later the best-selling Commodore 64, the hardware unit contained a number of internal analogue synthesizer and percussion sounds. In total, there are monophonic melody and bass sounds, a four-voice polyphonic chord voice and seven drum sounds. Each section could be addressed separately making this the first multitimbral desktop instrument. The synthesizer sounds offered very little in the way of customisation with a sustain and decay control at most although the drums are taken from the TR-606 and have the punch and clarity of that instrument. The unit is also equipped with eight CV/Gate outputs for controlling other suitably equipped instruments. By connecting their audio output back into the CMU-800, the sounds of the external instruments could be mixed with the internal sounds to create even fuller compositions.

While the CMU-800 was the first instrument in the Compu Music range, Roland also released a handful of other devices including the DXY-100 plotter which, among other things allowed the printing of musical notation. Perhaps most interesting for our discussion, however, is the CMU-810 which is a self-contained analogue monophonic synthesizer designed as a more complex and feature-rich accompaniment to the basic CMU-800 sounds. Housed in the same beige metal wedge shaped desktop unit as the CMU-800, the 810's front panel is dominated by a row of sliders and, although their plastic caps are different, they map out a signal flow that is very familiar. The CMU-810 repackages the sound generation components of the MC-202. It has the same controls, including the omissions and additions inherited from the SH-101, except here it is recast as the external sound source for the computer's sequencer.

Although the Compu Music series was not a commercial success, it marks a significant development in both the Roland's history and the trajectory of sequencing, external sound modules and the computer-based studio. The MC-202, albeit in slightly different form, is once again at the centre of another innovation and perhaps this suggests another way of conceiving of, and using, the instrument. Certainly, by the 1990s, the MC-202 was typically understood as a synthesizer with a built-in sequencer rather than as Roland had originally described it. Carpenter (1995) is illustrative, 'The 202 could be described as a keyboardless expander version of the SH101 (assuming you ignore the small, rubbery, Casio VL Tone-like attempt at a keyboard), but unlike the 101, the MC202 includes a two-track sequencer.'

Given the popularity of synthesizer expander models, simply ignoring the MicroComposer element and focusing on the built-in synthesizer could breathe new life into the instrument. However, even here the CMU-810 has some distinct advantages over the MC-202 because it possesses CV and Gate inputs. Initially designed to allow it to be controlled form the CMU-800's outputs, they are eminently usable today particularly following the resurgence of analogue connections that accompanied the Eurorack modular boom throughout the 2000s. Surely, though, the MC-202 also has CV and Gate inputs so could we not just use that instrument in the same way? Unfortunately, the MC-202's circuit means that the CV and Gate signals are intercepted by the CPU in order to process them and record them into the sequencer. This introduces noticeable latency which renders live playing all but impossible. This might seem like an oversight, but as we have seen, the MC-202's CV and Gate inputs were never intended to allow the built-in synthesizer to be played live. Instead, they were to allow pitches to be entered from an external keyboard rather than the rubber buttons of the instrument itself. That the processing lag that accompanies note input was not initially considered an issue speaks eloquently to a design intention that envisioned the MC-202 as an instrument for offline transcription and programming rather than real time performance. In the tradition of the MC-8 and MC-4, the MC-202 was a music processor and its sequencer was central. Unfortunately, that processing took time and so being a computer controlled MicroComposer demonstrably impacts the MC-202's instrument's live performance affordances.

Mods and Hackers

We might be forgiven for thinking that time has finally caught up with the MC-202 with its lag-inducing processor, powerful but convoluted linear sequencer, precariously volatile memory and reliance on cassettes for storage. However, two recent modifications are especially noteworthy here as they turn the MC-202 into a thoroughly modern instrument.

The Tubbutec MC-2oh2 is one of the most transformative of any electronic musical instrument modifications. The company's understated description of the project as an 'upgrade and feature extension' hardly does justice to the magnitude of the changes. Essentially, the MC-2oh2 modification replaces almost all of the digital components of the original instrument by giving it a new CPU, keyboard controller and even a thoroughly modern high resolution OLED display. Most obviously, by adding MIDI input and output alongside the existing CV and Gate, the update enables compatibility and interoperability with a wider range of external devices. MIDI also adds another flavour of synchronisation to the existing Din Sync which means that the MC-2oh2 can clock or be clocked from other instruments or modern computer-based systems.

But let us not think that this is about finding new ways to synchronise an old machine. Custom software implements an entirely new sequencer that does away with the cyclable linear arrangement of the original and replaces it with a pattern-based version that is simultaneously more modern and more xOx. Thirty-two patterns can be arbitrarily triggered or chained into Songs. Sequences and Songs can be edited while the instrument is still playing adding a level of improvisational potential absent from the original. And if that weren't sufficiently performance oriented, sequenced notes can be played in a variety of different orders, individual steps have probability settings to alter their modifiers such as Gate Time, and there are new controls for modifying arrangements in real time. The Arpeggiator is significantly expanded with multiple play modes and an expanded octave range and, like the sequencer, it can be synchronised to the original analogue Din Sync, analogue clock pulses or MIDI. The Arpeggiator and sequencer can also interact with the former inheriting motifs from the latter to create yet more live, performable opportunities. Crucially, sequences and songs can be saved without the need for batteries or cassette backup making the new instrument eminently more suitable for portable use.

While the core analogue sound circuits are scrupulously retained and the instrument can be returned to its original state if desired, the digital upgrade allows for additional sound sculpting features. There are additional software-generated envelopes for the Filter and Pulse Width modulation, while a second, independent LFO with a range of waveshapes is dedicated to the Filter. The MC-2oh2 also reinstates the Noise generator that was originally lost in the transition from the SH-101 and adds an envelope to that for good measure.

Cementing the lineage between the MC-202 and the earlier MCs, a third independent sequencer lane is added. This can be used to drive external instruments connected via MIDI. This feature alone might be enough to warrant the instrument claiming the name MC-303. But to connect the MC-202 to that next generation grooveboxes that actually began with the MC-303 in 1996, this third sequencer lane can also be used to trigger the array of built-in 8-Bit drum samples. The provision of customisable drums truly takes the MC-202 from prototype to fully fledged groovebox. And, unlike any other Roland groovebox instrument including the most recent MC-101 and MC-707, the MC-2oh2 is an analogue synthesizer. Given the scale of the upgrade, it probably goes without saying that the original CPU latency is addressed so the enhanced MC-202 can be played live just like a CMU-810.

Although the Tubbutec mod is quite astonishing in its scope, ambition and execution, Defective Records' 'MC-202 Hack' application offers a different but equally ingenious solution to the problem of the original instrument's fragile memory and backup procedures. While compact cassette tape was not the most reliable storage medium in the 1980s, subsequent years have seen them all but forgotten with very few cassettes, players or recorders currently in production. However, Roland's backup protocol does not actually require tape at all. In fact, what emerges from the Tape Out socket on the back of the MC-202 is actually digital data encoded as an audio file. After all, all that was required to interface with the Tape Out was a standard cassette recorder with an equally standard cassette tape exactly as one might use to record a favourite MC-202 laden track from the radio.

As the data stream is encoded as audio, this means that the Tape Out signal can be captured and recorded by any device capable of recording analogue audio. While this certainly includes cassette decks, it also includes computers, tablets, phones and even samplers. Once recorded, simply playing back the audio file into the MC-202's Tape In loads the data into memory. There is no way to speed up this process and it happens at the same rate as if recording to and playing back from tape but the near-obsolete media and recording device are now replaced by more reliable and readily available modern alternatives.

This is already a huge boon for the MC-202 composer who is saved from investing in libraries of cassettes and drawers of batteries. However, the MC-202 Hack software takes things a stage further as the creators have not merely provided a useful tool for archiving audio backups but have also reverse engineered the data format encoded in the audio stream and created a software application to decode and display it. This means that MC-202 sequence data can be visualised and edited on a computer as though it were a standard MIDI file. In fact, because the software operates bidirectionally, a standard MIDI file can be converted and transmitted to the MC-202. Obviously, the software does not add any new features but it simultaneously enables the use of editing tools outside those available in the MC-202's own operating system and allows those sequences to be saved on modern storage media. It also allows sequences created outside the MC-202 to have their playback verified on a different synthesizer. Perhaps the most fitting would be a clone or software emulation of the SH-101.

What is particularly fascinating about the innovation of the MC-202 Hack software, is that it is made possible in part by Roland's decision to use a tape interface in the original instrument. Had they opted for a built-in storage device such as the floppy drive that was included in the later MC-500, such an elegant solution would not now be possible. Of course, there are hardware floppy drive emulators but there is something undeniably neat about working with audio files and finding new ways to interrogate the data in their original format as they stream from the back of the instrument just as they have for four decades. Thanks to modern audio recording devices and the ingenuity of Defective Records' software developers, the MC-202's data files remain unchanged but are now more usable and can be more reliably backed up than when they were designed. Ultimately, we might argue that the MC-2oh2 upgrade and MC-202 Hack software unlock latent affordances in the instrument whilst adding a host of new opportunities that have enabled the MC-202 to catch up with the modern music technology world. Simultaneously, however, we should remember that the 'rediscovery' of analogue CV and Gate and Din Sync connectivity that accompanied the analogue resurgence and the renewed interested modular synthesizers, has allowed the music technology world to finally catch up with the MC-202.

Marching Backwards into the Future

What, then, should we make of the MC-202? Perhaps we could simply dismiss it as a failed instrument that was ill-conceived and subsequently resurrected and reinvigorated by the industrious work of modders who have turned it into the device it should or could have been from the outset? Alternatively, we could view it is a product of its time and of the very specific moment in 1983 just on the cusp of the establishing of the MIDI specification that promised stability and standardisation in communication between instruments. And we must remember that this moment is shortly to witness a wave of polyphonic synthesizers like the Yamaha DX7 that brought their shockingly new palettes and timbres that had never been heard before. Finally, it is hard to consider the MC-202 outside the backdrop of a new generation of computers and music sequencing software that was already beginning to transform the studio, the role of the musician and their relationship to technology. In a sense, each of these interpretations has merit and it is reasonable to argue that, in addition to being an underappreciated instrument in its own time that has been rediscovered and rejuvenated to become highly relevant today, the MC-202 also serves as an invaluable historical artefact that provides a link between different periods of music technology, production and practice. In particular, it tells us much about the turbulence and uncertainty of early 1980s electronic musical instrument production and consumption. For a synthesizer enthusiast, this was an exciting period with not just new models appearing but entirely new forms of synthesis and new connections between instruments and computers. For a synthesizer designer, however, it is clear that this was a time of great uncertainty. The MC-202, CMU-810 and the whole Compu Music range make it clear that Roland's engineers and research and development teams appreciated the growing importance of the computer and the centring of composition around the computer-based sequencer. It is also clear, however, that they were not at all sure how this would play out. Would the computer sit inside a dedicated musical hardware like the MicroComposer or in a more general purpose tower or desktop casing with mouse, keyboard and monitor and a host of remotely controlled sound modules? One answer comes via Michigan Synth Works who in 2022 created the MSW-810 which is described as a 'faithful recreation of the CMU-810 Monosynth' and supplied with a scan of the original Roland CMU manual as documentation. With the heart of the MC-202, the MSW-810 re-presents the instrument as an external sound module and takes advantage of the renewed currency of CV and Gate connections and the availability of external sequencing. By contrast, there is also a buoyant market for new grooveboxes that combine sound generation with sequencing in a compact, transportable form factor. It is easy to sympathise with the uncertainty of Roland's engineers as they contemplated possible futures. Would self-contained musical hardware instruments dominate or would the more modular computer and sound module system take over? As we now know, these weren't alternate visions and both have come to pass.

It is rather fitting then, that the impact of this two-part instrument should be felt widely. There is no doubt that Roland's post-1990s grooveboxes owe more than just their name to the MC-202 and these, in turn, set in train a series of developments that led to a slew of all-in-one instruments such as those produced by Elektron, Novation, Polyend and Teenage Engineering among many others. Even though they often include or are even entirely based around samples rather than analogue sound circuits, the Digitakt, Circuit Tracks, Tracker and OP-1 certainly owe a debt of gratitude to the MC-202. And, of course, instruments such as the Elektron Digitone, Syntakt and Analog Four eschew samples for various forms of analogue and digital synthesis in combination with their powerful sequencers and external instrument control. For their part, Roland have continued to develop their own line of grooveboxes and have begun filing in the unused MC-x0x numbers with the entirely digital MC-101 and MC-707 joining the ranks in 2019.

In equal measure, the MC-202 was ahead of its time while being utterly reflective of the moment of its creation and release. In combining sequencing, sound generation and external instrument control, it predicted the future for Roland and mapped out a direction for musical instrument design more broadly. Yet, at the same time, the MC-202 was resolutely stuck in the past. Changing fashions and functionality saw to it that the MC-202's analogue sound circuits and connectivity were all-but-superseded when the instruments were released while its sequencer was considered to be convoluted and lacking the storage and backup facilities that would seem essential to make sense (and use) of its portability.

Decades later, the rich tonality of its CEM3340 oscillator and IR3109 filter circuits are very much back in vogue. The instrument is the sound of Alessandro's Cortini's 'Sonno' (2014) album, for example. Recorded in a variety of locations while the artist was on tour, the entire album is produced with just an MC-202 processed with some effects pedals. 'I had an old Roland MC-202 and a few delay pedals. I would set them up on a table in the hotel room and connect it to a little speaker system, and I would set up something that would just run. Longer sequences would repeat and then I would just tweak things and then lay down.' (interviewed in Frosty 2017). The resulting album is a virtual showcase of the MC-202 and it is interesting to note that Cortini's use is centred around approaching the instrument on its own terms rather than being bound by the expectations of its prior use.

> It was an instrument new to me that I got excited about, that I had no background with whatsoever and I approached it in a way that was unfamiliar to how it was used before ... the MC 202 was more connected to techno and acid, which I love, but it is not the way that I ended up playing the instrument itself.
>
> (interviewed in Ilevičius n.d.)

Clearly, the distinctive tonality and portability are important features here but it is notable that the volatility of the sequencer memory and the qualities of its connectivity similarly shape the encounter. Decades after its design and release, the inability to save sequences or straightforwardly back up data via MIDI or USB, for instance, emphasise the MC-202's liveness and create a focus on real time tweaking and capturing the instrument's output as audio before it is lost.

Of course, the analogue connectivity and synchronisation protocols of the MC-202 have been given a new lease of life in a world of modular instruments in which control voltages and analogue clock signals are the lifeblood that course through the veins of Eurorack patch cables. Moreover, the analogue and digital modifications such as those designed by Tubbutec, comprehensively recast the MC-202 and allow it to be encountered as a thoroughly modern yet vintage instrument whose features and live performability could barely have been imagined when it was new and, in truth, now eclipse many instruments released in the decades since.

By inspiring the self-contained groovebox and the computer-controlled desktop sound module, the MC-202 acts as a pivot between two contrasting approaches to music production, composition and performance. It was designed on the cusp of two eras in music technology yet, as it has been modified and reimagined over the years, its combination of analogue and digital synthesis and sequencing, its array of interfacing options, its internal synthesis and external control and its dual focus on music processing and live performance has seen those divergent technologies and practices seamlessly converge. Ultimately, even in its original, unmodified form, the MC-202 is truly one of the most complex instruments (in every sense) that Roland has ever made. Because it is so prescient, it is also possible to argue that, despite its rather lacklustre sales and the fact that it remains overshadowed by the more commonly revered TB-303, TR-808 and TR-909, the MC-202 is not only one of the most important of the xOx instruments, but one of the most important of Roland's entire catalogue.

8 TR-909 Rhythm Composer

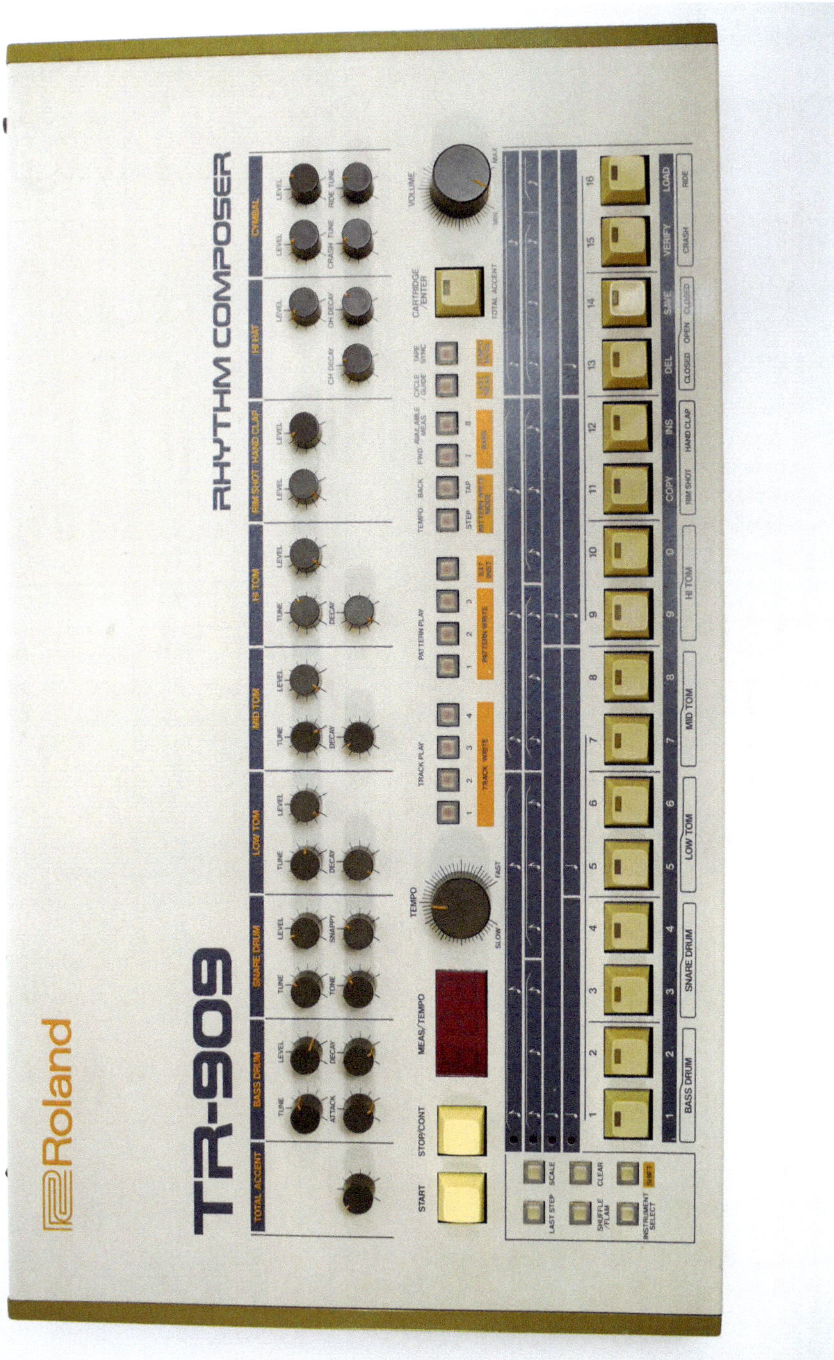

Figure 8.1 Roland TR-909 Rhythm Composer (image credit: implant4).

DOI: 10.4324/9781003374510-8

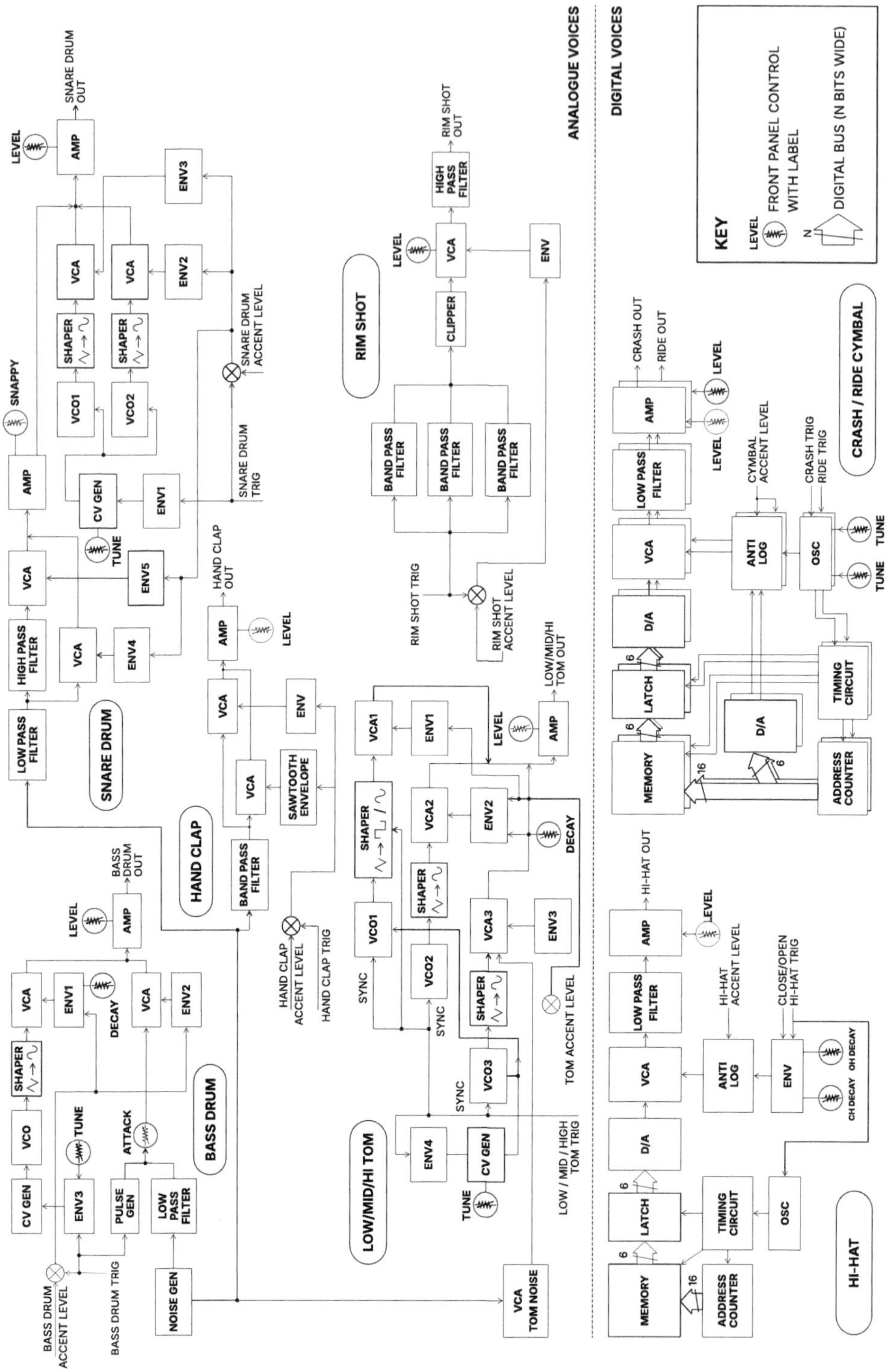

Figure 8.2 Roland TR-909 Simplified Block Diagram (adapted from Roland 1984 TR-909 Service Notes).

TR-909 Rhythm Composer

Released
1983 (discontinued 1985)

Sound source
Hybrid Analogue and Digital Circuits
Analogue: Bass Drum; Snare Drum; Low, Mid and High Toms; Rim Shot; Hand Clap
Digital (6-Bit PCM samples) Open and Closed Hi Hats; Crash Cymbal; Ride Cymbal

Real time controls
Individual Level Controls for all sounds
Bass Drum: Tune, Attack and Decay
Snare Drum: Tune, Tone and Snappy
Low, Mid and High Toms: Tune and Decay
Open Hi Hat: Decay
Closed Hi Hat: Decay
Crash Cymbal: Tune
Ride Cymbal: Tune
Total Accent Level

Tempo
Approximately 37 to 290 BPM
Control is labelled 'Slow to Fast' but BPM can be displayed in the 'Meas/Tempo' visual display

Memory
48 Rhythm Patterns (16 x 3 Pattern Groups) x 2 (Bank I and II)
4 Tracks (1 to 4: Total Maximum Measures 896) x 2 (Banks I and II)

Connectivity
Main Audio Output (L, R/Mono)
Multiple Outputs: Bass Drum, Snare, Low Tom, Mid Tom, Rim Shot, Clap, Hi Hat, Crash, Ride
Trigger Out: Rim Shot
Start/Stop Pedal Input
DIN Sync Input
MIDI In; MIDI Out (x2)
Tape Memory/Sync (Save/Out; Load/In)

Dimensions
486 (W) x 105 (H) x 300 (D) mm
19-1/8 (W) x 4-1/8 (H) x 11-13/16 (D) inches

Weight
4.5kg/9lb 15oz

Notes
According to Roland, fewer than 10,000 TR-909 were manufactured.

Suggested listening
Skinny Puppy 'Smothered' (Nettwerk, 1984)
Jeff Mills 'The Bells' (Purpose Maker, 1997)
Inner City 'Good Life' (10 Records, 1988)
Madonna 'Vogue' (Sire, Warner Bros.1990
Daft Punk 'Revolution 909' (Virgin, 1998)
Björk 'Hunter' (One Little Indian, 1998)

A Different Drummer

Looking back, it is clear that the TR-909 is a truly pioneering product for Roland. It is the company's first rhythm instrument to make use of digital samples in its arsenal of sounds. As we have seen, in the wake of the Linn LM-1, drum machine design had decisively, and almost immediately, shifted to the use of digital sound samples rather than the analogue circuitry whose 'unrealistic' sound had been blamed for the early indifference felt towards some of Roland's previous instruments including the TR-808. The TR-909 is also the Roland's first rhythm instrument to incorporate MIDI (Musical Instrument Digital Interface). Brand new in 1983, this remarkably long-lasting digital protocol allows instruments to communicate and synchronise with one another. The inclusion of MIDI is what makes it trivially easy to integrate the TR-909 into today's music studios and clearly signalled the instrument as cutting edge four decades ago. Without doubt, it is perfectly possible to read the TR-909 as an example of Roland's forward-looking approach to instrument design and, in particular, as an exemplar of the way the company embraced – and even defined – the emerging digital technologies and opportunities. After all, Roland did not simply adopt MIDI like so many other manufacturers. Rather, along with Dave Smith from Sequential Circuits, Roland co-developed the standard with xOx lead Tadao Kikumoto drafting the initial protocol. Seen in this light, if we were to search for an instrument to embody Roland's 'We Design The Future' corporate slogan, the TR-909 would seem to fit the bill very nicely.

Yet, for all this, the TR-909 was not immediately recognised as unequivocally futuristic or forward-facing. In fact, for many who encountered it on its initial release, it was seen as being stuck in the past. It was an instrument that did not move on significantly from its predecessors and rather unconvincingly edged the Roland Rhythm Composer towards the 'realistic sounding' future already ably mapped out by its competitors. Indeed, for some, the TR-909 even marked a step backward from the TR-808. While impressed with the refinements to the workflow and ease of use, Jon Lewin's 1984 piece in the UK published *One Two Testing* magazine is a case in point. Citing both the unrealistic nature of the sounds and the similarities with some of the TR-808's tones, the review disparagingly suggests that Roland should have named the TR-909 the '808½'.

Given that the TR-808 did not use samples, we might be forgiven for being surprised to find any comparisons between its unabashedly analogue sound and that of its successor. However, the truth is that while the TR-909 did use samples, it employed them only for some of its sounds. As such, although the Hi Hats, Ride Cymbal and Crash Cymbal are recreated digitally from PCM sample data, the remainder of the kit from the Bass Drum, through the Snare, High, Mid and Low Toms, Hand Clap and Rim Shot, are all the products of analogue synthesizer circuitry. The TR-909 then, is truly a hybrid machine. Of course, the TR-808, TR-606 and other 'Computer Controlled' instruments had made use of digital technology to control their sequencers and to offer pattern storage and recall. However, where those instruments had exclusively used analogue circuits to create their drum and percussion tones, the sound of the TR-909 draws on a mix of analogue and digital sources. It is a mix of synthesis and sample playback, of old and new, although by no means in equal measure. And, despite Roland's marketing discourse that claimed to bring together the best of both worlds, many commentators appeared to hear more of the old than the new and the upshot was an instrument often positioned as speaking yesterday's musical language. Indeed, Lewin's review goes on to criticise the sounds of the TR-909 as lacking 'the authenticity of real sounds for studio work' (where 'real sounds' is shorthand for digital samples) and ultimately concludes that the instrument is 'a sore disappointment.'

(Lewin 1984: 60). It is reasonable to assume that Roland must have shared a sense of disappointment too as, with the instrument failing to capture the imagination or find an audience, just 10,000 units were manufactured during the year that it was in production (Tsuboi 2020).

With the benefit of hindsight, we might more accurately look back at the TR-909 as a transitional instrument much like the CR-78. Just as the CR-78 had ushered in programmability and set the stage for the x0x series, so the TR-909 brought samples into Roland's sonic lexicon and mapped out the next stage of the journey for the series. More specifically, we might see the TR-909 as a manifestation of a complex of challenges being wrestled with within Roland about how to respond to rapidly changing market forces and the impact of digitisation while also recognising and drawing on the depth of design and technological expertise and the creative and engineering vision of its teams. Roland's promotional literature at launch sought to position the TR-909 as a 'Different Drummer' by emphasising the controllability of its analogue circuits compared with the inflexibility of samples, while also promoting the realistic nature of the samples it did include. In truth, it was a confusing message that attempted to tread the line between analogue and digital, past and future, old and new technologies. As such, for some at least, the TR-909 was not sufficiently different from its analogue predecessors yet too different to its all-digital competitors.

The idea that the sounds of Roland's instruments swam against the tide of the mainstream aesthetic only to be rediscovered and appreciated at a later date by a new generation of musicians has become a standard, almost unchallenged, part of the x0x instrument mythology. Certainly, if we follow the broad trajectories of electronic dance music from the late 1980s to the present, it is hard to underestimate the role of the TR-909 in providing the foundation of genres such as House and Techno. Indeed today, its popularity is such that the TR-909 ranks among one of the most revered and sought after of the x0x instruments – or any instrument, for that matter. Perhaps second only to the TR-808, the TR-909's status as a fetish object is, at least in part, a function of scarcity, but is also due to its sounds and their use in a host of influential records. However, while its characteristically punchy tones are undeniably iconic and have been recreated, reimagined and recirculated in a plethora of hardware and software emulations, sample packs and even separated out into discrete Eurorack modules, one of the defining features of the TR-909 is found in its sequencer. The innocuously labelled 'Shuffle/Flam' button pushes and pulls the TR-909 off the grid and takes the instrument to places that the metronomic, quantised accuracy of the TR-808 and TR-606 simply could not. If the combination of analogue and digital circuits gives the instrument its sound, it is the rolling, syncopated swing of its 'Shuffle' that gives the TR-909 its feel. And it is this combination of sound and feel that help to shape the sound of Techno and House music to this day.

Don't Call it a Drum Machine

After the exploration of comparatively inexpensive, lightweight, battery-powered music on the move that began with the TR-606 Drumatix, the TR-909 marked the turning of Roland's attentions back to the premium end of the market. Accordingly, promotional materials, and the case of the instrument itself, the TR-909 sees the return of the 'Rhythm Composer' title so notably absent from the TR-606. Indeed, Roland's marketing materials made great play of the name, albeit in a somewhat circuitous manner and with more than a slight dig at other unnamed manufacturers' similarly unnamed products along the way. The full-page advert for the TR-909 that featured in, among other places, the 1984 (Volume 2 Number 4) issue of the US-published 'Roland Users Group' magazine, implicitly recognises the competitive nature of the marketplace for rhythm units and uses Roland's soubriquet as a means of differentiating the TR-909 from its putatively inferior competitors. Under the headline 'Different Drummer', it proudly states,

> We don't call the TR-909 a drum machine for some very good reasons. True, it's a machine that makes drum sounds, but that's the end of any similarities between run-of-the-mill drum machines and the TR-909 Rhythm Composer. In fact, playing with the TR-909 is more like playing with a real drummer than anything else.
>
> (Roland Corporation US 1984: inside front cover)

Perhaps history unfairly colours the reading of this marketing discourse, but four decades on, there is more than a little defensiveness evident here. Accepting the hyperbole, the attempt to distance the TR-909 from the myriad other 'drum machines' (and perhaps even its own earlier instruments) is understandable given the growing number of digital, sample-based drum machines being developed by Linn, Oberheim, E-Mu and others. Yet Roland's allusion to their new instrument better replicating a 'real drummer' seems motivated as much by a desire to shift the focus away from the presence or absence of digital samples and the 'real drums' to which the Linn LM-1 had so loudly laid claim, than to any specific sequencing functionality or 'humanisation' features that might be employed to render an instrument less machine-like in its timing and performance.

In fact, elsewhere any nervousness on Roland's part makes way to even more aggressive marketing that centres on the idea that samples alone are simply not sufficient for the creation of 'real drums' and that the hybrid analogue/digital sound circuitry employed in the TR-909 is actually superior to the purely digital approach of its competition.

> We start with digital recordings of real drums, then through a 3-D waveform analysis, re-create the sounds through a hybrid digital/analog process. Not only does this provide the best drum sounds, but also the most flexible. Change the snap of the snare, the decay of the bass, you call it. The sounds you get are the sounds you really want.
>
> (Roland Corporation US 1984: inside front cover)

Here, while the word 'sample' is notably absent, digital recordings are presented as the starting point for Roland's reassuringly high-tech yet vaguely described approach. While the '3-D waveform analysis' may very well refer to the customised CP/M-80 computer upon which the TR-909's Hi Hat and Cymbal samples were edited (Tsuboi 2020), the importance of this text is more about its evocation of Roland's use of cutting-edge technology in the design of the TR-909. In addition to any non-specific allusions to space age science fiction technologies, the invocation of 3-D waveforms must surely be read as references to high end instruments such as the Fairlight CMI (Computer Musical Instrument).

While there were other sampling instruments and keyboards including the comparatively low-cost E-Mu Systems Emulator and ferociously expensive New England Digital Synclavier II, the Fairlight CMI had quickly become a shorthand for the cutting edge of musical instrument technologies (see Harkins 2020). As it was, to all intents and purposes, a computer tuned for audio processing, the hardware interface of the CMI comprised, among other things, a video monitor and light pen that enabled the user to make selections and perform editing on the screen itself as if they were directly manipulating the data. This was futuristic enough when the tasks involved loading and saving samples and other file management duties, in part because the light pen was, even in the late 1970s and early 1980s, an underused and esoteric device as likely to be seen plotting Rebel and Imperial attack trajectories in 'Star Wars' as in any domestic computing setting. However, beyond the hardware, it was the waveform visualisation that particularly communicated the CMI's high end digital credentials.

The Operating System of the CMI is split into a number of 'Pages' dedicated to specific functions with Page 6 offering a 2-D waveform display of any currently loaded sample in the instrument's memory. This view will be familiar to users of modern Digital Audio Workstations and presents sample data on a horizontal timeline. Page D, however, offers an alternative visualisation of the same data. Separating a sample into a number of discrete frames, Page D layers a series of 2-D slices in front of one another, essentially creating a pseudo-3-D scene in the manner of a toy theatre or the terrain models used in geographical information systems. The actual utility of the display in Page D is a matter of debate, but as an indicator of high end technology, it remains unusual today and second to none in the early 1980s. Accordingly, with Page D being a near-permanent fixture on the monitors of artists performing live with the CMI, '3-D waveform analysis' had been firmly cemented as an almost unimaginably complex technological process in the minds and imaginations of audiences and musicians alike.

Of course, what is so fascinating about Roland's references to 3-D waveforms and advanced sound analysis is that digital sampling technology is simultaneously presented as almost agonisingly innovative and futuristic but also musically inadequate. The references to 'flexibility' and being able to create 'the sounds you really want' aim to point to the comparatively static nature of the samples offered in the LM-1 and its immediate successors as well as to the limited tonal control over the sounds. For sample-based drum machines of this era, this might extend to volume controls and the ability to alter the pitch of a sample albeit with the associated change in playback speed lengthening or shortening the sound. Certainly, with single samples for each drum sound there was no opportunity to alter the balance between the body and wires of a Snare Drum or the clickiness of a Bass Drum. In truth, whether the sound of the specific Snare Drum or Toms in the LinnDrum or Oberheim DMX are any more distinctive or creatively limiting than their equivalent in the TR-909 is debatable, but there can be no doubt that Roland's marketing materials sought to promote the idea of drum synthesis as providing flexibility and customisation.

The advertorial commentary in the Roland UK Newslink feature published in the June 1984 issue of *Electronic Soundmaker & Computer Music* puts the case even more forcefully.

> The advantage of the TR-909 is that every drum sound can be 'tailored' to your music. Bass drum has tune, level, attack, and decay; snare drum has tune, level, tone, and 'snappy', all toms have tune, level and decay and even the Hi Hat has Open Decay, Closed Decay and Level while the cymbals have level controls and Crash and Ride tune controls. Being analogue when the toms are tuned higher, for example, no other parameter changes. In a digital system this would not be possible: higher tuning would give a shorter note.
>
> (Roland Newslink 1984b: 5)

The promotional text does seek to have its cake and eat it here because, while it is true that sample-based instruments of this era did not offer timestretching algorithms so that sounds played back at a higher pitch were, indeed, shorter in duration and vice versa, this is precisely the technology that powers the Hi Hats, Ride and Crash Cymbals in the TR-909 with precisely the same effect. As such, even though aspects of the analogue percussion sounds can be modified independently allowing them to be 'tailored', the TR-909's complement of digital sounds behave exactly in the manner Roland disparage when their Tune knobs are used to alter their pitches.

Given the speed with which the marketplace and cultural intermediaries such as reviewers and journalists had contributed to the creation of a deep and very real schism between analogue and digital drum machines, even a hybrid machine found itself swimming against the tide of an entire industry by not fully embracing the aesthetic of realism through sampling. As such, it is hard not to read Roland's introduction and initial framing of the TR-909 as an attempt to tread carefully between the reassuring use of digital terminology without truly committing to the technology, per se.

If Fairlight Made Drum Machines

Of course, attempting to differentiate the TR-909 from the increasingly large number of drum machines (run of the mill or otherwise) produced by competing manufacturers is one thing, but it is clear that Roland also sought to mark a break between this new instrument and its own forebears. Dwarfing the TR-606, the TR-909 is strikingly large. In fact, it is even a little larger than the positively giant TR-808. The TR-909 also returns to a predominantly metal construction after the detour of lighter plastic for the Drumatix. If the similarly plastic construction of the TB-303, SH-101 and MC-202 had been partly motivated by Roland privileging portability, the TR-909 returned to a size and form factor that could, at best, be described as transportable. This was a heavyweight, as well as undeniably heavy, instrument. However, for all its size, heft and literal weightiness, the TR-909 looks altogether different from the TR-808.

Immediately noticeable is the colour of its metallic case. Where the TR-808 was black, the TR-909 is a creamy, 'off white' beige. Surely there could not be a less subtle means of separating the two instruments. In terms of marketing and branding, the colour schemes present the two instruments as polar opposites. The TR-808 looks like the TR-909's negative. Similarly, with many of the competing drum machines from E-Mu Systems, Oberheim and Yamaha dressed in darker tones, the lighter hue allowed the TR-909 to easily stand out from the crowd. Perhaps it is not too fanciful to suggest that for Roland themselves, the bright TR-909 might have represented the hope of a new light dawning from the darkness of the TR-808.

There is further significance to the colour scheme, however. Just as the design of the TR-808 had signalled its departure from the era of home organ accompaniment units by discarding their wood-grained cabinetry, the TR-909 clearly spoke the visual language of the computer. It might not resonate in quite the same way today after Apple turned home computers multicoloured with the introduction of the translucent 'Bondi Blue' Mac in 1998, but before this point, computers were beige. Desktop cases, towers, monitors, keyboards, mice were all beige. Moreover, even though the TR-909, like the TR-808 before, is equipped with a large number of buttons and knobs to program its sequencer as well as real time controls for sculpting the analogue and digital sounds, the top third of its front panel is empty. Additionally, unlike Roland's previous rhythm instruments, the case of the TR-909 is angled to make it easier to use while seated at a desk. Combined with the large amount of unadorned beige metal at the top of the instrument, the sloping front panel further connects the TR-909 to the aesthetic and industrial design of home computers such as the Commodore 64 and Apple II.

The allusions to the personal computer do not end with the colour scheme. The flattened switches filling the lower half of the front panel of the TR-808 are replaced by what are essentially computer keycaps. While the upper section of the TR-909 is empty, the lower portion of the instrument is dominated by a row of 16 tall mechanical keys that could have been prized directly from a QWERTY keyboard (or perhaps even the Fairlight CMI which was a whiter shade than beige but nonetheless similarly stood out from the dark metal, plastic and wood of other instruments). The TR-909's computer keyboard keys, with their integrated red status LEDs, are also resplendent in glorious 1980s computing-beige. Gone are the playful gradients of reds and yellows seen on the TR-808 or the futuristic silver of the TR-606. By contrast, even though it retains some flashes of Roland's corporate orange to add subtle accents to the interface, the TR-909 is a rather more sombre affair. One splash of red is found in the three character, seven-segment LED display that can be set to show, among other things, the tempo of the current Pattern. Where previous Roland instruments had utilised combinations of rotary dials and switches to adjust settings with feedback being presented by the position of the physical controls themselves, here the LED display gives a consolidated, more immediate form of visual indication. Crucially, while it might have been some way from a 3-D waveform display, the TR-909's LED was an altogether more modern visual design cue. The result of the colour scheme, case design and hardware controls, is that the TR-909 is unmistakably businesslike, serious and grown up in its restraint. There are certainly no gimmicky 'Drumatix' names to be found here.

In fact, although the TR-909 revives the title of Rhythm Composer, it is missing the 'Computer Controlled' legend so proudly worn by both the TR-606 and TR-808 before it. This reflects no change in the design of the instrument which retains a microprocessor at its heart to govern the operation of the sequencer and to manage the storage and retrieval of Patterns and Tracks. Perhaps we might argue that such a label had been rendered unnecessary in an age in which digitisation and computerisation were such prevalent forces. Alternatively, we might suggest that the removal of the legend was a more deliberate attempt to imply a far greater degree of computerisation at work than simply 'control'. Taken in light of the industrial design, aesthetics and even the incorporation of mechanical QWERTY-style keyboard keys into the instrument, perhaps the intention was that the TR-909 was seen not as a computer-controlled instrument but rather as a computer. In one sense, the TR-909's industrial design seems to invite speculation about what would happen if Fairlight and Apple collaborated on a drum machine.

Of course, while the intention might have been to ensure that the TR-909 looked every inch the modern digital computer, the truth was that, inside the beige case, the instrument was brimming with analogue components and circuitry.

Kick, Punch

Regardless of whether its industrial design, aesthetics and hardware layout led to it been seen as different to the TR-606 and TR-808, the TR-909 certainly sounded unlike previous Roland instruments. If the tone of the TR-808 could be characterised as 'warm' with 'booming' bass frequencies and a sparkling fizziness in the high end that connects it to the earlier generations of analogue rhythm accompaniment instruments as well as the CR-78, with the TR-606 being charitably painted as abrasive, metallic and even harsh, the overall sound of the TR-909 is almost always described as powerful and punchy (Brewster 2023; Robehmed 2023; Stokes 2022b). There is a palpable attack and presence to the sound of the TR-909's kit with a pronounced attack transient and very little of the ringing decay that typified the TR-808.

Indeed, speaking of the development of the TR-909, Tadao Kikumoto notes that, once again the original intention had been to return to the flexible, complex drum synthesizer that had been the impetus for the TR-808. However, as before, the more ambitious aspects of this larger project were curtailed due to the cost of analogue components and the need to more quickly bring a product to market. As we have noted, while the sounds of the TR-808 are perhaps the most iconic of any drum machine (or Rhythm Composer) created before or since and continue to be used, reused, and emulated to this day, Kikumoto had been dissatisfied with many of the circuits. Speaking of the TR-808 Hand Clap, Kikumoto notes reveal his disappointment as well as the pragmatic nature of product development and the negotiation of design, cost and the schedules of product releases and marketing.

> it didn't sound like popping enough, but more like a bunch of bamboo sticks being hit, however we were running out of time. After we released the TR-808, I succeeded to improve it to a better sounding circuitry complete with popping sound generation, but the 808 handclap sound was already gaining a lot of popularity, and it was too late to replace it with the new circuit, nor launch the new one.
>
> (Kikumoto 2020)

It is clear that Kikumoto and the other engineers and designers at Roland had continued to critique and refine their work and this ongoing research and development into analogue drum synthesis informed the sonic fingerprint of the TR-909. The new instrument was an attempt to explore new territory but also to remedy some of the perceived failings and inadequacies of earlier instruments. Interestingly, while a Hand Clap does make its way into the roster of TR-909 sounds along with a Rim Shot, many of the TR-808's percussion sounds have no equivalent in the new instrument. Congas, Maracas, Clave and even the somewhat divisive Cowbell are all absent. Instead, the TR-909 focuses on a comparatively narrow range of tones. There is a Bass Drum; Snare Drum; Low, Mid and High Toms; Rim Shot; Hand Clap; Open and Closed Hi Hats; Crash and Ride Cymbals. This is an interesting choice that, again, appears out of step with the marketplace. Even the earliest Linn instruments had offered plenty of percussion samples alongside the staples of the rock and pop kit and other manufacturers followed suit with congas, bongos and even vocal snippets eventually creeping into the sample sets of instruments in Yamaha's RX range. Clearly the TR-909 palette is not quite as stripped back as the TR-606, but it still offers only ten sounds that, like the Drumatix, do not offer much embellishment beyond the bread and butter sounds of the rock drum kit.

Or perhaps that should be the bread-and-butter names of the sounds of the rock drum kit because, as Anniss notes, these were 'heavy and distinctive, but very obviously electronic and synthesized' sounds that 'weren't a particularly accurate reflection of the acoustic percussion instruments they were imitating' (Anniss 2016). Indeed, reflecting on the development of the distinctive palette for the TR-909, Kikumoto observes that, 'I gave up the idea of implementing a universal synthesizer to the 909. Instead, we pursued making attack with impact that were not available on the TR-808, and this evolved into a compressor-like sound.' (Kikumoto 2020).

In this context, the effect of an audio compressor is to control the dynamics, or variations in perceived loudness, of sound over time. Whether working with acoustic or electronic material, compression along with EQ (equalisation) are among the most widely-used and powerful tools in the sound engineer's arsenal. Used in moderation, compression can even out putatively undesirable variations in the volume of specific sounds over the course of a performance thereby bringing more consistency in levels. However, because audio compressors allow fine control over the balance between the attack and release portions of sound passing through their processing circuits, extreme settings can exert a high level of tonal shaping. In the context of drums, this can mean accentuating or diminishing the initial transient to create deeper, fuller sounds with sustain and body or aggressive tones with more bite and punch.

With their array of outputs that allow the sound of each drum to be separately processed with external effects units, instruments such as the TR-808 are particularly well-suited for studio use in this way. This is one further way in which the Rhythm Composer is distinguished from the TR-606 with its single audio monaural output combining all the sounds and disallowing individual processing (without modification). It is no surprise to find that the sounds of the TR-808 were frequently processed with compression in order to add sharpness to the otherwise rather softer sounds and, in particular, to add yet more sustain to the booming Bass Drum. It is interesting to note that, even though the TR-909 is similarly equipped with separate audio outputs for each of its sounds, the intention of Kikumoto and the engineering team developing the instrument was to create a tonality that built the aesthetic of external processing into the kit itself. To be clear, the TR-909 does not include any audio compression circuitry per se, so these are not sounds passed through an internal effects unit. Indeed, it would be many years before drum machines (or any other electronic instrument, for that matter) would include any form of built-in effects processing. Rather, the analogue circuitry in the TR-909 is designed to create sounds that mimic the effect of externally processed drums. As such, the sound design not only makes decisions about the core tonality of the instrument but also incorporates specific technical and aesthetic approaches and practices to the treatment of drums in the studio.

Snappy Circuits

Even though the larger project of a 'universal drum synthesizer' had been abandoned for the TR-909, the desire for more intricate sounds and the ongoing motivation to explore analogue synthesis techniques ensured that the sound generation circuits in the TR-909 were not simply repackaged versions of those from earlier instruments. Although Ian Gilby's early 1984 preview of the instrument is more than a little taken aback by the continued centrality of analogue synthesis, it is suitably reassured by the newness of the sounds as well as their origins in digital analysis and computer simulation.

> What does come as a surprise, though, is the discovery that all but the cymbal sounds on the 909, are synthetically generated (not sampled sounds). However, the percussion sounds available from the machine are totally 'new', not simply copies of the 808's sounds, and are the result of computer simulation tests used to analyse the changing character of drum sounds, in order to recreate authentic 'voices'.
>
> (Gilby 1984: 38)

The addition of samples might be the headline differentiator between the TR-909 and its predecessors as we shall see below, but we should not overlook the analogue design of the instrument which is considerably more complex than that of the TR-808, or any previous Roland instrument for that matter. Where the TR-808 had made extensive, and ingenious, use of Bridged-T networks to create sounds with a comparatively small number of components, the design of the drum sounds in the TR-909 took a different, and in many ways more traditional, approach. Analysing Roland's TR-909 Service Notes (1984, First Edition) which include schematics and some description of the operation of the circuits and which are adapted in Figure 8.2, we note that the Bass Drum, for instance, comprises two discrete elements that are mixed to create the composite sound.

The first part of the sound simulates the body, sustain and resonance of the drum while the second provides the attack portion that replicates the sound of a beater hitting the drum skin. As shown in the Bass Drum section of Figure 8.2, a VCO (Voltage Controlled Oscillator) generates a Triangle waveform which is fed into a wave shaper to transform it into something approaching a Sine wave. This resulting tone contains only the fundamental frequency with no complex harmonic overtones and is perfect for simulating the sound of the rich, deep Bass Drum voice. Tonally, this is very similar to the core of the TR-808 Bass Drum. However, where the TR-808 had utilised Bridged-T networks and exploited the natural attack and decay profiles dictated by the values of the components in the circuit, the use of a continuously running VCO in the TR-909 requires another means contouring to shape its tail to silence. As such, replicating the patching of an analogue subtractive synthesizer such as the Roland System 700 that Kikumoto and colleagues had used to model their earliest drum synthesis experiments, the overall output level of the TR-909 Bass Drum VCO is fed into a VCA (Voltage Controlled Amplifier) which is, in turn, contoured over time by an Envelope Generator.

Thus far, the combination of VCO, VCA and Envelope produce a tone of constant pitch that has an instantaneous attack and a gradual decay to silence. In fact, with short attack and comparatively slow release settings applied to the Envelope, the deep, slowly decaying Sine wave created by this element of the TR-909 Bass Drum circuit could passably emulate the character of a TR-808 Bass Drum or even a modern '808' bass tone. However, the design of the TR-909 circuit makes an important addition as the pitch of the VCO is modulated by yet another Envelope ('ENV3' in Figure 8.2) which causes it to quickly descend from high to low. It is this that adds the punchy transient to the TR-909 Bass Drum and unmissably differentiates it from the TR-808. If this decaying pitch sweep is too slow, the result would sound like a siren or spacey 1970s 'Disco Tom' effect. However, if set at just the right rate, this downward fall is no longer heard as descending pitch at all but rather as a burst of energy that adds a powerful thump to the beginning of the sound. To accentuate this effect yet further, the second element of the sound draws a signal from the instrument's common White Noise generator. As we saw with the TR-808 and TR-606, Noise had been used to create various effects including the simulation of a reverberation to give Toms as sense of space. For the TR-909 Bass Drum, the sound is notably dry and the Noise is used to augment the attack of the sound. In fact, once filtered to remove some of the hissy high frequencies, it is combined with a 'click' from a Pulse Generator and fed into another VCA with its own Envelope whose extremely short attack and decay times create a burst that simulates the beater hitting the skin of the drum. Mixed together, the elements of the sound create a Bass Drum circuit with a much-improved overall and the potential for a significantly more pronounced attack than that of the TR-808.

Four front panel knobs offer control of aspects of the TR-909 Bass Drum circuit. Level is the simplest and alters the volume of the sound from maximum to silence and allows a custom mix to be created or the sound to be muted with a deft flick of the wrist. The Attack knob gives separate control of the level of the Noise and Pulse Generator element of the sound. With this control, the 'click' can be dramatically accentuated which is partly responsible for creating the characteristically punchy 'compressed' TR-909 Bass Drum sound although it can be removed altogether to allow the circuit to generate a softer, more muffled tone. The Decay control allows adjustment of the VCA envelope that contours the body element of the sound. Interestingly, although the decay of the TR-808 Bass Drum had been elongated to an unrealistic length by Kikumoto in recognition of the sound design potential of the drum synthesizer ideal, the TR-909 Bass Drum circuit creates an altogether more abrupt, staccato and punchy sound even with its Decay setting at maximum. Without internal modification, the TR-909 Bass Drum is not capable of booming or ringing out in the style of the TR-808 and this difference in circuit design and aesthetic choice greatly impacts the flavour of the instrument and the ways it is subsequently used.

Although the final TR-909 Bass Drum front panel control is labelled Tune, it does not work in quite the manner that we might initially expect. Given that the design of the circuit includes a free-running VCO generating a continuous waveform, we might imagine the Tune knob simply adjusts its fundamental frequency. If this were the case, such a control might even make it possible, with plenty of practice, to perform pitched '808' bass lines with the TR-909 Bass Drum sound, albeit with less boom thanks to the shorter decay. However, as Figure 8.2 shows, the Tune control is not connected to the VCO at all but rather impacts the intensity of the Envelope that rapidly sweeps its pitch from high to low to (partially) create the transient of the sound. As such, the fundamental pitch of the Bass Drum actually stays the same with the Tune knob setting the point from which the downward pitch sweep begins. As such, it is not possible to tune the Bass Drum per se and certainly not possible to create distinctly different pitched notes.

The TR-909 Snare Drum and Bass Drum circuits are actually remarkably similar designs although the resulting sounds are, obviously, quite different. Again, the sound comprises two core elements with the resonating body of the drum combined with a burst of initial energy that simulates the striking of the drum membrane along with the rattle of the wire snares. Roland's Service Notes identify these two constituents of the Snare Drum sound, 'Drum' and 'Snappy'. In fact, each are actually comprised of two elements. Here, the 'Drum' is made from the combined output of two VCOs which, by virtue of different charging capacitor values, generate waveforms at slightly different frequencies. With VCO1 running slightly low in pitch, this creates a thicker, more complex overall tone. As with the Bass Drum, the outputs of the VCOs pass through waveshapers that reduce overtones before each signal is fed into its own VCA controlled by a dedicated Envelope. The inclusion of separate VCAs and Envelopes adds complexity and movement to the sound as each tonal element of the Drum has its own volume contour.

The 'Snappy' element of the Snare Drum sound draws from the TR-909's common Noise generator with the output being sent to a Low Pass Filter to remove the very high end frequencies. The filtered Noise is then split in two with one stream passing through a High Pass Filter to remove low frequencies before both are sent to separate VCAs controlled by dedicated Envelope Generators. The result is a sound that simulates elements of the initial crack of the stick on the drum head along with the rattle of the snare wires. By separately filtering two discrete streams of Noise there is noticeable interplay between the frequency ranges and variation over time that belies the name 'Snappy'. Overall, the Snare Drum circuit is a comparatively complex one, certainly when compared with those in the previous TR instruments. By involving four separate sonic elements, each of which are differently processed to make a distinctive contribution, the TR-909 Snare Drum has body and the punchy attack that Kikumoto so missed in the earlier instruments.

Like the Bass Drum, the Snare Drum also offers up control of its analogue circuitry on the front panel. In addition to the overall Level of the composite sound, there are knobs labelled 'Tune', 'Tone' and 'Snappy'. Most straightforwardly, 'Snappy' adjusts the level of the clicky Noise element and approximates the effect of tightening and loosening the drum head and snare wires. Slightly less intuitively, 'Tone' alters the duration of the Snappy sound which has the effect of altering the balance between the tonal and atonal element of the composite sound, while Tune allows the pitch of the oscillators in the 'Drum' element to be raised or lowered. In combination, these three front panel controls can transform the Snare Drum from an abrupt, staccato burst, through a decaying noisy sweep, all the way to a dull and somewhat lifeless thud with the Snappy element completely removed.

The circuits generating the High, Mid and Low Toms in the TR-909 follow a very similar pattern to those of the Bass Drum and Snare Drum. Indeed, the Service Notes claim that, 'SNARE, BASS, TOM TOM operate basically in the same manner and share the same Noise generator'. However, the circuits grow in complexity with each individual Tom comprising the output of three VCOs mixed with Noise. The three oscillators are differently tuned to create an immediately thicker, more complex sound and each is passed through a waveshaper to remove overtones before being sent to their own dedicated VCA which is contoured by its Envelope Generator. The output of VCO3 is mixed with Noise derived, once again, from the instrument's common source. That the amplifiers controlling the levels of the three oscillators each have their own Envelope shapes again gives the sound a complexity as different frequency bands decay over time at different rates.

While the circuit might increase in complexity, the front panel control of the Toms is slightly reduced compared with the Bass Drum and Snare Drum and runs to just overall Level, Tune and Decay. As Figure 8.2 shows, the Decay control actually only alters the decay rate of Envelope/VCO2 which means that, with it set to maximum, the tail of the sound comprises just that single oscillator. Tune alters the basic pitch but, as the three oscillators are synchronised, their relative tunings retain the core tonality of the sound. Although the frequency intervals of the three oscillators making up each Tom sound cannot be altered, the front panel Tune knobs allow the pitch of each Tom sound to be controlled much like an acoustic drum kit. As Roland's TR-909 Owner's Manual suggests, 'Controlling the pitch of these three Tom sounds, you can use them as Melody Tom.' (Roland 1983e: 8). As with the other TR-909 sounds, the Toms are aggressive and punchy with an almost woody tonality to them. Even with the Decay knob at maximum, these Toms create a staccato effect with an emphasis on their attack rather than body. Compared with the TR-808 Toms whose individual pitches are far more noticeable, it is harder to create a 'Melody Tom' effect with the altogether more abrupt and 'compressed' TR-909 variants.

The remaining two analogue circuits bear much in common with their namesakes in the TR-808 and, accordingly, are much simpler, and efficient, in their design. The Hand Clap is almost identical to the design in the TR-808. Given that Kikumoto has discussed his disappointment with the original sound, it might come as something of a surprise to see so few changes. The basic principle is identical with a two-part sound made up of filtered Noise sent through different shaping processes. One stream of Noise is fed through a sawtooth Envelope Generator that simulates the effect of multiple hands (or bamboo sticks) clapping in succession by repeatedly triggering the sound. Alongside this, a separate Noise stream is contoured by an Envelope with a longer Decay that seeks to simulate the atmospheric ambience of a room or hall. The resulting sound in the TR-909 is subtly different from its earlier incarnation in the TR-808, however, because the means of generating Noise differs between the two instruments. While the earlier instrument used the mythical out-of-specification 2SC828-R transistor to create truly random noise with its characteristics high end sizzle, the Service Notes for the TR-909 reveal that the instrument implements a 'quasi-random noise generator' based around two shift registers (Roland 1984: 6). In other words, the noise source in the TR-909 is not truly random and is generated digitally. In fact, the pattern of pseudo-random noise repeats every few seconds. Interestingly, because this common Noise source is tapped and utilised by many of the circuits, when multiple sounds high in Noise (or with their Noise elements set to maximum) are triggered simultaneously there is an audible phasing effect that can appear to thin out the soundscape.

The last, and simplest, of the TR-909's analogue circuits is the Rim Shot which is based around the same Bridged-T network technology that formed the basis of many of the sounds in the TR-808 and TR-606. In the TR-909, the output of three Bridged-T band pass filters excited by a sequencer trigger pass through a simple diode clipper before being mixed and contoured by a single Envelope-controlled VCA. The combined output is subsequently sent to a High Pass Filter to remove any unwanted low frequency information. Although it uses a very similar circuit to the TR-808, where that sound was soft and almost delicate, the TR-909 Rim Shot is weighty, unmistakable and almost brash thanks in no small part to the addition of the clipper in the circuit. Like the other sounds in the TR-909 palette, the Rim Shot sounds like it has already been compressed, perhaps even over compressed. As with the TR-808, and perhaps betraying their connection, neither the TR-909 Rim Shot nor Hand Clap sounds offer any front panel controls beyond a simple Level knob to control their volume in the mix.

In using VCOs, VCAs, Filters and Waveshapers, the circuit design of the TR-909 is based around far more traditional analogue synthesis technologies and structures than the TR-808 and TR-606. One effect of this change is that, in general, there is far less sonic variation between different TR-909s as their circuits are more stable. There is an exception, however as both the Hand Clap and Rim Shot makes use of the same Bridged-T network designs as those earlier TR instruments. Because the tonality and decay of Bridged-T oscillators is partly governed by the specification and tolerances of each discrete component, there is potential for considerable sonic variation in these circuits. This is especially notable in the case of the TR-909 Rim Shot where the three Bridged-T oscillators are the only sound generating components in the circuit. As such, variations between individual instruments can be very pronounced. Indeed, the Service Notes suggest that there was awareness of this early on with Roland making formal modifications to the design of the Rim Shot circuit on at least one occasion after the instrument's initial release.

In fact, alongside component lists, schematics and block diagrams, the TR-909 Service Notes make it possible to identify at least five revisions of the instrument by their Serial Number range. In addition to some more cosmetic changes such as the replacement of the rubber battery compartment cover with a metal one, the document details a number of more sonically impactful alterations. In particular, changes to components in the Bass Drum, Rim Shot, and Hi Hat circuits are described along with the serial number from which the alternations became standard. While the technical specifications are, unsurprisingly, thorough so that we know that the value of Resistor R417 is changed from 12k to 3.3k in the Rim Shot circuit for units with serial numbers higher than 415300, the musical consequence and rationale is rather more vague with the Notes stating this is, 'For giving the voice more realistic sound.' (Roland 1984: 7). Elsewhere, the document outlines a comparatively early alteration that extends the tuning range of the Bass Drum control while the final revision, 'will emphasize attack of TOM TOMs.' (Roland 1984: 7). As we shall see below, although some updates such as those fixing glitches in the Operating System are welcome, revisions to the sound generating circuits are not universally appreciated and many of the after-market modifications offered by third parties are motivated as much by seeking to return later instruments to the putatively superior earlier versions as they are to explore the latent synthesis capabilities lurking inside the metal case.

Metal Beats

Of course, although the majority of the sounds of the TR-909 are generated by analogue circuits, it is the digital samples that are the most distinctive and widely promoted feature of the instrument. Given the palpable excitement and enthusiasm for sampling that had swept the drum machine marketplace, it is instructive to remember Kikumoto's attitude towards the technology. Considering sampling to be 'too easy' and lacking 'craftsmanship' (Kikumoto 2020), this attitude had clearly motivated the continued use and development of analogue synthesis beyond the TR-808. Moreover, recent interviews have suggested that even with the TR-909, Kikumoto was reluctant and even resistant to adopting samples. Reflecting on the need to quickly deliver a product, Kikumoto notes that, 'As the time was up, we were forced to introduce PCM sampling technology for metal percussion sounds such as cymbal.' (Kikumoto 2020).

Indeed, Kikumoto recalls that the need to quickly bring a new, competitive product to the market very nearly exerted an even greater impact on the TR-909, 'Kakehashi-san, being worried about the R&D delay of the 909, came to our lab and advised us to switch to PCM sampling on drums as well.' (Kikumoto 2020). Ultimately, Kikumoto managed to convince Kakehashi to let the team pursue the hybrid approach and continue refining their analogue synthesis circuits alongside digital samples. Atsushi Hoshiai, who was responsible for the design and programming of the digital elements of the instrument, recounts a somewhat similar story which, among other things, confirms that the initial plan was that the TR-909 would follow in the all-analogue tradition of its predecessors. However, Hoshiai's account points to a slightly different motivation for the eventual use of samples that recognises the limitations of analogue synthesis in satisfying the desire for realism.

> At the time we started working on the TR-909, we were still planning to make the sounds analog. For the snare and kick drum, this didn't pose any problems as we could easily make great sounds with analogue circuits. But the cymbals we just couldn't get right, no matter how hard we tried. That's why we decided that, just for the cymbals, we would use digital samples.
>
> (Hoshiai 2014)

Whether the inclusion of samples in Roland's new instrument was really as begrudging as Kikumoto claims and whether they were included in order to compete with digital instruments such as the LM-1 or to compensate for the perceived limitations of analogue synthesis in the simulation of metal percussion, their implementation in the TR-909 reveals a technical and creative inventiveness that reminds us that the task was neither easy or lacking in craft.

In fact, while we have already noted that analogue synthesis is not especially well-suited to the recreation of cymbals, even with Roland's distinctive approach to the use of a swarm of square waves, filtering and distortion, sampling did not provide a straightforward solution. This is not only because of the amount of high frequency information in metallic percussion typically requiring high sample rates and Bit depths for accurate reproduction, but also because, as they ring out, cymbals generally have a far longer decay than other drum and percussion sounds. This means that, compared with the relatively short transients of Bass Drums and Snares, storing samples of cymbals requires a good deal of memory which, in the early 1980s, was extremely expensive. Today, with data storage routinely measured in Gigabytes or even Terabytes, it might be hard to imagine that just a few tens of kilobytes would be sufficient for an entire instrument. Indeed, although it was widely lauded (and widely used) as the first sample-based rhythm instrument and boasted Toms, Congas, Cabasa and Handclaps alongside its Bass Drum and Snare, the Linn LM-1 had not included Ride or Crash cymbals. Even the instrument's Hi Hats used an ingenious technique that eked out both open and closed sounds from a very small amount of sample data. With all these creative solutions to minimise memory usage and the absence of lengthy cymbal samples, the LM-1 still debuted at $5500.

In order to keep the price closer to the $1000 that Kakehashi aimed for, it was clear that there would need to be yet more efficiencies and lateral thinking. Seen in this light, while the TR-909 can certainly be read as a continued expression of Kikumoto's creative vision of a drum synthesizer, the predominance of analogue sound circuitry and minimal use of samples could be understood as pragmatic cost-effectiveness. However, Roland's engineers went further still. Where Roger Linn had opted for 8-Bit samples in the LM-1 and the later LinnDrum (which did include cymbal samples but still retailed for $3500), Atsushi Hoshiai's innovation was to reduce the Bit depth. The 6-Bit samples used in the TR-909 significantly reduced the cost of the required digital storage as LeRoy (2023: 229) notes. The lower Bit depth also lent the samples a noisy, almost distorted tonality. In other circumstances, this reduction in the resolution of reproduction might have been heard simply as low audio quality. Yet, in the context of the TR-909, these crunchy metallic sounds were actually a remarkably good fit. As the design of the analogue circuits had focused on creating hard-hitting, aggressive sounds, the slight but noticeable crispness in the cymbal samples gave them the bite to cut through a mix. Crucially, their lower fidelity and slightly distorted tone was entirely consistent with the 'punchy', 'compressed' aesthetic of the instrument.

Hoshiai's contribution to the sound of the TR-909 extended considerably further than the choice of memory chips and data format, however. Despite his primary role on the project being a technical one, Hoshiai was also a drummer. As such, he not only had insight into the musicianship of rhythm, tone and performance, he also had a drum collection. Of course, this was invaluable as Roland began the shift away from the creation of original sounds through electrical components and analogue circuit design and towards the collection, curation and recording of new sounds. 'So we chose to use one of my cymbals which had a sound I really liked. If I had to make a new rhythm machine today, I would still use that very same cymbal.' (Hoshiai 2014).

In fact, all the sampled metallic percussion sounds in the TR-909 derive from Hoshiai's personal cymbal collection. The Crash and Ride come from a Paiste 18" Formula 602 Thin Crash while the Hi Hats are the product of a mismatched pair comprising a Zildjian on the bottom and Paiste on the top. Interestingly, although there were recording studio facilities at the Roland headquarters, Tsuboi notes that Hoshiai was unsatisfied with the results obtained in this environment. In search of the perfect tone, he waited until after hours and tried recording in the office. Experimenting with different microphone positions, he eventually alighted on a spot by a co-worker's desk.

The somewhat ad hoc origins of these sounds are utterly fascinating but, in this regard, they are by no means unique. Although there has been much speculation about the identity of drummers who performed the various hits and strikes that made their way into the sample memory of the LM-1, Roger Linn has noted that, "The truth of it is that most of the sounds came from a Los Angeles session drummer named Art Wood. There were a few other things; the hand claps came from Tom Petty and the Heartbreakers!" (Roger Linn in Horkins 1985: 28).

While this provenance might seem grander than Hoshiai's nighttime recording sessions in an empty office, there are remarkable parallels. The sounds of the LM-1 were not recorded in a lavish studio but rather were created at Linn's home. Speaking in 2016, Linn recalls the recording sessions. 'Art came over to my house … the sound recording booth was a closet so I put Art in this little closet with one drum at a time.' (Linn 2016). Given that the sounds of the LM-1 and TR-909 would go on to achieve iconic status and would continue to be used across multiple genres of music many decades after their initial release, their rather humble origins are truly fascinating. One might expect these sounds to be the product of lengthy, and costly, sessions in high end recording studios with multiple drum sets, libraries of microphones, expensive mixing desks, post processing equipment and teams of creative technicians, producers, engineers and performers. Certainly, since the 1980s, we have seen the emergence of an entire industry dedicated to the creation of sample libraries that employ exactly these technologies, venues and personnel, yet in these earliest days, it is important to note that none of this infrastructure existed. As such, sample creation was an altogether less formalised, though in no way less creative, process.

Further evidence of creative problem-solving is found in the way the samples are handled in the TR-909. We might be forgiven for thinking that, while analogue circuit design seems inherently complex, especially in relation to the creation of percussion sounds, sampling is an altogether simpler undertaking. Certainly, Kikumoto had seemed to suggest this but his critique appears to be aimed more squarely at the aesthetic and creative dimensions of sampling rather than the technical, per se. In terms of process, it is tempting to imagine that the digital recordings are simply made, stored on ROM chips and played back in their entirety when a key is pressed or a sequencer note is triggered. However, while this might be closer to the case in an era of plentiful storage and high-quality sampling, the approach taken in the design of the TR-909 presented a real challenge. Quite simply, although they help to tackle the problem of prohibitively expensive storage, 6-Bit samples suffer from an extremely unfavourable signal-to-noise ratio. This is less of a problem with short, punchy sounds that have little sustain but if we think about the anatomy of a cymbal, we find an altogether different profile. The volume contour of a cymbal comprises an initial transient that gradually gives way to a slowly decaying tone that rings out to silence. Except, with 6-Bit sampling, there is no silence, just noise and plenty of it.

Attempting to record and play back a cymbal strike in its entirety at 6-Bit resolution yields an appropriate attack (perhaps even with some pleasing distortion and digital clipping to add character and bite) but quickly descends into a muddy mess in which the decaying cymbal is drowned out by the noise floor and the limits of the sampling technology.

True to its status as a hybrid instrument, and reflective of the expertise and heritage of Roland's design teams, the solution to this digital problem is reassuringly analogue. Essentially, the samples are treated as sound sources to be modified rather than forming complete sounds to be played back in their entirety. They are material to be shaped and further processed by the synthesis engine. As such, the bursts of sampled cymbals stored in the ROM chips of the TR-909 are contoured by VCAs and Envelope generators just like the clusters of analogue square waves in the TR-808 and TR-606. Rather than create a lengthy sample that includes the entirety of the cymbal's decay, the level of a comparatively short sample is contoured by an envelope that adds the fade to silence. The novelty of this arrangement is such that it is called out in a discussion in the TR-909 Service Notes documentation.

> [the sound] has an envelope somewhat different form that of actual Hi Hat sound. This is because the Hi Hat sounds have been compressed before digitalized and Pulse Code Modulated (PCM) in order to have greater S/N ratio and higher decay resolution.
> (Roland 1984: 5)

To be clear, the sample is not looped to extend its length. Rather, the original recordings were made with high levels of audio compression which add sustain by raising the level of the decaying portion. As such, the initial burst of transient energy is retained but the natural decay of the cymbal is not heard and, instead, the compressed, sustained sound is contoured by the analogue synthesis circuitry before any unpleasant noise is heard. Of course, even with the creative use of compression, such a synthesized envelope has to curtail the sound prematurely which gives the resulting cymbal a somewhat choked quality. However, it retains the quality of the sound and, crucially, preserves the initial transients that Kikumoto's research and development had discovered to be so important in defining sonic character.

The use of analogue VCAs and Envelopes both responds to and minimises the impact of cost-saving, comparatively noisy 6-Bit samples and by eliminating the need to digitally record the full decay of the sound, it reduces the required sample time and the memory chips for storage. All three of the sampled cymbal sounds stored in the TR-909's memory operate in this way with control of the Decay of both the Open and Closed Hi Hats being available as variable controls on the front panel. As can be seen in Figure 8.2, the Hi Hat Decay controls have no direct impact on the sample data but rather alter the decay rate of the Envelope routed to the VCA in the circuit.

For the Ride and Crash cymbals, the envelope times cannot be altered from the front panel but control is given over the tuning of both sounds. In the absence of modern time-stretching algorithms, these variations in pitch are achieved by changing the playback speed of the samples, though, unlike the purely digital implementation of sampling in other instruments, the presence of synthesis circuitry in the audio path ensures that the decay of those digital sounds remains under the control of the analogue VCA and envelope with no unwanted noise creeping in.

Looking beyond the individual circuits, we find a further fascinating, and often-overlooked facet, of the TR-909 that directly arises from the use of samples in the instrument.

Both the Main audio outputs that combine all of the sounds in accordance with the front panel 'Level' controls, and the separate outputs that allow each sound to be mixed and processed independently, have Low-Pass Filters in their signal paths. As their name suggests, while they do not affect frequencies below their cutoff setting, these filters reduce the amount of high frequency information that leaves the instrument. What is surprising, however, is that these sets of filters do not have the same cutoff frequencies. This means that sound that comes from the Main outputs is different from that appearing at each of the individual outputs. For each of the separate outs, the cutoff is set to 15KHz which, while within the range of human hearing, is high enough so as to not significantly impact on the tonality of the drums. On the Main outputs, however, the filters have their cutoff frequency set to 7KHz which most certainly does impact on the sound. It is for this reason that the TR-909 sounds palpably different when heard through its different sets of outputs. The sound creation circuits remain exactly the same, but the analogue components in the final stages of the signal path apply markedly different degrees of high frequency shelving leading to an appreciably airier or duller tone.

The reason for such a drastic filtering is a consequence of the instrument's digital circuitry, CPU and the inclusion of samples, and in particular the relatively low Bit depth and sampling rate. As Reid (2002a) notes, the implementation of samples in this format can introduce harsh digital quantisation noise, aliasing and other unwanted high frequency artefacts into the audio signal path. This is especially problematic when each sound is cumulatively mixed together at the Main outputs. In a related way, one of the later revisions to the circuits that is documented in Roland's Service Notes, changes the value of a Capacitor to roll off further low end frequencies in the Hi Hat circuit. While operating at the other end of the spectrum, the object of this change is, again, to remove excessive noise generated in the instrument's digital sound generation.

As such, while it is usual to describe the TR-909 as a hybrid instrument by virtue of its mix of synthesized and sampled sounds, digging further into the circuits, we see that even its samples are essentially analogue/digital hybrids. It is hard not to look at the schematics of the TR-909 and not see Roland leveraging their investment and expertise in analogue synthesis technologies both in order to augment the instrument's palette but also to help remedy the shortcomings of the emerging digital technologies of sound generation, especially in the context of relatively cost-efficient implementations with such limited memory. In fact, the more one looks inside its beige, computer-styled case, the more analogue the TR-909 becomes.

Going Off Grid

Another area in which Roland clearly leveraged their expertise and heritage was in the sequencer of the TR-909. The fundamental principles are carried over from both the TR-606 and TR-808 although the row of 16 Step Buttons now become 'Main Keys' in a manner more redolent of a computer interface. Regardless of their name change, these keys still do dual service as the means of selecting Patterns and editing individual Steps. As this might suggest, the instrument also inherits the separate 'Write' and 'Play' modes of its forebears although, curiously, the TR-909 reverts to the TR-808 implementation in being truly modal. Where the TR-606 had allowed for seamless switching between Pattern Write and Pattern Play modes, the TR-909 requires the sequencer to be stopped before the mode can be changed and playback restarted. On the face of it, this appears a retrograde step from the TR-606 as it disallows the continuous operation of the instrument or the making of on-the-fly edits to Patterns. However, as we shall see later, despite the design of the instrument suggesting wholly discrete modes of operation for creation and playback, a number of performers have subsequently developed techniques that not only work around this implied workflow but that comprehensively transform the TR-909 into a dynamic and flexible live instrument. These later practices and improvisational techniques notwithstanding, it is certainly the case that the TR-909 Owner's Manual maps out of a workflow that will be familiar from the TR-808 documentation. New though it might be, the TR-909 remains conceptually based around the transcription of existing rhythmic material into the machine language and operational procedures of the instrument.

That said, the underlying design and architecture of the TR-909 does seem to take a number of steps back in various areas of its sequencing capabilities compared with the TR-808. Gone, for instance, are the two Parts and A-B Variations that extended the duration and flexibility of the TR-808 Pattern beyond a single measure. Similarly absent are the Intro and Fill options that allowed yet further Pattern variations to be programmed and then triggered live in order to create song-like arrangements without delving into the Rhythm Track or having to fix a particular arrangement in memory. Prima facie, the removal of these various sequencing and performance features appears to leave the TR-909 significantly pared down compared with the TR-808. Certainly, it appears to be more oriented around the single sixteen-step Pattern which apparently rules out higher resolutions of thirty-second notes or odd time signatures like 12/8 or 5/4, for instance. The changes also appear to make a much sharper division between Pattern and Track modes by firmly locating the creation of longer arrangements and song structures in the latter.

By way of compensation, however, TR-909 inherits the Pattern Chain functionality of the TR-606 that allows the creation of ad hoc loops of Patterns in adjacent memory locations. The implementation of Pattern Chaining in the TR-606 came with some interesting limitations that meant that the 16 memory locations in each of the instrument's two Groups were split into four Blocks. While Chains of up to two, three or four Patterns within any given Block were possible, they could not straddle Blocks meaning that careful programming and planning was required. The TR-909 does away with these technical and creative limitations and allows arbitrary Chaining of any Patterns in adjacent memory locations. Indeed, all 16 Patterns may be Chained together to create a continuous loop. Moreover, the implementation of Chaining in the TR-909 allows the connections between adjacent Patterns to be memorised in Pattern Write mode so that they become permanently set and selectable with a single key stroke from Pattern Play mode. One consequence of this is the virtual reinstatement of the 1st and 2nd Part feature of the TR-808 which makes possible the creation of Patterns with time signatures that exceed 16 steps. Used in combination with the Scale features that are imported from both the TR-808 and TR-606, the TR-909 is, on closer inspection, perfectly capable of creating complex rhythm Patterns beyond 4/4 or with a step resolution higher than sixteenth notes. It does not broadcast this fact particularly effectively outside the Owner's Manual, however.

As such, although the TR-909 sequencer might initially appear to lack much of the flexibility of the TR-808, the expansion of the accessible, performance-oriented Pattern Chain feature introduced in the TR-606 not only offers the same functionality but actually goes some way further. Of course, as a Rhythm Composer, the TR-909 includes the same ability to create Rhythm Tracks as its predecessors. However, the memory capacity of the instrument now allows programming and storage of eight completely separate Rhythm Tracks containing up to 896 measures. Interestingly, 'separate', here does not simply refer to the distinctiveness of the particular arrangement of Patterns but rather to another quirk of Roland's memory allocation and organisation. Pattern memory is split into two Banks each containing three Pattern Groups with 48 individual Patterns each, for a total of 96. Given the complex and divided nature of the memory, and once again, implying a workflow that has the requisite Patterns pre-programmed and later arranged into Tracks, the TR-909 Owner's Manual cautions,

> Later when programming a Rhythm Track (one whole music piece) by combining the Rhythm Patterns you have written, you are not allowed to use the Rhythm Patterns in different Banks. Please remember this, and try writing the Rhythm Patterns you may need for programming a Track into the same Bank.
>
> (Roland 1983e: 13)

Obviously, Pattern Chaining does not offer the same flexibility as the Rhythm Track and is limited to looping consecutive Patterns in a single Group, yet the removal of the Block restrictions imposed in the TR-606 coupled with the ability to create longer Chains does serve to blur the lines between Pattern and Track modes. With thoughtful programming, two loops of eight measures containing seven basic (even identical) Patterns followed by a variation or fill can be created, for instance. Creating such loops is made even easier thanks to the inclusion of data copying procedures that allow Patterns to be easily duplicated between memory locations and even across Groups and Banks. These newly implemented, and conspicuously computer-inspired, digital data management features along with the enhanced Pattern Chain functionality in the TR-909 creates the possibility of creating and performing fills, intros and outros in the more improvisational manner of the TR-808.

The move towards liveness is felt elsewhere in the design of the TR-909 sequencer. Where the TR-808 and TR-606 both included Real Time writing modes alongside their Step Time methods of note entry, the use of the 'Instrument Select' control and Tap button meant that Patterns were still constructed one sound at a time. The TR-909 does away with both the Instrument Select and Tap controls and, in its Real Time Pattern Write mode, each of the instrument's analogue and digital sounds are mapped to the 16 Main Keys. As such, combinations of Bass Drum, Snare Drum and Hi Hats can be Tapped in live in a single pass. Of course, the sequencer allows for additional real time overdubs to add complexity. We might see this live performance mode as a partial nod to the LM-1 and LinnDrum which were based around interfaces of simultaneously playable Pads. Liveness might also further explain the choice of keys utilised in the TR-909 interface. While perhaps not as playable as the Linn's Pads which would help define subsequent sample-based instruments such as the Akai MPC series (which were also designed by Roger Linn), the TR-909's computer keys certainly offered a more tactile feel than either the TR-606 or TR-808. That said, the feel more closely approximated musical typing rather than striking a drum but perhaps the association with computing was not one Roland found unwelcome.

As with the earlier Roland TR instruments, Real Time and Step Time Writing are closely integrated in the TR-909 with notes tapped in live being available for subsequent editing by switching modes. However, Step Write is also significantly enhanced with a host of new features not seen on previous instruments. 'Flam' brings the ability to simulate the effect of a drum being hit with two sticks slightly out of time. This effect, commonly performed by drummers to provide emphasis is implemented in the TR-909 by playing the first strike on the beat with the second triggered later. The interval of the Flam can be varied with eight possible settings ranging from none to a noticeable double strike. Flam can be applied per step but is not available for all sounds and is restricted to just the Bass Drum, Snare Drum and the three Toms. Although there is a dedicated Flam button which is used to set the interval, the means of applying the effect to a given sound is rather less intuitive and involves pressing two of the Main Keys simultaneously. The slight opacity of the interface notwithstanding, it is interesting to observe Roland following the trend of adding features that directly mimic the performance of a drummer in order to minimise the machine-like qualities of the instrument.

On that note, given Kakehashi's commitment to the importance of the Accent feature, it will come as little surprise to find this is also expanded in the TR-909. Two distinct flavours of Accent are possible in the instrument with what was implemented in the TR-808 and TR-606 now referred to as 'Total Accent'. This describes an Accent that, when set on a particular Step, affects all sounds that are triggered. As with the earlier instruments, the TR-909 has a rotary control than can be used to set the level of this Accent so its intensity can be adjusted. In addition to this global 'Total Accent', however, the TR-909 allows individual steps to be Accented for specific instruments. In this way, the Bass Drum, Snare Drum, three Toms and Open Hi Hat can each have their own dedicated Accents written into their respective sequencer lanes. This is achieved through the Main Keys which, rather than acting as simple binary switches per step, now cycle between Off, Normal volume and Accented. As the Owner's Manual notes, the Main Keys' integrated LEDs indicate the status of the steps as they cycle between unlit, dimly lit and brightly lit accordingly. Unexpectedly, while the instrument does not have touch sensitive keys, individual Accents may also be written in Real Time by using the pairs of Main Keys allocated to the relevant sound. The right Key plays a note at normal level with the left inputting an Accented step. Although the individual Accents are not available on the Rim Shot, Hand Clap or Cymbals, the combination of the Total Accent and individual Accents for the other sounds does offer considerably enhanced potential for replicating the complex dynamics of a drummer's performance.

It is in the TR-909's Shuffle feature, however, that we see possibly the most significant and impactful of Roland's attempts to emulate the feel of a human performer. In truth, perhaps Roland was as much seeking to emulate the feel of the Linn LM-1 whose Shuffle feature had been much-lauded (and much debated as the nuances of its timing are analysed and pored over, see Scarth and Linn 2013). As we shall see below, this feature goes on to define the TR-909 every bit as much as its iconic sounds, yet it is introduced in the instrument's Owner's Manual with extraordinary understatement. 'You may enjoy Shuffle Rhythm arrangement.' (Roland 1983e: 26). It is fair to say that a good many people did enjoy Shuffle.

Essentially, Shuffle is a simple alteration of the timing of the instrument's sequencer. By delaying alternate eighth or sixteenth notes, Shuffle effectively moves some beats off the grid to create a swung feel. Indeed, other drum machine manufacturers often refer to the feature as 'Swing' rather than 'Shuffle'. However, irrespective of their name, these alterations in timing are often presented as key to overcoming any unwanted machine-like qualities of the sequencer's playback. Whether they necessarily make the performance less metronomic and more 'human' is debatable as Swing and Shuffle are typically applied in a strictly regulated and predictable manner. Roland's implementation in the TR-909 offers seven degrees of variation selected by using the Shuffle/Flam modifier key with the intensity chosen by the appropriate Main Key. It is worth reiterating that these settings offset the alternate rhythmic divisions by a set amount across the entire measure and, thus, across the entire performance by precisely the same degree. There is no subtly shifting variation as we might find in modern 'humanisation' algorithms. Similarly, as there are no micro timing variations within the measure, there is none of the rushing or dragging that one might expect from a human performer and the Shuffle is as mechanically regimented as any other aspect of the machine. Whether it truly provided the key to unlock the 'full variety of drumming' as Roland's (1984: 2) TR-909 product brochure suggested, Shuffle, along with Flam and the greatly expanded Accent features, certainly marked out the TR-909 as the next generation of Rhythm Composer.

MIDI

It was the inclusion of MIDI, however, that really marked out the TR-909 as a forward-looking piece of music technology. Indeed, the TR-909 was not only the first Roland rhythm instrument to adopt the standard, it was the very first drum machine in the world to be equipped with MIDI ports. The importance of this is hard to overstate. As we have seen, the Musical Instrument Digital Interface was co-developed by Roland and, among other things, it created a standard communications protocol that allowed instruments from any manufacturer to communicate with one another. Prior to this, a number of analogue methods had existed including Control Voltages (CV), Gates and Triggers and proprietary systems such as DIN SYNC and Roland's early Digital Communications Bus (DCB). However, while there was some standardisation, the specification of the voltages in analogue systems could not be guaranteed across manufacturers. Roland, like many other instrument makers used a system based around 1 Volt/Octave signals to define the pitches of notes, for instance, while both Korg and Yamaha used an incompatible Hz/Volt system meaning that instruments were not interoperable. Even the voltages of pulses like the triggers and clocks that were used to provide rhythmic timing information varied across manufacturers and instruments meaning they were not automatically interoperable.

MIDI changed this by offering a standardised specification that allowed a single cable to transfer pitches, note durations, dynamics, timing and clock information as well as a host of real time control information that could be used to alter aspects of sounds and performances. As such, as long as it was equipped with MIDI In and Out ports, any instrument from any manufacturer could, in principle, be straightforwardly attached to any other to allow at least some degree of synchronisation and control depending on the specific implementation. This meant that MIDI-equipped controllers such as keyboards could be used to control the sound sources of multiple synthesizers simultaneously while sequencers could record the musical performance and controller information, play it back and allow it to be edited. For rhythmic instruments like drum machines, the most obvious application of MIDI is to provide a standardised synchronisation protocol. As each instrument has its own built-in sequencer, MIDI clock allows each to run in perfect synchronicity as well as providing virtual transport controls to start and stop in unison. Certainly, the TR-909 MIDI implementation facilitated this and, with both In and Out ports, the instrument could either receive clock data from another device or send data so that it acted as the primary source of timing information for a larger system.

However, while timing and synchronisation are the most obvious, and perhaps most immediately useful, aspects of the MIDI specification for a drum machine, the TR-909 went altogether further. First, its sounds could be controlled by an external instrument or sequencer, effectively transforming it into a source of sounds. This predicted the marketplace for dedicated keyboard-less sound modules that would come to dominate the late

1980s and 1990s. Before long, dedicated MIDI controller keyboards that sent data to other devices, but made no sound themselves, along with computer-based, multitrack sequencing software, meant that instruments did not require their own controls, keyboards or sequencers. Roland, in keeping with many other manufacturers, would soon (re)package many of their analogue and digital instruments as 'MIDI sound modules'. Indeed, by 1990, Roland's R8M placed a library of digital drum sounds in a rackmounted unit for use with external MIDI sequencing hardware and software thereby completely dissociating sound creation and sequencing into two processes and devices.

Controlling the sounds of the TR-909 via MIDI offered more than just a potentially more flexible sequencer, though. It also enabled the sounds to be played in different ways and with more dynamics than the instrument alone made possible. As well as defining musical notes and duration, the MIDI specification includes, among other things, a 'velocity' parameter. The name refers to the speed with which a key is pressed (or perhaps a drum is hit) which is broadly analogous to the strength with which it is struck. Although it can, in theory, be mapped to any parameter that an instrument makes available (and potentially multiple parameters simultaneously), MIDI velocity is very often tied to the loudness of a given note. With an appropriate velocity-sensitive controller, this means that individual notes can be triggered at different volumes thereby giving an approximation of the behaviour of an acoustic instrument. We have seen already that the implementation of the Accent in the TR-909 was significantly more sophisticated than in previous Roland instruments and that Kakehashi had placed great importance on this feature in order to add dynamism. However, well-developed though it was, the TR-909 Accent still only offers a comparatively limited number of possible levels ranging from none through to a step having an individual sound and Total Accent applied. Controlling the TR-909's sounds remotely over MIDI, however, they are considerably more dynamic with a potential 128 levels of loudness.

As such, not only did the implementation of MIDI in the TR-909 predict the emergence of the sound module and the concentration of sequencing around computer-based software, it also drew attention to the inherent limitations of the sequencer built into the instrument itself. Self-evidently, the analogue and digital sound circuits can respond in a considerably more dynamic way than the TR-909's sequencer can exploit. Again, this is likely due to a combination of software and hardware constraints including the difficulty of programming such fine-grained resolution with a limited array of hardware controls and no visual display beyond a three-character, seven segment LED. In short, the sounds of the TR-909 are more versatile when they are controlled from something other than the instrument's own sequencer. Indeed, both the promotional materials and Owner's Manual give example cabling diagrams that show the TR-909 being played live from keyboards including the HP-400 Electronic Piano. This instrument, clad in faux-wood finish and looking every inch the Home Piano its 'HP' monicker suggests, makes for a somewhat jarring companion to the TR-909 especially if the intention had been for the new Rhythm Composer to be a fixture of high-end recording studios. Its place is better understood, though, when we consider the comparatively limited range of MIDI-equipped keyboard instruments Roland had in the product line at this stage. As the company's first MIDI-equipped instrument, the Jupiter-6 might have been suitably 'professional' yet it neither transmitted nor responded to MIDI velocity so was not fit for the job. Although it may have been the only choice, the sight of the HP-400 on the back page of the TR-909 product brochure speaks to the immediate challenge of describing as well as taking advantage of these new features.

If hindsight encourages us to think that the designers of the TR-909 implicitly recognised that MIDI would sound the death knell for the hardware sequencer, the instrument has one more surprise in its specification sheet and on its front panel. Sitting innocuously alongside the three Pattern Group selection switches is a button labelled EXT INST. As if to underline its importance, it is one of the few buttons that does not have an alternative function. Yet, the features it unlocks are among the least understood of the instrument. In fairness, this might be because, aside from the EXT INST button itself, there is nothing on the instrument's panel to explain the functionality. Moreover, the documentation in the Owner's Manual is as brief as it is cryptic. What it does say, however, is that, 'This mode is provided for potential control of external sound source through MIDI.' (Roland 1983e: 37).

Deciphering the accompanying diagrams, we learn that, in addition to functioning as a sound source for other sequencers and controllers, the sequencer in the TR-909 can be used to drive other sound sources. This goes beyond simply sending MIDI clock or transport information to control the synchronised running, starting and stopping of another instrument's sequencer. Rather, by enabling EXT INST mode, the TR-909 can send MIDI note data to another sound source at the same time as it triggers its own internal sounds. Perhaps we might be forgiven for thinking that this is a simple matter of the TR-909 sequencer being set to trigger the internal drum sounds while the same data is converted into MIDI messages so that notes can be transmitted to the outside world. Indeed, as late as 1990, the MIDI Guidebook that Roland distributed with its instruments long after the specification had been introduced and adopted by almost all manufacturers, suggests this as a possible use for a MIDI-equipped drum machine. With the MIDI Out of the drum machine connected to the MIDI In of a digital piano, the guide notes the following. '...if the drum machine is playing a pre-programmed rhythm, the piano will play the same part ... whenever the drum machine plays a bass drum, the piano will play a 'C' etc.' (Roland 1990: 11).

While translating a drum machine pattern into a musical phrase by mapping each sound to a pitched note might yield some interesting results, this is not the implementation in the TR-909. Rather, the TR-909's EXT INST mode is much more flexible and operates in addition to its internal sounds and with an entirely separate and fully programmable sequencing extension dedicated to MIDI output. A Pattern stored in the TR-909 memory can actually comprise note and Accent data controlling the internal analogue and digital sounds as well as a full 16 lanes of note and Accent data that are transmitted via the MIDI Out. These sequences are saved together in a single memory and, aside from sharing some settings including the Last Step value and Shuffle amount, they are totally independent.

To program MIDI sequences, the EXT INST key activates a mode in which each of the 16 Main Keys are mapped to a MIDI note ranging from the 'C' two octaves below Middle C through to the E flat below Middle C. This note range makes it possible to create simple melodic phrases or basslines and, although the range of notes is fixed, the EXT INST patterns can be extended by Pattern Chaining or Track mode in the same way as those controlling the internal sounds. While the Owner's Manual is almost unhelpfully succinct in its description on the EXT INST functionality, Roland's promotional materials made great play of the feature particularly in terms of using the TR-909 to trigger the drum and percussion sounds of other instruments. Whilst dutifully reassuring the reader that the TR-909 already comes equipped with great sounds, it promises yet more. 'Even better – in addition to the 11 internal drum sounds, add up to 16 more drum sounds with no major surgery' (Roland Corporation US 1984: inside front cover).

The reference to major surgery is surely a veiled dig at the facility to alter the sounds in instruments such as the LinnDrum which was undertaken by replacing the EPROM chips on which the digital sound data was stored. As a hardware precursor to the software-based sample libraries that have become staples of contemporary music production and retail, collections of new chips with different drum and percussion sounds were available from Linn and other companies. However, while this added flexibility to the instrument, it came at a cost and required opening the casing and physically removing and replacing chips from the circuit board. In Roland's promotional literature, though, while we can detect the recognition that the built-in sounds might not chime with the dominant aesthetic of realism, there is no option to replace the sounds no matter how involved or invasive the process might be. As such, EXT INST is sold as offering the possibility to augment rather than replace. Certainly, much is made of the complexity and flexibility of the instrument's sequencer as a means of controlling other (digital) sound sources even if features like Flam are not available over MIDI. Nonetheless, the TR-909 EXT INST mode is extremely capable. Indeed, while the 11 internal sounds are more responsive when controlled externally, the instrument's ability to output 16 discrete MIDI notes in EXT INST mode means that the TR-909 is, in this respect at least, a more wide-ranging instrument when its sequencer is paired with an external sound source.

A Sore Disappointment?

For all the creative and technical inventiveness and despite the effort expended on extolling the virtues of a hybrid instrument with different types of sound source and an expanded sequencer, the TR-909 was not especially well-received upon its initial release. Removing some of the complexity of the TR-808's multi-function rotary controls used to select the constituent Parts of a Pattern or the active sound to be programmed, was welcomed by some. However, while the direct selection of sounds from the 16 Main Keys was praised in principle, it brought some unintuitive operation.

> The good news here is that the TR909 has dispensed with the plethora of multifunction rotary selectors that made programming and chaining song patterns such a drawn-out process on the 808. … This isn't quite as foolproof as it sounds, however. To select the open hi-hat, for example, you have to push both the accented and non-accented closed hi-hat selectors, which makes real-time programming of that voice more than a mite tricky.
>
> (Goldstein and Twigg 1984: 53)

At least part of the reason for the lukewarm response is surely a product of the attempt to add such a rich variety of features and functions to a comparatively limited hardware interface. Although the Open and Closed Hi Hats are treated as different sounds, with their own circuits and separate front panel controls for their Level and Decay, we might expect them to be selectable by their own dedicated Main Key. However, the implementation of the new Accent feature means that the two Main Keys allocated to the Hi Hats control the 'Normal' volume and 'Accented' version of the Closed Hi Hat. In this way, the interface is consistent across the instrument as pairs of Main Keys are allocated to other sounds such as the Bass Drum, Snare and Toms. This leaves the Open Hi Hat with no dedicated Main Key. Of course, Roland could have reduced the number of sounds to more comfortably fit the 16 keys or broken the consistent design language of the x0x series by adding a seventeenth key to accommodate the expanded sound palette. As it is, there is something of a mismatch between the ambition and actuality of the hardware and software.

These issues notwithstanding, the operation and flexibility of the TR-909 was broadly praised by contemporary reviewers but there was near unanimous agreement on one thing. Compared with the Oberheim DX, E-Mu Drumulator, and Sequential Drumtraks, whose 'digital' credentials were presented as bringing an almost automatic superiority, the sound of the TR-909 fell short.

> It's what it sounds like that's important ... the TR909 is a drum-machine, and it sounds like it. Apart from the cymbals, it uses little pieces of fiddly electrical stuff to make noises that sound like drums. It's not a Linn, nor even a Drumulator, and would not be mistaken for such ... It doesn't give the same degree of expression as these, even if special provision has been made for 'shuffle' and 'flams', and it certainly sounds less 'real', which is an audible deficiency.
>
> (Lewin 1984: 60–61)

Interestingly, given the comparative lack of control over the samples in the competing and putatively superior instruments from E-Mu, Linn et al which often extended no further than Level and Pitch, there is also criticism of the degree of sound shaping in the TR-909.

> The TR909 is perhaps a little lacking when it comes to providing the user with parameters for adjusting the preset sounds. As on the 808, the bass drum has user-variable tuning and decay, and in addition, the 909 goes one better by offering adjustable attack as well, though the effect of turning this control to maximum isn't particularly realistic or musical.
>
> (Goldstein and Twigg 1984: 52)

It is notable here that part of the criticism is not that the sounds are analogue rather than digital per se but that, compared with the flexibility offered in Roland's previous instruments, the TR-909 is seen as more limited. By exposing too little of its circuitry to the user, the TR-909 is understood to be underexploiting its analogue synthesis foundation. Ultimately, it is not a sufficiently well specified drum synthesizer.

> The noise that let the 909 down, to my ears, was the snare. Even with full fiddling with the tune, tone, and snappy controls, it still sounded like a synthesised thud with a splash of white noise, which is not what one would expect from a £1000 drum-machine.
>
> (Lewin 1984: 61)

That the sound of the TR-909 was unappreciated at first and fared unfavourably alongside fully sample-based instruments is a well-rehearsed cornerstone of the history of the instrument and neatly fits the dominant discourse of the xOx instruments more broadly. However, Lewin's comment about price also points to another important, and certainly less well-documented, issue. The TR-909 was not marketed with a clear audience or context in mind. This might seem an unlikely claim given that Roland's UK 'Newslink' promotional literature apparently sets out a very clear case.

> Roland have moved slightly out of the current trend for all-digital machines, but for good reason. The TR-909 is likely to be used by professional musicians and studios to whom absolute control is paramount and who need MIDI versatility, large memory, and impeccable sound.
>
> (Roland UK 1984a: 5)

However, elsewhere, a somewhat less emphatic tone is detectable.

> When I asked Alan Townsend, who is Roland's chief demonstrator, what market they were aiming for, he replied that it was hoped that the 909 would find its own level. This absence of a clear marketing policy could well turn this machine into Roland's first white (well, pale grey plastic) elephant.
>
> (Lewin 1984: 60)

One reading of the TR-909's early failure to find a market is that, while its flexibility and programmability were well-suited to commercial studios and producers, its sounds lacked the realism and currency that mainstream production required. And for the burgeoning home studio market, the instrument was too expensive to be competitive as Lewin notes. In retrospective celebrations of the TR-909 (originally published on 9 September or '909 Day') insulated by the knowledge of the instrument's later critical appraisal and impact, even Roland have noted this lack of clarity of purpose and audience.

> In hindsight, instead of targeting professional studios, the TR-909 was probably more suited to 'bedroom' producers who needed drum sounds and a versatile sequencer that could power additional equipment – so that they could create an entire piece of music with a minimal setup.
>
> (Roland 2019b)

Ultimately, this lack of clarity in the positioning of the TR-909 as a commercial proposition, coupled with the unfashionable nature of its sounds, meant the instrument was roundly criticised by many on arrival. Even the lure of new technologies such as MIDI and the potential this connectivity offered to augment the built in sounds by synchronising the instrument with other drum machines or even to ignore the internal sounds altogether and use the TR-909 as a MIDI sequencer to compose rhythms with external sound sources, proved insufficient.

> Even with its wonderful 'state-of-the-art' (this means 'modern') micro-chippery (tape dump, copy, tape sync, MIDI etc), the TR909 still sounds like a drum-machine, instead of a machine playing drums.
>
> (Lewin 1984: 60–61)

As such, with an estimated 10,000 units produced according to Roland (see Tsuboi 2020), Roland discontinued the TR-909 after just a year. With a considerably shorter production lifespan and fewer instruments manufactured, this meant that the TR-909 was actually a less successful commercial product than the TR-808 had been. In fact, according to Richard Katz, a product manager at Roland US in the early 1980s, many of the TR-909s that were sold were heavily discounted with Roland soon reducing the asking price to just under $300 in an attempt to clear stock (Stokes 2022b). Of course, the TR-909 would certainly find an enthusiastic audience, many of whom would be drawn to the instrument precisely because, like the TR-808, the TR-909 sounded like a drum machine rather than a machine playing drums. However, the appeal of the TR-909 was complex. Its syncopated Shuffle imparted a feel and groove that the TR-808's rigid sequencer simply could not yet its palette was almost unbelievably clipped, aggressive and ... compressed.

Techno Rebels

Today, it is hard to imagine that the TR-909 was not expressly made with the club sound system in mind. Even though we know that the engineers and designers at Roland were unaware of the widespread uses of their instruments until many years later and undertook their developments in a studio that had left them unaware of the true depths their Bass Drums generated, the TR-909's punchy sound, rich in transients and punch, is remarkably well-suited to large PAs. However, although it has become almost synonymous with forms of dance music, the first commercial recording featuring the instrument is usually attributed to Canadian electro-industrial band Skinny Puppy. The band's 'Remission' six-track EP was released in 1984 just a few months after the launch of the TR-909 and features the instrument in an extremely prominent role. Interestingly, producer David Ogilvie's introduction to the instrument was not an especially auspicious one. 'At first we turned on the 909 and it was far too clean. We were like, "What have we got here?" I struggled and was very stretched for ideas.' (Ogilvie 2019).

It was only with external processing in the studio, aided by the instrument's separate audio outputs, that the band found a sound they were happy with. Curiously, this is a sentiment expressed by a number of musicians, DJs and producers. Juan Atkins, for instance, disliked the sound of the instrument so much that he returned to the shop he had purchased it from and exchanged it for a sample-based Sequential Drumtraks.

> Actually, I had a 909 when they first came out, but I took it back and exchanged it for the DrumTraks because the 909 wouldn't sync to the Sequential Circuits system we were using in Cybotron. Also, at that time the kick drum on the 909 sounded horrible to me, but recently it's come into vogue so everyone's giving an arm and a leg to get hold of one. To think that I had one and I took it back – for the DrumTraks, of all machines.
>
> (Trask 1988: 42)

Given what we know of the lukewarm introduction of the TR-909, the idea of the instrument being returned in favour of a more apparently modern digital replacement does not come as much of a surprise. However, that Juan Atkins was one of the founders of what became known as Detroit Techno makes this rather more intriguing. In combination with Derrick May and Kevin Saunderson (collectively known as The Belleville Three), Eddie Fowlkes, Blake Baxter and James Pennington, he would prove influential in the development of a genre of electronic dance music built upon the near-constant foundation of the TR-909. Inspired by European electronic music produced by artists including Kraftwerk, Tangerine Dream, Gary Numan and Giorgio Moroder as well as Japan's Yellow Magic Orchestra and the electronic funk of George Clinton's Parliament/Funkadelic, Atkins et al expanded on the synthetic aesthetic to create a pounding, mechanised sound strongly rooted in the metronomic precision of the drum machine, with the TR-909 sitting alongside and taking over from the TR-808 and offering a very different sonic flavour. As Derrick May notes,

> Unlike the 808, which is really colored – it has this mechanical sound, and it's soft and sexy but powerful and fierce – the 909 was *dry*. It was more of a challenge. It had a raw blankness to it. You could create your own painting on this canvas.
>
> (May in LeRoy 2023: 234)

The key here was, as Ogilvie had found, processing the individual sounds of the TR-909 with effects units such as guitar pedals. Ironically, this was exactly the same technique that Prince had used to create the signature sound of his Linn LM-1 as engineer Susan Rogers has noted (Rogers 2016). As Eddie Fowlkes recalls of the TR-909, "We learnt that machine, and EQing was a big part of the art. I know the people in Chicago would EQ their 909s differently to us in Detroit." (Jenkins 2019).

It is no exaggeration to say that the TR-909 is practically inseparable from Techno. The instrument brought an immediately fresh tonality that was unlike anything Roland, or any other instrument maker, had created before. Of course, many new sounds are shocking simply because they are unfamiliar but the TR-909 was different. Even once the novelty wears away, the sound remains utterly arresting. Regardless of any imitative intentions, what emerged from the instrument's outputs sounds wilfully synthetic but more than this, it is forceful, even aggressive. There is nothing reassuring or enveloping here. The Bass Drum is not deep like the TR-808, nor is it clicky like the Linn but it has incredible bite and punch. The Snare is a brutal stab of energy matched only by the Rim Shot whose smack is almost violent. While still full of punch, the Toms do not possess the same strident qualities and seem better suited to creating continuous rolling, running patterns that weave in and out of the Bass Drum rather than occasional fills or flourishes. Perhaps more than any sound, however, it is the metallic cymbals that bring Techno alive. The slightly distorted, clipped Hi Hats have an insistent quality while the Ride cymbal provides the perfect counterpoint to the Bass Drum and Tom. All of this arrives directly because of Kikumoto's ongoing research and development and focus at Roland on the importance of transients in defining the specific quality of a sound. This concentration on finding ways to add the attack that was seen to be missing from the TR-808 creates a compressed, almost confrontational, sound in the TR-909. One effect of its well-defined attacking punchy transients is to let the TR-909 cut through a PA system. Moreover, its combination of Bass Drum and Ride cymbal create interesting interplay especially when pushed at high levels through speaker systems. Exploiting the built in characteristics and protection circuits of PA systems, fading the Bass Drum in an out with its Level control can effectively suck the Ride cymbals in and out of a mix creating an intense change in dynamics.

As we saw earlier in the discussion of the TR-808, Roland's 16-Step sequencing interface affords the creation of rhythms that are not tied to the (l)imitation of the performances of human drummers. Like the earlier TR-606, the TR-909 continues this tradition by allowing the triggering of multiple sounds on single Steps in combinations that materially defy the sticking ability of the most able performer. While this could be coded as 'unrealistic' if imitation were the goal, and there is certainly nothing to stop programmers creating patterns that sit within the bounds of human performance, for the pioneers of Techno, the TR-909 soon became a critical platform allowing the exploration and expression of a machinic aesthetic. Not only were the TR-909's sounds clearly synthetic, but its sequencer afforded the creation of rhythms that could only be performed by machines.

Crucially, although the synthetic qualities of the TR-909 are immediately evident, Techno was not simply the sound of machines. Rather this was the sound of music being made by people with machines (that had been made by people). This was a music in which people explored the potentialities of machines in a relationship that was simultaneously guided by the technological possibilities that were marshalled and shaped by exploratory compositional and improvisational practice and a deeply-rooted sensibility. At the heart of Techno, then, is a complex but harmonious relationship with technology. Through the creative acts of making, composition and sound design, people become deeply connected to technology and channel their expressivity through it. Moreover, as performances are shared in public either live or on record, those foundational technologies and practices connect people with one another. The distinctiveness of the Detroit sound is notable and, while the influence of Kraftwerk and others can be heard in the early works of the Bellville Three, there is a politically-distinctive quality that reflects the geographic, cultural and socio-economic context of its origin. Atkins has spoken widely of his introduction to the work of theorist Alvin Toffler in a 'Future Studies' class at high school and the influence this had on the conceptual development of Techno (Sicko 1994). In works such as 'Future Shock' and 'The Third Wave', Toffler explores the relationship between technology and social change in the shift from an 'Industrial' to an 'Information Age'. The influence of Toffler is suffused throughout this new musical culture and is explicit in use of the phrase 'techno rebels' which, along with 'cybotron' and 'metroplex' form part of the language of early Techno, being both the name of Atkins' group and subsequent record label respectively. However, there is also a broader commentary in Toffler's work that recognises the increasing pace of technological change but that outlines a relationship between humans and new technologies that is not simply overwhelming or alienating. Instead, computers and by implication computer controlled Rhythm Composers can be put to work in the creation of new forms of expression.

It is also impossible to appreciate the growth of Techno without consideration of the history of transition and transformation in Detroit, itself, which is so closely tied to the trajectories of the growth and decline of the automotive industry as well as Motown records which had relocated to Los Angeles in the early 1970s. Indeed, as Deforrest Brown, Jr (2022) notes in *Assembling a Black Counter Culture*, the emergence of Techno in Detroit must be understood as a complex consequence of the impact of industrialised labour and the automation which had transformed what had been the heart of the US motor industry. Underscoring this point, in a 1988 interview that discusses the difficulty of getting his music heard given that he had no UK distribution, Atkins reveals that, 'I had people come up to me and say they thought Cybotron was some white guys from Europe. People couldn't believe we were actually Detroit musicians. We sold between 7000 and 10,000 copies of that record in Detroit.' (Trask 1988: 39).

As Mike Rubin has suggested, while pioneers such as Atkins, May, Saunderson along with the 'second wave' of producers including Carl Craig, Stacey Pullen and Kenny Larkin, found a receptive audience in Europe, Techno soon become portrayed as 'white' music.

> As techno's digitized sounds crisscrossed Planet Rock, most of the producers and consumers who tuned in were white and the identity of the music's funky forefathers got lost in the transmission. Meanwhile, black audiences and musicians ceased to see techno as their own art form, leaving the electronic pioneers of Detroit not only feeling spurned by indifferent white-owned record companies, but estranged from their own community as well.
>
> (Rubin 2000: 112)

Crucially, in recovering the origins of Techno as a unique form of working class, Black musical and cultural production, Brown's work offers an important and timely response to the cultural assimilation of the genre within predominantly white, European electronic music contexts and discourses.

If it was the punchiness of the TR-909 that underpinned Techno in Detroit, it was the instrument's Shuffle that shaped the distinctive syncopation of House music. Pioneered by DJs and producers such as Frankie Knuckles, Ron Hardy, Jesse Saunders, Chip E., Joe Smooth, Steve 'Silk' Hurley, Farley 'Jackmaster' Funk, Marshall Jefferson as well as the group Phuture, in Chicago's underground club culture, House inherited much from Disco both musically and culturally. Disco had steadily grown in popularity with movies like 'Saturday Night Fever' (1977) and its soundtrack created by the Bee Gees at least partly responsible for bringing the genre into the mainstream. Such was the popularity of Disco that considerable opposition was mobilised against it, apparently protesting its musical derivativeness and prevalence. Perhaps the most significant and contentious incident was the 'Disco Demolition Night'. This promotional event was held at Chicago's Comiskey Park, the home of the Chicago White Sox baseball team, on 12 July 1979. Enlisting the support of Steve Dahl, a radio DJ who had been publicly campaigning against Disco for some time, the event offered tickets at a reduced rate of 98 cents to anyone who arrived with a Disco record. The accumulated vinyl was to be ceremonially destroyed in an explosion on the pitch. With at least 50,000 people attending, the event ended in a riot as records were destroyed to chants of 'Disco Sucks'. Writing an overview of the year in 'Rolling Stone' magazine in December 1979, Dave Marsh noted that,

> The antidisco movement, which has been publicized by such FM personalities as notorious Chicago DJ Steve Dahl, is simply another programming device. White males, eighteen to thirty-four, are the most likely to see disco as the product of homosexuals, blacks and Latins, and therefore they're most likely to respond to appeals to wipe out such threats to their security. It goes almost without saying that such appeals are racist and sexist, but broadcasting has never been an especially civil-libertarian medium.
>
> (Marsh 1979)

While Dahl denied any such motivation, more recent commentaries such as Gillian Frank's (2007) 'Discophobia: Antigay Prejudice and the 1979 Backlash against Disco' have similarly foreground an underlying homophobia and racism in the movement and many of its acolytes. The event had significant consequences and impacted sales of Disco records as well as mainstream labels' desire to promote the genre. Regardless, the ethos of Disco did not disappear and DJs such as Frankie Knuckles drew inspiration from the music while transforming its sound into an altogether more electronic one. As Rietveld (2011: 6) notes of this underground resurgence, 'In its renewed marginalised cultural position, dance culture returned to the secrecy of night culture, a creative space outside the spectacle of the major music industry, enabling it to develop relatively unfettered in new directions.'

One of the ways House reimagined Disco was through the mechanisation and automation of its sound. While House relied on drum machines rather than live drummers in a manner that might have interested Toffler, the programmed patterns betrayed the rhythmic DNA. The tempos of House tracks ranged between 120–130bpm and were to similar Disco (and tended to be a little slower than Techno of the same era). Continuing the Disco influence, the four-on-the-floor Bass Drum marking out every beat was similarly punctuated with a Hi Hat pattern that put the Open Hat on the off-beat, falling between the pulsing Bass Drums. The combination of sharp, punchy Bass Drum and crisp sampled Hi Hats makes the TR-909 particularly well-suited and creates a driving beat that occupies both the high and low ends of the frequency range while leaving sonic space in the middle of the mix. A more distinctive feature of the typical TR-909 House beat arises from the use of the Hand Clap and Snare Drum. We might expect the Snare to fall on the second and fourth beats however, it is the Hand Clap that very often performs this role making it the third of the core sounds in the kit. The Snare Drum, instead, is employed for longer syncopated sequences and rolls of sixteenth notes that often nearly fill a whole measure. In this way, the role of the Snare Drum transcends simple timekeeping to become a more complex rhythmic feature. For each of the sounds, their naming or the conventional means of using their putative reference points in the acoustic drum kit need not impact their use in the TR-909. Sounds are encountered on their own terms in combination with whatever pattern of triggers maximises their effect.

In this regard, excepting variations and differences in processing and EQ as we saw above, the fundamental basis of TR-909 Techno and House patterns have much in common. Indeed, while tempos have increased, the essential building blocks of these foundational drum patterns remain in place today However, where Techno typically focuses on the pounding sound of the Bass Drum, almost relentless repetition and a wilfully machine-like solidity in timing, House beats take extensive advantage of the TR-909's Shuffle feature. In truth, Techno producers often bake in a subtle shuffle to create a running or rolling effect, yet it is in House that we typically hear some of the most overtly swung backbeats in electronic music. So, although Techno can be Shuffled, the effect is typically more subtly applied whereas, in House, it is a definable feature of the groove with the lengthy sequences of Snare Drums we noted above serving to emphasise the syncopation. While we have noted that Roland was actually following Roger Linn's lead, that the TR-909 was the first Rhythm Composer to include the feature made it particularly desirable.

> The 909 was the first drum machine where you could put swing on the drums ... Even the drums we loved from other drum machines or sampled, were triggered off the 909, because of its MIDI capabilities.
>
> (Jazzy Jeff in Jenkins 2019)

It is hard to overstate the impact of Shuffle in the TR-909 and it is perhaps not too fanciful to see it as a key reason for the instrument's status as it brings an unexpected feel to a machine with such resolutely synthetic sounds.

> 'I was excited about creating something with a bit of swing to it,' he remembers. 'Because, everything that I'd been programming up to then had basically been coming from the 808, with a straight 16th-note pattern. There was one track by Mantronix called Ladies, and it just had this swing. I was like, "How the fuck did he do that?" Then I realised it was the 909. That was the excitement of wanting to use the swing from the 909. That's how the bassline for Pacific State came about.'
>
> (A Guy Called Gerald in Stokes 2022b)

It is also interesting to note that the TR-909's ability to function as a sequencer as well as a sound source was central to its appeal for some. Indeed, just as A Guy Called Gerald had used the TR-808 to synchronise the SH-101 and MC-202 using its analogue DIN Sync connection, the inclusion of MIDI in the TR-909 gave the instrument additional value.

Once Atkins had overcome the initial dislike of the instrument, for instance, he created the drums to the influential 'No UFOs' (1985) by connecting the Sequential Drumtraks to the TR-909 he had subsequently re-acquired (Brown 2021).

One particularly notable feature of the uses of the drum machine in the emergent context of House and Techno was that it was seen both as a recording and live performance instrument. As Frankie Knuckles notes of the TR-909,

> And I would use it live in the club. I would program different patterns into it throughout the week, and then use it throughout the course of a night, running it live, depending on the song and playing it underneath, or using it to segue between some things.
>
> (Knuckles interview in Broughton 2018)

In fact, despite its many credentials as a studio synchronisation and sequencing tool the TR-909 has not only continued to be used as a live performance instrument, but has been transformed by some artists into a solo instrument. Perhaps the best-known and most accomplished exponent is Jeff Mills whose 'Exhibitionist' series of performances capture a rare virtuosity. Interviewed in 2022, Mills recalls his initial approach was to identify ways of segueing into and out of records by using pre-programmed patterns on the TR-909 to provide elongated bridge sections. However, after performing with other musicians including drummer Tony Allen, Mills began to approach the TR-909 as a standalone instrument. Central to Mills' use of the TR-909 as an instrument is the harnessing of the full range of its capabilities.

> I came up with this idea of using the stop and start buttons to play it as an instrument in itself. Playing the first beat or the first two beats of a pattern, or the first three beats, or inverting it by hitting the continue button first and then the start button. Just finding new ways to be more creative with the machine.
>
> (Mills in Payne 2022)

Of course, given the strict modality of the TR-909's Write and Play modes, we might be forgiven for wondering whether the 'Exhibitionist' performances do ultimately involve the triggering of pre-programmed patterns. However, closer study reveals that Mills actually creates the entire performance with a single, ever-changing pattern created in Write mode. Any starting or stopping of the sequencer is for rhythmic effect rather than to switch the instrument's operating mode. Instead, Mills makes use of the various level controls to mute specific sounds while tapping or step programming the material. Through the combination of programming, muting and unmuting with the Level controls along with manipulation of the various sound sculpting parameters, Mills comprehensively transforms the TR-909 from studio tool to solo instrument. Locked permanently in Pattern Write mode and with such virtuosic use of the sequencer, the manipulation of the sound controls makes this perhaps the closest the TR-909 comes to being the drum synthesizer Kikumoto had dreamed of.

While the TR-909 remains inexorably connected with House and Techno and continues to drive these genres today, the distinctive tone and feel of the instrument has been used across a remarkably wide range of genres. Madonna's 'Vogue' (1990) takes the Shuffled House beat firmly into the world of commercial pop replete with syncopated snare fills which build into and out of sections of the song. Here, the Shuffle so characteristic of House is centrestage with the exaggerated groove accentuated yet further by carefully programmed notes on the final sixteenth note of a beat both emphasising the Shuffle and solidifying the downbeat. Bjork's 'Hunter' (1997) takes a rather different approach and creates a double time pattern that simultaneously evokes military drumming yet is inescapably machine like in its precision. This unique pattern programmed and performed in a single take by Mark Bell particularly takes advantage of the instrument's Scale feature to alter the time signature and provide greater resolution to the sequencer. Like Mills' Exhibitionist series, Bell's use of the TR-909 here combines programming and live performance with real time manipulation of the front panel sound controls to take full advantage of the analogue synthesis. What is so fascinating, however, is that the Pattern uses only two sounds, the Bass Drum and Snare Drum. Most obviously, interest and variation comes from the extensive use of the instrument's multi levels Accent to create an incessant, urgent rolling effect. There can be no mistaking that these Patterns come from a machine but by varying the level of Accents and taking advantage of the ability to apply them to individual sounds and not just the whole instrument, the Bass Drum and Snare play off each other and the whole sequence alternates between building and holding back. As the TR-909 Accent feature is not simply a level control but rather interacts with the synthesis circuits, the result of such a varied sequence of Accents dramatically alters the sound. Some strikes are strident with others seemingly muted as if damped by a hand. More than this, Flams and the Scale mode are used to create Patterns that extend the apparent resolution of the sequencer and allow notes to sit outside the typical note grid. Crucially, by creating numerous Patterns with odd numbers of beats in different Scales, an entire Pattern Chain can be created that shifts in and out of sixteenth notes and triplets. While it does not use the full range of the instrument's sounds, or perhaps because of it, 'Hunter' must rank as one of the most accomplished showcases of the full range of the TR-909's sequencing capabilities.

If these tracks might be read as virtual celebrations of the instrument, Daft Punk's 'Revolution 909' (1998) leaves little doubt as to the nature of its homage. Given that the track is also a political critique on the French government's position on rave events or 'teknivals', there could hardly be a better instrument to head the rebellion.

Evolution 909

The centrality of the TR-909 to such a range of musical genres has ensured that the instrument and, in particular, its dry, punishing sounds have remained important parts of contemporary music production. In terms of the instrument itself, with an estimated 10,000 units produced (Tsuboi 2020), original TR-909s are not easy to come by. This scarcity coupled with the desirability of the instrument means that they command high prices on the second-hand market. Although the TR-808 is generally the most consistently highly valued of the two xOx Rhythm Composers (and of the xOx instruments more broadly), recent years have seen the TR-909 edge ever closer. Both the TR-808 and TR-909 now occupy a similar price bracket and routinely change hands for between three to five times their original retail 1980s prices and continue to rise.

As with the TR-808 and TB-303, there is a bitter irony in this situation that is hard to overlook. As we have seen, the growth of the reputation, legacy and desirability of the TR-909 can be traced back its comparative affordability in the early second-hand market following its discontinuation. The status of the TR-909 as an instrument that had been largely discarded, or even ignored, by the professional recording studios and home users with whom Roland had hoped it would find an enthusiastic audience, plays a crucial role in allowing it to pass into the hands of the musicians, producers and DJs who turned it into the creative centrepiece of their art, identity and expression. As Paul Hartnoll of Orbital whose 1989 'Chime' features the TR-909 that the group continues to perform live with today, notes,

> You couldn't afford a DX7, so you bought a DX100; you couldn't afford a 707, so you bought a 909 – you were getting all the cheaper versions, but actually those machines created the sound of techno and house music.
>
> (Hartnoll in Buskin 2006)

Of course, the result of the transformation of the instrument from a technological experiment and retail failure, as Kikumoto puts it (LeRoy 2023), into a musical and cultural object of desire, is that the original hardware instrument is now available for purchase only by those with considerable economic means. For most, it remains an unobtainable object that may never been seen in the flesh let along played. Clearly, there are some contexts in which the instrument continues to be used, recorded and performed with as we have seen, yet it is an unavoidable truth that at least some instruments exist as collector's items and markers of taste, distinction and wealth.

The high value and scarcity of the TR-909 means that, like the TR-808, instruments are very often kept in a relatively unaltered state. However, given the number of hardware, firmware and software revisions that Roland issued in the course of the instrument's brief retail lifespan, there are opportunities for owners to update their machines. Interestingly, 'update' in this context does not necessarily mean implementing any or all of the subsequent revisions in an earlier instrument. For many connoisseurs of the TR-909, some of the later circuit revisions are considered less desirable than those that shipped with earlier instruments. The Rim Shot is a case in point as we noted above. While Roland's explanation of the change was to provide a 'more realistic sound' (Roland 1984: 7), the most noticeable difference is a change in pitch with the newer, higher pitched version frequently being described as sonically inferior to the original. The revision to the Bass Drum circuit, by contrast, which was described as 'expanding the TUNE range' (Roland 1984: 7) is typically described as adding more 'punch' to the sound. Reading the circuit in Figure 8.2, this makes sense as we recall that the 'Tune' control actually affects the rapid pitch sweep at the very start of the sound rather than altering the fundamental frequency of the body of the drum. For many enthusiasts and owners, the additional attack and thump makes this revision the preference over the earlier circuit.

As Roland's Service Notes are so detailed and outline which ranges of serial numbers originally shipped with which circuit revisions and firmware, it is possible to identity the original 'version' of any given TR-909. Additionally, because the specific components, along with their location in the circuit, and the required changes in their values are listed for each element of each revision, updating or reverting is perfectly possible for those with the requisite skills and parts. It follows that, by implementing some and not others, or by downgrading previously applied updates, it is possible to create hybrid combinations of revisions that Roland never released in any single TR-909.

Interestingly, the impact of these revisions and the hybridised approach to their implementation is, however, codified in many of the clones, remakes and emulations that have emerged over the years. Most recently, both Roland and Behringer have released hardware instruments that explicitly seek to recreate not only the sound but also the workflow of the TR-909. Roland's TR-09 is part of the company's Boutique series and shares an identical form factor with all of the other instruments in that range. Like all bar one of the Boutique instruments, the TR-09 is a digital recreation that uses ACB (Analogue Circuit Behaviour) modelling algorithms to emulate the analogue circuitry of the original instrument. Despite its diminutive size, the TR-09 includes almost all of the hardware controls of the original instrument though the physical buttons and knobs are different with relatively simple switches replacing the computer-style QWERTY keys of the original instrument. Key combinations including the use of the array of 'Shift/Modifier' buttons for selecting Flam, Shuffle intensity as well as sound selection are all carried over, too.

There are some more notable omissions, however, chief of which is the removal of the EXT INST button perhaps reflecting its esotericism but certainly altering the functionality of the instrument. It should also be noted that, while it is broadly faithful in its layout and number of controls, the front panel of the TR-09 is rather cramped with the rotary controls for sound shaping and levels being very tightly packed. For setting values, the miniaturised interface might present fewer challenges, yet it dramatically impacts on the usability and live playability of the instrument and it is hard to see the kinds of performances pioneered by Jeff Mills being comfortably performed with controls so close together. Of course, MIDI is still featured but now extends beyond clock or note triggering and allows remote control of the virtual circuits. This might go some way to overcoming the cramped interface but does rather undermine the purpose of a hardware instrument. In addition to modelling the sounds of the TR-909, the TR-09 also implements the sequencing workflow albeit with a few alterations. The fundamental distinction between Patterns and Tracks remains with Pattern Chaining offering the same flexibilities and potentials it did in the original. However, shifting between Play and Write modes no longer requires the instrument's sequencer to be stopped and restarted. The implementation of the seamless behaviour found in the TR-606, does render the TR-09 conducive to being used as a live performance instrument without the workarounds we noted above, albeit with a miniaturised interface that makes this a somewhat less appealing prospect.

Although they share an aim in emulating the original instrument, Behringer's RD-9 takes a rather different approach to the TR-09. Most obviously, while there can be little doubt as to the reference point, nowhere in the documentation for the Behringer instrument is the number '909' mentioned. Instead, references abound to 'reviving a timeless analog design from one of the best classic drum machines of yesteryear' and the 'Authentic recreation of the influential and highly sought after legendary analog drum machine'. Just like its unspoken muse, Behringer's instrument is large. It is still not quite as big as the TR-909, but it is imposing, nonetheless. Sonically, the RD-9 takes the same hybrid approach to sound creation as the original with a mix of analogue circuits and digital sample playback. As with Behringer's other x0x recreations (and their recreations of other 'vintage' instruments inspired by Roland and other manufacturers), the circuits replicate those from their points of reference with modern replacements for unavailable components. Perhaps what is most interesting about Behringer's RD-9 is the way it takes part in the canonisation of specific revisions of the TR-909. As such, the Bass Drum and Rim Shot appear to be modelled on Roland's second revision with its additional 'Tune' punch, while the other sounds either revert to the first revision or favour the further updated circuits found in later instruments with higher serial numbers. One key departure from the fundamental design that spanned all revisions of the instrument, however, is the absence of the low pass filters on the Main and individual audio outputs. In head-to-head comparisons with TR-909s of different revisions, this contributes to the noticeably bright, and perhaps even punchier, sound of the RD-9.

Unlike some of the company's other recreations, the RD-9 deviates considerably from the hardware interface of the TR-909. The sound sculpting knobs are similar to the TR-909's in both their shape and spacing and there are QWERTY-style Main Keys, although these are not mechanical and are caps for membrane switches. The remainder of the interface is rather different, however, and reflects significant alterations in the operation of the sequencer which includes features such as probability, randomness, polymeters and note repeat that will be familiar from more recent sequencers in instruments such as the Akai MPC and Elektron series.

While authenticity in the analogue recreation or digital modelling of the instrument's circuits appears to be at the heart of much of the marketing and promotional discourse supporting the TR-09 and RD-9, both also include a number of additional features and effects. Although the RD-9 dispenses with the corrective filters in the TR-909 audio signal path, it adds an analogue multimode filter with dynamic control of cutoff frequency and resonance along with a 'Wave Designer' circuit, for instance. These imaginative musical manipulation features expand the palette of the instrument and offer new ways of hearing the TR-909. Interestingly, Roland's approach here is different as its Boutique extensions seek to replicate techniques such as compression for the Bass Drum and Snare Drum that have become more commonly associated with the processing of the TR-909 over the years. Moreover, delving into its menus, the TR-09 offers further control of its virtual circuits with extended tuning ranges that, as with the Boutique TR-08, go some way to mimicking Kikumoto's drum synthesizer vision. More recently, the TR-1000 sees Roland taking what initially looks to be a different approach again in offering the company's first analogue sound generation circuits since the original TR-909. However, on closer inspection, as with the circuits derived from the TR-808, the TR-1000's analogue TR-909 circuits are similarly expanded in keeping with the way the instrument's sounds have been understood and used over the decades. Additional parameters such as an overall Pitch control for the Bass Drum, along with extended ranges for existing controls such as tuning, each expand the sonic potential of the original circuits to reflect the ways in which the TR-909 has been processed, expanded and modified.

As a flagship instrument with a retail price considerably in excess of £2000/$2000, it remains to be seen whether the TR-1000 becomes a platform for the hardware modifications and circuitbending that are emulated in the instrument's digital ACB models of the TR-808, TR-909 et al. However, at around a tenth of the price, the comparative low cost of the Behringer RD-9 and the similarity of its analogue circuits to those in its 1980s counterpart has given rise to a vibrant community of modders. Some modifications offer fine tweaks to the sound by altering the degree of waveshaping that transforms the Triangle wave in order to change the character of the Bass Drum, or to alter the amount of Tom Noise to soften or sharpen their attack. In addition, some rather more complex alterations exist that take advantage of the synthesis architecture of the TR-909. One particularly notable modification is the implementation of audio-rate amplitude modulation which is elsewhere described as 'X-Mod' on Roland synthesizers such as the JX-8P. Regardless of the sonic utility of such a mod, its availability on both the RD-9 (and TR-909) speaks to the fact that, unlike Roland's earlier TR-808 which used ingenious but far simpler Bridged-T networks, the TR-909 with its VCOS and VCAs, has far more in common with the architecture of a traditional subtractive synthesizer.

Although the TR-09 and RD-9 are the most widely available recreations at present, there have been many other hardware recreations. Along with more partial remakes from companies such as Jomox, by far the most attentively detailed instrument is the Din Sync

RE-909 which, like the company's RE-303, is described as a replica rather than a recreation or clone. More than any other instrument, the RE-909 extends its attention to the sounds, sequencer and even the physical form factor, look and feel of the TR-909. That said, while full kits are available with companies offering building services for those not wishing to undertake the DIY project themselves, the RE-909 remains a relatively expensive and, therefore, esoteric and specialist product.

Over the years, markedly different approaches have been taken. Novation, for instance, released the Drumstation/D-Station in 1996, making it one of the first mainstream products that sought to simulate the sounds of the TR-909 along with the TR-808. The 19″ rackmount module was described as being, 'ideal for producing the kind of classic synthesized drum sounds which have returned to popularity in recent years in various styles of dance music.' (Novation 1996: 1). In addition to using an early form of software modelling which Novation referred to as 'Analogue Sound Modelling' (ASM), the Drumstation is notable for completely eschewing the sequencer of the TR-909 (and TR-808) and focusing solely on the sounds. In this way, the instrument predicts the strategy taken by Tiptop Audio whose range of Eurorack-compatible modules recreate the various TR-909 sounds, albeit here in analogue format. Like the Novation instrument, however, the Tiptop modules offer a variety of extensions to the original circuits along with additional front panel controls for manual manipulation and CV inputs for automated modulation with a modular system.

As with the TR-808, the dissociation of the TR-909 sounds and sequencer in these modules has important consequences not the least of which are the removal of the inherent restrictions and opportunities that are scripted into the method of programming the original instrument. Indeed, outside the TR-09 and RD-9 (and far scarcer RE-909), banks of samples either in other drum machines, keyboard workstations or sound libraries remain the most prevalent source of TR-909 sounds. However, while the replacement for the TR-808 sequencer might be seen to break the close connection of sounds and programmability that were encapsulated in the original instrument, the design of the TR-909 already speaks to a future in which sounds are not so tied to the specific behaviours of the triggers and Accents. The incorporation of MIDI had begun breaking apart an instrument's sounds and the means of controlling them into separate processes and products. Viewed in this way, we could argue that the TR-909 has always been a sound module and that its contemporary existence as a soundset to be used in novel contexts and with different sequencing tools is a logical, perhaps even inevitable, outcome of its design and a reflection of the period of its inception.

9 One After 909

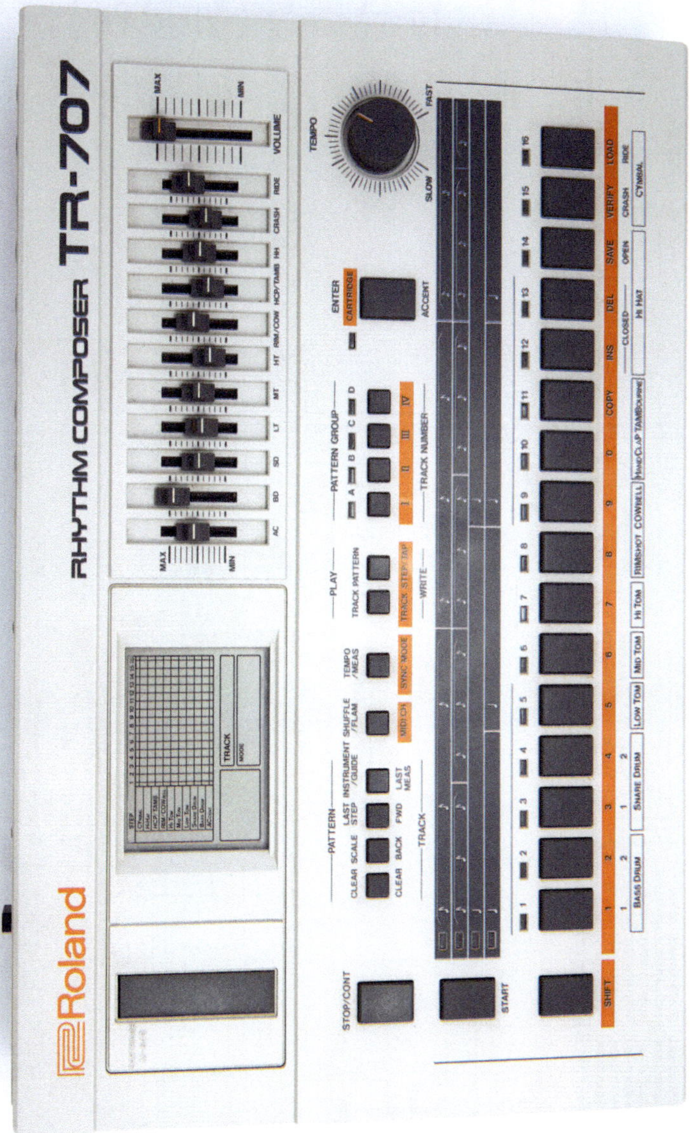

Figure 9.1 Roland TR-707 Rhythm Composer (image credit: James Newman).

DOI: 10.4324/9781003374510-9

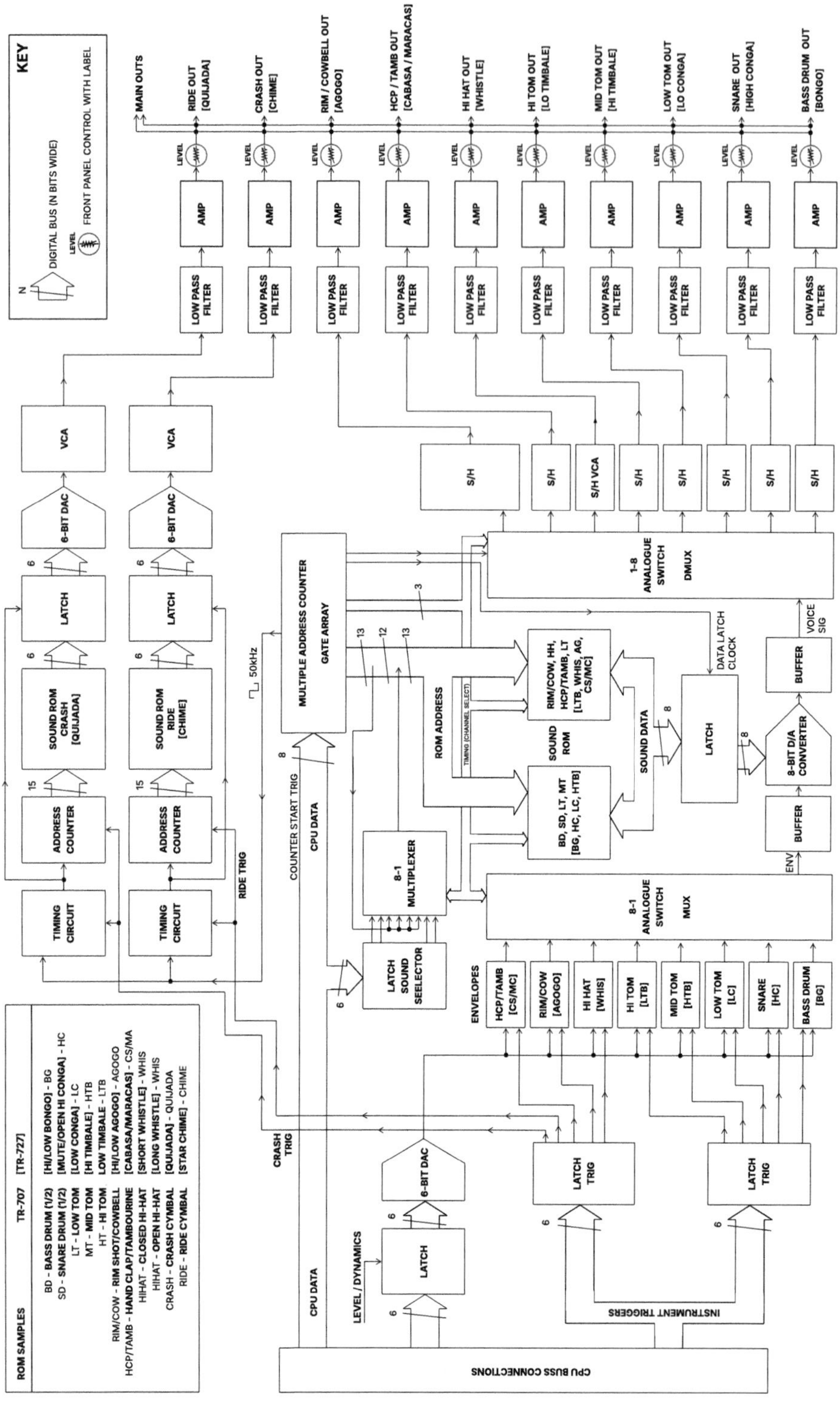

Figure 9.2 Roland TR-707 [and TR-727] Simplified Block Diagram (adapted from Roland 1985 TR-707 Service Notes).

What's in a Name?

Throughout this book, we have seen an underlying discourse rehearsed in the marketplace, in reviews, and in instrument design that embeds an expectation of innovation. The TR-909 was born into an era in which it was not sufficient for a new drum machine, or even a Rhythm Composer, merely to offer new sounds. To be a truly new instrument, it should be unlike that which came before, either by dramatically expanding upon it or upending the presumed direction of travel altogether. Decades later, with at least one analogue revival having reignited interest in the circuits that underpin the TR-909 and the other x0x instruments, it is hard to imagine just how obsolete analogue synthesis had become.

> By 1984, the infant technology of digital sampling was beginning to produce useful results and MIDI, the newfangled 'musical instrument digital interface' that allowed synthesizers to communicate with each other, was on the cusp of changing the way everything worked, both on stage and in the studio. You couldn't give away an analog synthesizer – and that's where the story gets interesting.
>
> (Berk 2000: 192)

How, then, should we make sense of an instrument like the TR-909 given that it seems to move swiftly from a transitional, pragmatic design object, through unwanted, ill-judged commercial failure to become a celebrated and essential foundation of numerous forms of contemporary electronic music, cultural identity and creative expression? As Wilson (2016) notes, 'The TR-909 may have been a stopgap before Roland finally released the drum machine it felt it needed to beat the likes of the Linndrum, but it unwittingly created the pulse of the modern nightclub.'

With the benefit of hindsight, it is easy to look back at the TR-909 and slot it into the dominant x0x discourse. The wrong instrument at the wrong time, its sounds were unfashionable and unappreciated in a world apparently obsessed with the realism and authenticity that only samples could deliver. Reflecting on the TR-909 after its discontinuation and replacement with the all-digital TR-707 in 1985, Julian Colbeck (1985: 38) noted that, 'It's always struck me as surprising that Roland – who, after all, pretty well inaugurated serious drum machines with the TR-808 – fell behind when it came to producing low-cost digital devices.'.

We might be tempted to think of the TR-909 as being hamstrung by its own past, or rather by Roland's past. Does the TR-909 speak to a company unwilling or unable to let go of its legacy? Do we find Roland so committed to its belief in the power of its analogue circuits and the concept of a flexible drum synthesizer that it is unable to see the present, let alone design the future? Perhaps in the TR-909 we see the living embodiment of one of Marshall McLuhan's most famous aphorisms. 'We look at the present through a rear-view mirror. We march backwards into the future.' (McLuhan and Fiore 2001: 75).

Were Roland unable to commit to a fully digital instrument because they were guided by the course they had previously set for themselves? Certainly, if we take McLuhan's point, we might see a company that had, throughout the 1970s and early 1980s, fixated on, refined and perfected its approaches to analogue sound synthesis and that was destined to continually remake its old instruments rather than truly embrace the change and disruption of the new. Poetic though this might be, we must remember the pragmatic reality of research, development and production processes that mean companies invest in technologies and teams and build experience and expertise over time. And, of course, we also know that Roland is the company that just a few years later would launch the D-50, their first fully digital synthesizer whose Linear Arithmetic (LA) Synthesis built on the development and sound analysis work on transients undertaken during the development of the analogue circuits in the TR-808 and TR-909.

> Through my development experience of TR-808 and 909, I found that 90% of all sound identities occur during the quick transformation of overtones arising from the initial touch to around 20 or 30 milliseconds. The sound after the attack is not so important. Actually, this way of thinking led to developing the D-50.
>
> (Kikumoto 2017)

Perhaps we might instead think that the TR-909 was an instrument ahead of its time. Combining analogue circuits and digital sampling technology, it predicts a plethora of hybrid instruments that harness the strengths of different approaches to sound creation and control.

> 'In those days there were many sample-based drum machines,' Nishi and Hoshiai continue. 'For example, Linn Drum, Oberheim DMX and the E-mu Drumulator. At the time, the drum machine trend was that of the "real sound", so we feel perhaps the market was not yet ready for the synthesised drum sound that we pursued.'
>
> (Jenkins 2019)

Of course, there is some merit in these ideas as we have seen even if it is sometimes difficult to pick a path through the post-rationalisation and legacy-building that has a tendency to privilege certain narratives and provide the reassuringly neat connective tissue of causation.

Given the indifference and even disappointment that greeted the instrument on its initial release, we might be tempted to define the TR-909 by what it was not rather than what it was. Even a brief survey of initial reviews provides a list of the instrument's deficiencies as they appeared in the context of the early 1980s. Fixed sounds, few samples, limited control, and possibly just as bad, a palpable sense that it reused yesterday's technology without sufficiently moving forward. And it is perhaps this point that leads us to a somewhat different reading of the TR-909 that is more complete and encounters it not just for what it was but also as a reflection of a turbulent and uncertain time in electronic musical instrument design.

Ultimately, the TR-909 was a transitionary instrument. Most obviously, we see this in the combination of analogue and digital sound synthesis but also in the digitally controlled sequencer that sends variable voltages around its circuits to create different intensities of Accent and the VCAs and Envelopes that shape sampled sounds while concealing digital noise. Yet, the TR-909 is also an instrument equipped with MIDI that is highly responsive to digital control and capable of sequencing other suitably equipped, forward-looking instruments. Alongside this, however, sit resolutely analogue synchronisation methods including DIN Sync and a voltage trigger output. It is in this way that the TR-909 reveals itself as an instrument bridging two eras. It sits between analogue and digital synthesis, analogue and digital communications and synchronisation protocols. It behaves like one of Roland's analogue Rhythm Composers yet it wears the garb of a home computer. The TR-909 stands as a record of an industry in flux and a company palpably struggling to make sense of the rapid speed of change.

According to the accounts of Kikumoto and others, the TR-909 was a product hurried to market yet is, simultaneously, an instrument that was part of a longer, ongoing research and development project exploring approaches to creating drums and percussion as well as broader synthesis and control. Despite their frequent designation of the x0x instruments as 'analogue', we have seen that digital systems steadily grow in importance and impact throughout the few short years since the design of the CR-78. Digitally controlled sequencers, storage and integrated circuits rightly allowed the TR-808 to bear the legend 'Computer Controlled' and, with sampling technology already available in, admittedly high end (and high price) instruments like the Fairlight CMI and E-Mu Emulator, the arrival of sample-based drum machines was inevitable. The uncertainty as to precisely how, and over what period of time, Roland could navigate this territory is effectively written into the design, and even the name, of the TR-909. Recalling Lewin's somewhat scathing (1984: 61) review that suggested a renaming to '808½', the instrument did bear a number of pre-release production names. As Atsushi Hoshiai reveals, 'During development, the team called the instrument the TR-880, then TR-880D (D for digital), then TR-909D, and finally the TR-909.' (Tsuboi 2020).

It almost goes without saying that any design project is a mass of twists and turns as issues, problems and unanticipated opportunities arise. Rare is the product that makes its way unaltered from initial concept to finished article. What we see in these codenames is a fascinating record of that journey that charts an instrument beginning its life as a derivative of the TR-808, digitally expanding its potential and becoming the TR-880D, beginning to acquire its own identity as the TR-909D, before eventually settling back as the TR-909 as its digital credentials become hybridised. These numeric shifts and the addition and removal of the 'D for digital' speak eloquently to the transitional nature of the instrument as it both becomes something new yet pulls back from fully declaring what it might have been.

Surveying Roland's promotional materials, it is interesting to note what was foregrounded in 1984.

> Think of any way to interface, and you'll find it on the TR-909. MIDI, Sync-24, Tape Memory Save/Load, RAM-Pak Program storage, they're all here. So what does this mean? It means that years from now, when other drum machines are sitting in the closet gathering dust, your TR-909 will still be on the job.
>
> (Roland 1984: inside front cover)

Perhaps it is not due to the ability to load and save patterns to cassette tape or RAM-Pak, and it is certainly not the case that all the other drum machines are gathering dust in closets, but there is absolutely no doubt that the TR-909 is still very much alive and kicking.

Digital Dynamite

On the face of it, the comparatively swift discontinuation of the TR-909 and its replacement with the all-digital TR-707 marked the end of an era for Roland. The new instrument is entirely based around samples with not a hint of analogue sound generating circuitry. In fact, offering no sound sculpting capabilities and even eschewing the pitch controls that allowed manipulation of the TR-909's sampled cymbals, the TR-707 comprehensively turns its back on Kikumoto's vision of a drum synthesizer. Certainly, for reviewers in 1985, the situation was clear. As Colbeck (1985: 38) notes, 'Quite apart from the oddity of 9 coming before 7, is this the drum machine that Roland should have launched instead of their TR-909? The TR-707, on the face of it, is more in line with present-day requirements for digital drum machines.'

Indeed, its shift from analogue sound circuitry to digital samples leaves it very often overlooked and until recently only begrudgingly recognised as part of the xOx family by virtue of its name and number. However, as new as it might initially seem, and as new as Roland might have wanted to ensure that it seemed in its contemporary marketing and promotion, the TR-707 owes a significant debt to the TR-909 as well as a number of the company's other earlier instruments. As such, while the main focus of this book falls on the (predominantly) analogue instruments, the relationship of the TR-707 to these designs and the way it both breaks away from and builds upon them warrants discussion.

Despite the promotional materials that proclaimed the instrument as 'Digital Dynamite', the TR-707 employs one of the same 'hybrid' approaches as the instrument it superseded. With 8-Bit drum samples and 6-Bit cymbals presenting the same challenge in terms of signal to noise ratios, the TR-707 reuses the approach pioneered for the TR-909. As Figure 9.2 shows, the TR-707 contours its digital drum and percussion samples with analogue envelopes and analogue VCAs (Voltage Controlled Amplifiers). In this sense, while the sounds are markedly different and reflect the trend for 'realistic' renditions of acoustic tones, we might see the TR-707 as part of the continuing development of Roland's technologies and approaches. Indeed, even at a sonic level, the TR-707 bears more than a passing resemblance to its predecessor. The Bass Drums, Snares and Toms are most definitely new and have an abrupt, clipped bite that is emblematic of the 1980s aesthetic and reflective of the extremely limited amount of sample memory available. The Bass Drum, in particular, has a clicky sound that is far more Linn than xOx. The TR-707 Hand Clap, Hi Hats, Crash and Ride Cymbals, however, are so similar to the sounds found on the TR-909 that many have speculated that they are either samples of that instrument or, in the case of the Hi Hats and Ride Cymbal especially, that they use the same underlying sample data. It is worth noting here that memory remained an expensive commodity and so, even though the TR-707 is entirely based around sampling, its design is remarkably efficient. While we might be used to measuring sample libraries of tens or even hundreds of gigbytes today, Roland's 'TR-707/727 Service Notes' (1985: 7) reveal that most of the TR-707 sounds are each allocated just 4 kilobytes (kB) of precious ROM space, with the longer decays of the Toms and Hi Hat necessitated 8kB each. If the TR-707 has a punchy, clipped sound, then surely these comparatively miniscule samples are part of the reason. There are certainly no booming TR-808 Bass Drums here. Even with the ingenious enveloping technique, the Crash and Ride Cymbals occupy 32kB each perhaps reminding us why such sounds were absent from the original Linn LM-1.

Visually, the TR-707 is also an instrument of contrasts. Its case is entirely plastic rather than metal, yet it has the same angled wedge shape, the same shade of 'Personal Computer beige', and reprises the same grey and orange colour scheme for its livery and labels from the TR-909. Its lightweight and noticeably smaller construction not only helped Roland achieve a lower price point than its previous Rhythm Composers, retailing at £499 on release, but also lent the instrument a more contemporary, compact feel compared with the bulk and heft of the earlier machines. For some contemporary reviewers, however, while the price was enthusiastically received, the lightweight construction now verged on 'flimsy' (Lewin 1984: 60–61).

However, alongside these similarities are some crucial differences. Although there are no sound shaping controls, a bank of 11 faders dominates the top right of the instrument's front panel. Along with an overall Volume control, they offer level control for each of the built-in, unalterable sounds. The shift from the rotary controls to faders puts the TR-707 in an interesting position in separating it from the TR-606, TR-808 and TR-909 and appearing to connect it with the SH-101 and MC-202 synthesizers. This is ironic given that the TR-707's faders offer precisely no control of any synthesis parameters. Instead, perhaps the row of densely-packed faders, speak to the audio mixing desk and serve to make a more positive feature of the control over the independent levels of sounds rather than pointing to the absence of synthesis controls. The faders certainly allow easier live muting and unmuting of individual sounds as they can be quickly flicked from top to bottom of their range in a way that rotary controls make harder.

If any feature of the TR-707 makes a statement about the novelty of the instrument and its departure from the x0x series, however, it must surely be the large LCD. Presented as a grid comprising 16 columns for each of the subdivisions in a bar, this approach will be familiar from the TR-808 onward. However, where earlier instruments mapped 16 steps to 16 physical buttons, the TR-707 used its LCD to show the sequence data for every instrument simultaneously. By visualising the steps for each of the instrument's sounds, the display becomes central and provides a complete overview of a measure of music at a glance. Individual drum hits are shown on the matrix with a circle graphic. The presence and absence of these circular indicators at any point on the grid mimics the behaviour of LEDs embedded in the earlier Step Buttons and makes it immediately obvious which sounds are playing – and which are not – at any moment in time. While they no longer contain LEDs, 16 Main Keys are carried over from the earlier x0x designs and may be used for Step or simultaneous Real Time 'Tap' programming as with the TR-909. In the TR-707, however, the presence of a separate display means that the Main Keys feel more like a computer keyboard than they did on the TR-909 even if they look more like the drum pads of a Linn or Yamaha instrument. The expansion of the x0x interface was much praised in contemporary reviews. Colbeck, for instance, notes how it retained the strengths of the x0x approach while adding that the display was,

> ...not only a useful feature to have when it comes to programming, but also an educational one, especially for those who find the rigours of step-time writing beyond their scope due to a lack of music notation skills.
>
> (Colbeck 1985: 38)

That said, although the display was unlike anything seen on an earlier x0x front panel, it was not entirely new. Indeed, two years earlier in 1983, the Boss DR-110 Doctor Rhythm had included a very similar grid-based LCD, albeit in slightly cut down form. While there were rows for the Bass Drum, Snare Drum, Open and Closed Hi Hats, the Cymbal, Hand Clap and Accent all shared a single row that could be cycled to show the individual data. Again, what we see in the TR-707 is a forward-looking instrument that continues to explore and exploit existing approaches and technologies. Indeed, as Henrit (1985: 79) noted upon the instrument's release, 'It incorporates features from the cheapest and most expensive Roland machines. It uses the wonderful and useful graphic display from the 110 and the same sort of assigning features and digital sounds to be found on the 909.'

In fact, the situation is actually more complex than this because the DR-110 was merely Roland's first commercially-available implementation of an idea which can be traced back even further. In terms of functionality, of course, the display bears much in common with the EKO Computerhythm and Roland's own research and development tool that had been used to program its earlier instruments. However, where the former had used a matrix of illuminated push buttons and the latter comprised an array of 16x24 plug points that were physically patched, the visual design DR-110 display drew its inspiration from a different source. Perusing the TR-808 Owner's Manual, the DR-110 display is right there staring back at us from the page. As the programming of the TR-808 is described, it is illustrated with a grid mapping out 1–16 steps and a row for each sound with individual hits represented with circular graphics in the appropriate locations on the grid. Indeed, the Owner's Manuals for the TR-606 and TR-909 use exactly the same grid notation throughout.

In one sense, although it had never been part of the instruments' physical forms, the underlying design and documentation of the xOx rhythm instruments had inculcated the concept of the grid. In another, we might view the DR-110 and TR-707 as finally bringing to life this core idea, that had been so central to the creation and explanation of the xOx rhythm concept. It is certainly striking to open the pages of a manual from 1980 and find an almost perfect facsimile of the defining interface feature of an instrument that would not come into production for another five years. But, of course, this provides yet further evidence that, even though the TR-707 initially appears to resolutely draw a line under analogue drum synthesizer experiments that came before, there is much continuity throughout the xOx family of instruments.

If we wanted further evidence of the connection between past and present, we find it writ large in the TR-727. Released shortly after the TR-707, this instrument is almost identical to its slightly lower numbered predecessor with only its name and the blue livery that replaces the orange immediately distinguishing them. However, pressing Start ushers forth a whole new sonic landscape as, where the TR-707 continues the tradition of the TR-606 and TR-909 in homing in on the sounds of a rock drum kit, the TR-727 is wholly dedicated to 'Latin' percussion. High and Low Bongos, High, Low and Muted Congas, High and Low Timbales, High and Low Agogos, Cabasa, Maracas, Short and Long Whistle, Quijada and Starchime make up the full suite of sounds. Using the same technology as the TR-707, all samples are shaped with the same analogue circuits to minimise extraneous noise. Indeed, Roland's Service Notes consider the instruments sufficiently similar that they are treated as one in circuit and block diagrams.

Perhaps the most niche and targeted of all the xOx instruments, the TR-727 palette is reflective of a trend in percussion instruments as well as a nod to the past. Instruments such as the LinnDrum as well as the later and comparatively inexpensive Yamaha RX-15 combined Bass Drums, Snares, Hi Hats along with Congas, Bongos and Cabasas, for instance. Indeed, we should remember that the TR-808 had synthesized Congas, Bongos, Maraca and Clave alongside its booming Bass Drum and snappy Snare before the TR-606 and TR-909 stripped things back. In this way, Roland's TR-7x7 instruments followed in the sonic footsteps of the earlier CR and TR rhythm units with their palettes of sounds mixing rock drums and 'Latin' percussion to perfectly match the Bossa Novas and Beguines in their banks of preset patterns. As such, the TR-7x7 instruments look back and forward for their sonic inspiration while using a mix of digital sampling and analogue shaping, along with a display technology grounded in Roland's analogue drum synthesizer development and documentation. Taking this on board, perhaps the TR-7x7 might actually be the real hybrid, transitional xOx rhythm instrument. At the very least, while they seem to mark a decisive departure from the instruments they apparently supersede, the TR-707 and TR-727 can be viewed as the culmination of many of the core approaches that underpinned the development of the xOx series.

Certainly, continuity was noted in the early reviews of the instruments though, unlike many of their x0x predecessors, the TR-707 was well-received on launch with reviewers recognising the innovation and currency of the new sound and the familiarity of the interface and methodology.

> It was the TR707, unveiled towards the end of 1984 but already a huge success across the globe, that signalled Roland's return to the drum machine fore. The reason for its success? A combination of some excellent digital sounds, and a programming system derived from that used in the 808 and 909 – and which was therefore familiar territory for an awful lot of musicians.
>
> (Goldstein 1985: 92)

Despite being widely lauded, reviewers noted a number of issues. Interestingly, some of these pointed to the ways in which the instrument fell short of the TR-909 as well as its competitors. Goldstein (1985) and Lewin (1985) among others noted the comparative step back in the implementation of the Accent. Where the TR-909 had a two-tiered system with some individual sounds having their own Accent in addition to the Total Accent affecting all sounds on a given step, the TR-707 offered just the instrument-wide settings, albeit one with a more variable setting. More pointedly, criticism was levelled at the static nature of the samples. Compared with the previous TR-909's pitch control of its Ride and Crash Cymbals and separate Decay settings for the Open and Closed Hi Hat samples, Lewin's (1984: 60–61) review is telling in suggesting that, 'not having tuneable drums is the main stumbling block.'

Roland would eventually remedy this in later Rhythm Composers although not before the TR-505 (1986) removed even more features and functions such as individual outputs and shrank the display to a sub-DR-110 style single row. After this, the curiously named TR-626 (which bears no obvious connection with the TR-606), retained the smaller display but offered twice as many 'Rock' and 'Latin' sounds as either the TR-707 or TR-727 which, crucially, could be tuned up and down seven semitones. Curiously, and unlike its analogue or digital x0x predecessors, the TR-626 also added a number of preset patterns that could not be erased, much to the chagrin of some reviewers.

> You might have noticed that I have not mentioned the preset rhythms which occupy half of the available pattern memories. This is because I find the notion of this kind of ROM memory abhorrent. It's not even as if the beats are anything to get excited by. For the most part, they are very conservative renditions of static and tedious rhythms, the sort of beats you'd find on a home organ, and to my mind a total waste of good memory space. I can think of many good reasons for having the machine come out of the box filled with standard drum parts, but none for having those patterns a permanent feature of the available facilities! I don't like being negative, especially about an otherwise superb gizmo, but I really feel that the designers have goofed here.
>
> (Badger 1987)

In another unexpected move, the sounds of the TR-707 were also added to Roland's MKS-7 'Super Quartet' instrument where they sat alongside a selection of preset 'Chord' and 'Bass' and 'Melody' sounds drawn from the Juno-106. As a rackmountable sound module with very few controls on its sparse front panel, the MKS-7 could only be controlled by an external keyboard or sequencer connected via MIDI. Built around a bank of easily-accessible preset synthesizer and drum sounds with very limited editing features, the all-in-one nature of the Super Quartet harks back to Roland's earlier forays into Compu Music with the CMU-800 and CMU-810, especially in its 'ivory' colour variation that perfectly coordinates with the beige PC towers and monitors of the time. Although it is something of a curio even among synthesizer enthusiasts, the MKS-7 along with the preset patterns built into the TR-626 perhaps provide some useful insight into the connection between Roland's designers and the wide range of uses to which their instruments were being put. Certainly, Kikumoto has noted that,

> We did not know anything about the "Mid-O fever" [editor's note: the worldwide popularity of the TR-808, TR-909, TB-303, etc.] with new music in the US and Europe until 1992, when a well-informed guy from an overseas department told us about it. Even Mr. K [editor's note: Roland creator, Ikutaro Kakehashi] and the overseas branches did not know about the underground market until that time.
>
> (McCabe 2020)

Nonetheless, even without tuneable samples, the TR-707 and TR-727 found an enthusiastic group of musicians, DJs and producers who adopted their characteristically tight sounds. For instance, although the TB-303 is the most characteristic element of the sound, both the TR-707 and TR-727 were widely used in early Acid House. Indeed, Phuture's seminal 'Acid Tracks' places the TR-707 front and centre while House tracks such as 'Washing Machine' (1986) and 'Move Your Body' (1986) demonstrate the centrality of the TR-707 and TR-727 in the work of Mr Fingers and Marshall Jefferson among others. Perhaps the most consistent use of the TR-707, however, is found in popular Egyptian 'Al Jeel' music. Mixing influences from Pop and Reggae along with classical Arabic instruments such as the oud and ney, the distinctive crispness of the TR-707's clipped drums continues to play a key, even leading, role.

Despite its use in the foundation of genres of electronic music and the continued centrality of the instrument in Jeel, the TR-7x7 did not go on to achieve the same amount of adulation as the TR-808 and TR-909. This is surely attributable to the limited degree of control over the sounds though this has been addressed by modders and circuitbenders who have embraced the instrument in part because it remains comparatively inexpensive and available. As such, many TR-707s are now found bristling with additional switches, knobs and faders that allow real time control over the playback of the instrument's samples. In addition to overdriving the outputs to create various flavours of distortion, modification of the clock speed that governs the rate of sample data playback yields pitch control as well as introducing copious amounts of now fashionable digital artefacts and aliasing with Snares and Bass Drums transformed into screeching high pitched bursts or booming rumbles of digital noise.

For those not wishing to drill holes in the cases of their instruments, other modifications, such as those created by UK-based HKA Design, maintain the functionality of the original instrument but augment or even replace its sample data. By fitting newly-designed custom ROM chips, the TR-707 user can access multiple banks of samples including sounds from the LinnDrum, Oberheim DMX and TR-808 and TR-909. Taking advantage of the digital nature of the sound sources and reverse engineering the data format in which samples are stored on its memory chips, these modifications are intriguing. On one hand, they could be considered to speak to a dissatisfaction with the TR-707 and its

TR-505 and TR-626 counterparts for which ROM expansions are also available. Perhaps these upgrades continue to venerate the earlier Roland machines and aim to add some of their cultural and sonic capital to the digital successors, albeit through sound that is frozen in comparatively static samples rather than dynamic, controllable synthesis circuits. However, these modifications might also speak to the flexibility of digital instruments. Not only do the multiple banks transform a single instrument into a veritable tour of influential drum machines, they also bring those sounds under the control of a highly developed hardware and visual interface that is the apotheosis of the 'TR-REC' programming method. The TR-707 is, after all, the physical embodiment of the means by which the TR-808 and TR-909 were explained in their own documentation.

Interestingly, although neither Roland nor any other manufacturer has created hardware clones of the TR-7x7 instruments, a software re-creation is available as part of Roland's 'Cloud' service. What is particularly notable about this software, however, is the way it not only replicates the behaviour and sounds of the original units (which are effectively treated as one) but also how it incorporates a host of after-market circuit bending modifications. In fact, great play is made by Roland of the new flexibility and sonic control that the Cloud plugin offers which nods towards yet does not fully recognise the work of the communities and engineers that have kept alive and reimagined the instrument over the decades. Of course, the software recreation of the various modifications and alterations lack the tactility and materiality of their physically circuit bent counterparts, yet they do ultimately transform the TR-707 into something approaching the drum synthesizer that it never was.

The software reinterpretation of the TR-707 makes it particularly evident that Roland has adopted and appropriated the playful improvisations of musicians and circuit benders who have explored and reworked its instruments. Of course, this is far from unique to the TR-707 and, in the recreations of Boutique hardware and Cloud software plugin recreations, we see Roland folding these new understandings, musical 'probings' and newly revealed affordances into their official canonical versions of the instruments. Viewed in this way, the TB-03, TR-08, SH-01a and Roland Cloud plugins, are not so much emulations of the original instruments so much as they are simulations that collect the accumulated knowledge, the transformative performance-oriented practices, and the myriad approaches to external processing and internal modification. In one sense, then, the various x0x synthesizers, Rhythm Composers and drum machines, are not really instruments at all but rather are ongoing stories and developments in technological, artistic and creative practice. Above all, the ongoing availability of the sounds and workflows in these recreations and simulacra ensures that the meanings and uses of 'x0x' will continue to be reimagined and reconfigured by new transformative practices of creation and play, recreation and replay.

Coda: Tomorrow Returns

Nowhere is this continual probing and reimagining more evident in Roland's own TR-1000 Rhythm Creator. Released at the end of 2025, the instrument is most immediately understood as a celebration of Roland's 1980s rhythm instrument lineage. Indeed, the marketing and promotion for the new instrument playfully reframes the 'TR' designation initially standing for 'Transistor Rhythm' as 'Tomorrow Returns'. Of course, we might also read the TR-1000 as a direct response to the criticism of Roland's sole focus on digital emulation in the face of other companies such as Behringer making analogue recreations of instruments like the CR-78, TR-808, TR-909 and TR-606. This criticism is compounded by the fact that these instruments have often arrived with larger, putatively easier to use interfaces, and lower price tags than Roland's diminutive, digital Boutiques. Yet, the TR-1000 is far from a simple remake or clone of a TR-808 or TR-909 despite offering circuits derived from both of those instruments. Nor is it a straightforward continuation of the company's revered TR-x0x product line. Most obviously, the 16 analogue tones offered in the TR-1000 actually constitute only a subset of the full TR-808 and TR-909 palettes.

> It features 16 of the most coveted circuits from the TR-808 and TR-909, faithfully recreated from the original designs with detailed engineering and carefully selected modern components.
>
> (Roland 2025a)

As we noted earlier, there is more to the TR-1000 than recreating the TR-808 and TR-909 circuits with contemporary electronics. With the TR-1000, Roland offers some telling extensions to the analogue sound generation circuits of the TR-808 and TR-909.

> Fusing vintage soul with expanded control, this engine delivers the authentic analog TR sound, boosted with new parameters and vastly expanded dynamic range.
>
> (Roland 2025a)

Clearly, there is a careful balance here between 'authenticity' and a recognition of the expectations of contemporary musicians and performers. In particular, these extensions and expanded ranges speak to the ways in which the TR-808 and TR-909 along with the other x0x instruments explored throughout this book have been reconfigured through the creative practice of musicians, performers, producers and DJs. As we saw in the detailed discussion of the TR-808, its 'booming' Bass Drum has become a staple of numerous genres of popular music yet this is a sound that has been transformed and mutated through hardware modification of the original instrument, external processing, and even through sampling, so as to extend its decay and to refashion it as a pitched bassline instrument. Seen in this light, the 'message from the developers' that prefaces the TR-1000 Owner's Manual is revealing.

> On the analog side, improvements such as pitch control, expanded parameter ranges and wider dynamics were added to the analog circuitry without compromising the original spirit and sound character.
>
> (Roland 2025b: 7)

Thus, the design ethos that underpins the creation of the analogue circuits present in the TR-1000 is not one motivated solely by the meticulous recreation of four decade old originals. Rather, these circuits and the real time controls offered over them are responses to what those original instruments and their circuitry have become over many decades of use and exploration. They are responses to what these circuits have come to mean musically and culturally, how they have been continually used, reused and reimagined in the hands of countless generations of players. And to make the case even more forcefully, it is notable that the TR-1000 also includes digital recreations of the analogue circuits from the TR-808, TR-909 (along with the TR-606, TR-707, TR-727 and CR-78). Interestingly, not only are there digital recreations of every sound from these instruments (including those present and absent in the analogue reimaginings of the TR-808 and TR-909), but these emulations offer yet more range, flexibility and scope than the analogue originals or remakes.

> On the digital side, you'll find a collection of reimagined Roland drum sounds that sound and behave like wild mutations of their origins.
>
> (Roland 2025b: 7)

Here, then, the digital Analog Circuit Behavior emulations dramatically extend parameter ranges or even add entirely new controls that are based on third-party, aftermarket modifications and hardware circuitbending of the analogue originals. In this way, while it bears more physical similarity to the TR-8 and TR-8s, albeit with a rather more sober colour scheme than the acidic 'Aira Green', the TR-1000 can be read as an instrument created in the mould of the TR-909 rather than a development of the sample-based TR-707, TR-505 and TR-626. The TR-1000 is, after all, a hybrid instrument that combines analogue and digital sound generation technologies.

Even here, though, there are some important differences as the TR-909 was the product of an era in which the future appeared to be unquestionably digital and in which analogue circuitry was coded as obsolete. The TR-1000, however, returns to analogue circuits but also places them alongside digital algorithms that emulate them and the extended controls that arose out of technical modifications and creative practice. Moreover, these analogue circuits and modelled emulations comprise just part of the TR-1000's palette which is rounded out by further digital sound generation techniques such as Frequency Modulation, sampling and sample manipulation such as slicing and timestretching.

As such, the TR-1000 neither uncomplicatedly looks back at an analogue past nor does it simply map out a digital future. Like the TR-909, though, it is most definitely a transitional instrument for Roland. Whether the TR-1000 is the first in a series of analogue or hybrid instruments is not yet clear as even the Roland FAQ tentatively shared by Kirn on the *Create Digital Music* blog indicates.

> WILL ROLAND BE RELEASING MORE ANALOG PRODUCTS BASED ON LEGACY GEAR?
>
> Future product decisions are made by Roland Japan, and there's no further information available at this time.
>
> (cited in Kirn 2025)

Whether it is the first of many or a singular experiment, it remains the case that the design and release of the TR-1000 marks an important moment in Roland's relationship with its history of influential pre-digital instruments. Prima facie, it might appear to mark a decisive departure from Kakehashi's doctrine of 'not chasing ghosts' yet its mix of extended analogue and digital sound generation and expanded sequencing features place it firmly on the x0x continuum. And, surely, if we wished for a means of exemplifying the ways in which the musical, social, cultural, economic, political and technological meanings of drum machines and synthesizers are shaped over time by the collision and interplay of designerly intention and the creative practice and probings of musicians, the TR-1000 is precisely that writ large.

References

808 (2015) Directed by Alexander Dunn. Available at: Apple TV+. https://tv.apple.com/gb/movie/808/umc.cmc.5tsjwlv888qfbelx9hcq8tufg.

Aitken, Stuart (2011) 'Charanjit Singh on how he invented acid house … by mistake', *The Guardian* (10 May). https://www.theguardian.com/music/2011/may/10/charanjit-singh-acid-house-ten-ragas.

Akrich, Madeleine (1992) 'The De-Scription of Technical Objects', in *Shaping Technology, Building Society: Studies in Sociotechnical Change*, Wiebe E. Bijker and John Law (Eds), Cambridge, MA: MIT Press, pp. 205–224.

Altice, Nathan (2016) 'The Playing Card Platform', *Analog Game Studies, 1*(IV). https://press.etc.cmu.edu/file/download/1006/9fb288de-5c88-41b0-b193-2b0b7cb86d77.

Anniss, Matt (2016) 'Instrumental Instruments: 909', *Red Bull Music Academy* (9 September). https://daily.redbullmusicacademy.com/2016/09/instrumental-instruments-909.

Anniss, Matt (2017) 'Instrumental Instruments: Roland SH-101', *Red Bull Music Academy* (29 September). https://daily.redbullmusicacademy.com/2017/09/roland-sh101-instrumental-instruments.

Armstrong, Stephen (2012) *The White Island: Two Thousand Years of Pleasure in Ibiza*. London: Transworld Digital.

Arnold, Jacob (2015) 'Ron Hardy at The Music Box', *Red Bull Music Academy* (18 May). https://daily.redbullmusicacademy.com/2015/05/ron-hardy-at-the-music-box.

Attack (2016) 'Hardware Focus: SH-101', *Attack Magazine* (2 November). https://www.attackmagazine.com/technique/hardware-focus/roland-sh-101/.

B.R. (2020) 'Yamaha's DX7 Synthesiser Changed Modern Music', *The Economist* (31 December). https://www.economist.com/prospero/2020/12/31/yamahas-dx7-synthesiser-changed-modern-music.

Bacon, Tony (1982a) 'Roland TR606', *Electronics & Music Maker* (February), pp. 18–19.

Bacon, Tony (1982b) 'Roland TB-303 Bass Line', *Electronics & Music Maker* (April), pp. 20.

Badger, Mark (1987) 'Roland TR626 Rhythm Composer', *Sound On Sound* (November), pp. 14–19.

Barker, Chris (2012) 'Interview: Juan Atkins on the Detroit Techno Sound', *MusicRadar*, 3 March 2012. Available at: https://www.musicradar.com/news/tech/interview-juan-atkins-on-the-detroit-techno-sound-532619.

Barnett, Scot and Boyle, Casey (2017) *Rhetoric, Through Everyday Things*. Tuscaloosa: The University of Alabama Press.

Batt, Nick (2019) 'SonicLAB: Behringer RD-8 Drum Machine – First Look', *Sonic State (*18 June 2019). https://www.youtube.com/watch?v=Y5xp1jXdBIw.

Beat on the Cheap (1984) *Electronic Soundmaker & Computer Music* (February), pp. 32–37.

Beaumont-Thomas, Ben (2014) 'The Roland TR-808: The drum machine that revolutionised music', *The Guardian*. https://www.theguardian.com/music/2014/mar/06/roland-tr-808-drum-machine-revolutionised-music.

Beecher, Mike (1983) 'The Bermuda Triangle: A Science Fiction Story in Sound by Isao Tomita', *Electronics & Music Maker* (February), pp. 50–52.

Behringer (2019) 'MS-101 Live Performance Tricks and the "Plastic?" Question', *Behringer YouTube Channel* (16 April). https://www.youtube.com/watch?v=fNUZgl45xZI.

Bein, Kat (2020) 'Sound Behind the Song: "Autriche" by Autechre', *Roland.com* (1 October). https://articles.roland.com/autriche-autechre/.

Berk, Mike (2000) 'Technology: Analog Fetishes and Digital Futures', in Peter Shapiro (Ed.) *Modulations: A History of Electronic Music: Throbbing Words on Sound*. New York: Caipirinha Productions, pp. 188–205.

Bjørn, Kim (2022) *Inspire the Music: 50 years of Roland History: Insights, Instruments, Innovators*. Frederiksberg: Bjooks.

Bogost, Ian and Montfort, Nick (2009) 'Platform Studies: Frequently Asked Questions'. *Proceedings of Digital Arts and Culture 2009*, University of California, Irvine. Retrieved from http://escholarship.org/uc/item/01r0k9br.

Brewster, Will (2023) 'Nine of the Best Tracks to use the Roland TR-909', *Mixdown* (13 January). https://mixdownmag.com.au/features/nine-of-the-best-tracks-to-use-the-roland-tr-909/.

Bristow, Dave (1982) 'Yamaha CS-01 Breath Control Synthesiser', *Electronics & Music Maker* (September), pp. 14–16.

Brotman, Dan (2022) 'Revisiting the Roland TB-303 Origin Story', *Future Music* (11 November). https://futuremusic.com/2022/11/revisiting-the-roland-tb-303-origin-story/.

Broughton, Frank (2018) 'Frankie Knuckles on the Birth of House Music', *Red Bull Music Academy* (interview originally conducted in 1995). https://daily.redbullmusicacademy.com/2018/02/frankie-knuckles-1995-interview.

Brown, Deforrest (2022) *Assembling a Black Counter Culture*. Brooklyn, NY: Primary Information.

Brown, Vanessa Jane (2021) 'Techno's Godfather Speaks: An Interview with Juan Atkins', *Reverb* (26 March). https://reverb.com/fr/news/interview-juan-atkins.

Buskin, Richard (2006) 'Classic Tracks: Orbital Chime', *Sound On Sound* (December). https://www.soundonsound.com/techniques/classic-tracks-orbital-chime.

Cameron, John (2022) 'Understanding the DJ Pierre/Mr. C Feud and Acid House's Origins', *Selector* (28 July). https://selector.news/2022/07/28/dj-pierre-mr-c-acid-house-feud/.

Cann, Warren (1981) 'The Linn LM-1 Drum Computer', *Electronics & Music Maker* (September), pp. 57–58.

Carlozo, Lou (2019) 'Prince's Drum Machine: How His Use of the Linn LM-1 Heralded a New Age of Pop Rhythm Creation', *Reverb* (24 June). https://reverb.com/uk/news/prince-and-the-linn-lm-1.

Carpenter, Tom (1995) 'MC202 Microcomposer (Retro)', *Sound On Sound* (August). https://www.soundonsound.com/reviews/roland-mc202.

Carter, Chris (1997a) 'Roland TR808', *Sound On Sound* (May). https://www.soundonsound.com/reviews/roland-tr808.

Carter, Chris (1997b) 'Roland MC8 Microcomposer (Retro)', *Sound On Sound* (March). https://www.soundonsound.com/reviews/roland-mc8.

Chatterjee, Suprateek (2015) 'RIP Charanjit Singh, The Most Influential Musician You May Have Never Heard Of', *The Huffington Post* (7 August). https://web.archive.org/web/20150709043323/http://www.huffingtonpost.in/suprateek-chatterjee/rip-charanjit-singh-the-m_b_7750646.html?1436343715.

Ciani, Suzanne (2016) 'Seven Waves', *Bandcamp*. https://suzanneciani.bandcamp.com/album/seven-waves.

Clark, John (2013) 'My Studio – Debukas', *Attack Magazine*. https://www.attackmagazine.com/features/my-studio/debukas/.

Cohen, S. (1972) *Folk Devils and Moral Panics: The Creation of the Mods and Rockers*. London: McGibbon and Kee.

Colbeck, Julian (1985) *Keyfax: The Guide to Electronic Keyboards*. London: Virgin Books.

Colbeck, Julian (1986) *Keyfax 2: The Guide to Electronic Keyboards*. London: Virgin Books.

Colbert, Paul (1983a) 'Cheap Synths', *One Two Testing* (December), pp. 67–74.

Colbert, Paul (1983b) 'Yamaha DX7', *One Two Testing* (November), pp. 10–13.

Colbert, Paul (1983c) 'Roland MC202 Microcomposer', *One Two Testing* (December), pp.10–13 and 88.

Collin, Matthew (1998) *Altered State: The Story of Ecstasy Culture and Acid House*. London: Serpent's Tail.

Computer Music (2014) 'Blast From the Past: Roland SH-101', *Music Radar* (originally published 17 April 2014). https://www.musicradar.com/news/tech/blast-from-the-past-roland-sh-101-598483.

Computer Music (2018) 'Blast From the Past: Roland MC-202 MicroComposer', *MusicRadar* (originally *Computer Music*, 9 May). https://www.musicradar.com/news/blast-from-the-past-roland-mc-202-microcomposer.

Crombie, Dave (1982) 'The Japanese Music Fair', *Electronics & Music Maker* (January), pp. 12–15.
Crombie, David (1985) *The Synthesizer and Electronic Keyboard Handbook*. London: Pan.
Crossley, Neil (2023) 'The MU's 130th Anniversary: We Look Back on the Union's Rich History and the Battles Won and Lost over the Decades', *Musicians' Union* (20 June). https://musiciansunion.org.uk/news/the-mu-s-130th-anniversary.
Davis, Dylan (2022) 'Acid Patterns: How People are Sharing a Visual Notation System for the Roland TB-303 to Create and Recreate Acid House Music', *Organised Sound* 27(1), pp. 7–19. doi:10.1017/S1355771822000164.
Dayal, Geeta (2010) 'Thoughts on "10 Ragas to a Disco Beat"', *The Original Soundtrack* (5 April). https://web.archive.org/web/20120305185502/http://www.theoriginalsoundtrack.com/2010/04/05/thoughts-on-10-ragas-to-a-disco-beat/.
Dias, Alexander (2020) 'Sound Behind the Song: "Flat Beat" by Mr. Oizo', *Roland.com* (1 October). https://articles.roland.com/flat-beat-mr-oizo/.
Doershuk, Bob and Milano, Dominic (1987) 'Inside Roland: Visionaries in Action', *Keyboard* (August), pp.14–16.
Doktor Avalanche (n.d.) https://www.the-sisters-of-mercy.com/tech/doktors.htm.
Douglas, Adam (2023) 'CR-78: The Whole Story', *Roland.com*. https://articles.roland.com/cr-78-the-whole-story/.
Doyle, Tom (2015) 'Classic Tracks: A Guy Called Gerald "Voodoo Ray"', *Sound On Sound* (July). https://www.soundonsound.com/techniques/classic-tracks-guy-called-gerald-voodoo-ray.
Edwards, Chris and Wilson, Jonathan (2009) 'Sines of Reality', *Engineering and Technology* (19 October). https://eandt.theiet.org/2009/10/19/sines-reality.
Ellis, David (1981) 'Casio VL-Tone', *Electronics & Music Maker* (August), p. 61.
Ellis, David (1982) 'Firstman SQ-01 Sequence Synthesizer', *Electronics & Music Maker* (March), pp. 12–13.
Epand, Len (1976) A Phantom Orchestra at your Fingertips', *Crawdaddy* (April).
Eshun, Kodwo (2000) 'House: The Reinvention of House', in Peter Shapiro (Ed.) *Modulations: A History of Electronic Music: Throbbing Words on Sound*. New York: Caipirinha Productions, pp. 72–87.
Frosty (2017) 'Alessandro Cortini on Falling in Love with Synthesizers', *Red Bull Music Academy*. (9 March). https://daily.redbullmusicacademy.com/2017/03/alessandro-cortini-interview.
Gilby, Ian (1984) 'Roland TR-909 Digital Rhythm Composer', *Electronics & Music Maker* (March), p.38.
Frank Gillian (2007) 'Discophobia: Antigay Prejudice and the 1979 Backlash Against Disco'. *Journal of the History of Sexuality*, 16(2), pp. 276–306. https://doi.org/10.1353/sex.2007.0050.
Goldstein, Dan (1985) 'Latin Lessons', *Electronics & Music Maker* (August), pp.92–93.
Goldstein, Dan and Twigg, Geoff (1984) 'Roland TR909 and MSQ-700 Rhythm Composer and Digital Keyboard Recorder', *Electronics & Music Maker* (April), pp.52–54.
Goodwin, Simon (2019) *Beep to Boom: The Development of Advanced Runtime Sound Systems for Games and Extended Reality*. Abingdon: Routledge.
Goodyer, Tim (1986) 'Beat Box Chic', *Music Technology* (November), pp.70–71.
Gordon, Jeremy (2023) 'The evolution of Steve Albini: 'If the dumbest person is on your side, you're on the wrong side', *The Guardian* (15 August). https://www.theguardian.com/music/2023/aug/15/the-evolution-of-steve-albini-if-the-dumbest-person-is-on-your-side-youre-on-the-wrong-side.
Hamill, Jasper (2014) 'The world's most famous electronic instrument is back. Will anyone buy the reissued TB-303?' *Forbes* (25 March). https://www.forbes.com/sites/jasperhamill/2014/03/25/one-synth-to-rule-them-all-roland-takes-on-clones-with-reissue-of-legendary-tb-303/.
Harkins, Paul (2020) *Digital Sampling: The Design and Use of Music Technologies*, Abingdon: Routledge.
Harkins, Paul and Prior, Nick (2021) '(Dis)Locating Democratization: Music Technologies in Practice', *Popular Music and Society* 45(1): pp. 84–103. https://www.tandfonline.com/doi/full/10.1080/03007766.2021.1984023.
Hass, Jeffrey (2020) *Introduction to Computer Music*. https://cmtext.indiana.edu/toc.php.
Hayes, Susan (2019) 'TR-808 The Downbeat of Innovation', *donlewismusic.com*. https://donlewismusic.com/blog/blog/6022562/tr-808.
Heermans, Andrew (2016) 'Rhetorical Synthesis: The Story of the Roland TB-303', *Digital Writing and Research Lab*. https://www.dwrl.utexas.edu/2016/11/22/rhetorical-synthesis-the-story-of-the-roland-tb-303/.

Henrit, Bob (1985) 'Roland TR-707', *International Musician & Recording World*, (January), pp. 79–80.
Heywood, Benjamin (2020) 'Sound Behind the Song: "I'm a Mess" by Big Black', *Roland.com* (1 October). https://articles.roland.com/im-a-mess-big-black/.
Hill, Andrew (2003) 'Acid House and Thatcherism: Contesting Spaces in Late 1980s Britain', *Space and Polity*, 7(3), pp. 219–232. https://doi.org/10.1080/1356257032000169695.
Hoffman, Atticus (2014) 'The Beat Goes On: The Roland TR-808, From Failure to Iconic Status', *Audio Kultur* (5), pp. 22–25.
Horkins, Tony (1985) '"I Murdered the Drummer!"', *International Musician & Recording World* (June), pp. 28–29.
Hoshiai, Atsushi (2014) 'AIRA – TR-909 Rhythm Composer', *RolandChannel* (21 January). https://www.youtube.com/watch?v=pXsMvTSCkuY&t=1s. https://www.attackmagazine.com/features/my-studio/skudge/.
Hub, Tabitha (2015) *R is for Roland: Selected Roland Synthesizers and Drum Machines from 1973 to 1987: An Illustrated Book*. Berlin: Electronic Beats.
Humberstone, Nigel (1994) 'Orbital: Recording "Chime", London Studio' *Sound On Sound* (April). https://www.soundonsound.com/people/orbital-recording-chime-london-studio.
Igoudin, Alex (1997) *Impact of MIDI on Electroacoustic Art Music*. CCRMA, Department of Music, Stanford University.
Ilevičius, Paulius (n.d.) 'Processes Behind the Analogue Lullabies of Alessandro Cortini', *Secret Thirteen*. https://secretthirteen.org/interview-with-alessandro-cortini/.
Iverson, Jennifer (2021) 'Jennifer Iverson: TR–808: Race, Groove, and Drum Machines', *University of Chicago News* (10 February). https://music.uchicago.edu/news/jennifer-iverson-tr-808-race-groove-and-drum-machines.
JaegerOslo (2017) 'An Unlikely Legacy: A Brief History of Roland's TB-303', *JaegerOslo*. https://jaegeroslo.no/an-unlikely-legacy-a-brief-history-of-rolands-tb-303/.
Jazzy Jeff (2024) 'All About That Bass', *Twenty Thousand Hertz (podcast)*. https://www.20k.org/episodes/allaboutthatbass.
Jenkins, Dave (2019) 'Roland TR-909: The History of the Influential Drum Machine', *DJ Mag* (1 February). https://djmag.com/content/rolands-iconic-tr-909-we-chart-history-influential-drum-machine.
Jenkins, Mark (1983) 'Yamaha DX7', *Electronic Soundmaker & Computer Music* (November), pp. 46–47.
Jenkins, Mark (1984) 'Personal Keyboards', *Electronic Soundmaker & Computer Music* (January), pp. 30–34.
Jenkins, Mark (2020) *Analog Synthesizers: Understanding, Performing, Buying*. Abingdon. Routledge.
John (2016) 'Electronica for Audiophiles: 303 Edition', *Darko.Audio* (3 March). https://web.archive.org/web/20231001051926/https://darko.audio/2016/03/electronica-for-audiophiles-303-edition/.
Johnson, Derek (1994) 'Roland SH101 [Retrozone]', *Sound On Sound* (February). https://www.soundonsound.com/reviews/roland-sh101-retrozone.
Johnstone, Robert (1994) *The Sound of One Chip Clapping: Yamaha and FM Synthesis*. MIT Japan Program, Massachusetts Institute of Technology. http://hdl.handle.net/1721.1/17123.
Jones, Andy (2024) '"The Union Passed a Motion to Ban the Use of Synths, Drum Machines and any Electronic Devices": The Day the "Loony" Musician's Union Tried to Kill the Synthesizer (which also happened to be Bob Moog's birthday)', *MusicRadar* (1 January). https://www.musicradar.com/news/the-union-passed-a-motion-to-ban-the-use-of-synths-drum-machines-and-any-electronic-devices-the-day-the-loony-musicians-union-tried-to-kill-the-synthesizer-which-also-happened-to-be-bob-moogs-birthday.
Jones, Brian (1984) 'Roland Drumatix Modifications', *Electronics & Music Maker* (February), p. 42.
Kakehashi, Ikutaro (1985) 'Designs on the Future (Roland Newslink)', *International Musician & Recording World* (April), pp. 79–80.
Kalepu, Akhil (2015) 'History of the TB-303: Roland's Accidental Legend', *DJ Techtools* (2 December). https://djtechtools.com/2015/12/02/history-tb-303-rolands-accidental-legend/.
Kikumoto, Tadao (2017) 'D-50 30th Anniversary', *Roland.com*. https://www.roland.com/uk/promos/d-50_30th_anniversary/.
Kikumoto, Tadao (2020) 'Episodes of the Mid-O Series', *RC-808*. https://rc-808.com/episodes-of-the-mid-o-series/.

Kikumoto, Tadao (2021) 'How was TR-808, TR-909, and TB-303 Born?', *DTM Station* (7 August). https://www.dtmstation.com/archives/909.html.
Kirn, Peter (2019) 'It's 606 Day – Remember When Roland Made a Drum Machine Like a 303?', *Create Digital Music* (6 June). https://cdm.link/2019/06/its-606-day-remember-when-roland-made-a-drum-machine-like-a-303/.
Kirn, Peter (2020) 'Drumatix: The Perpetual Appeal of the TR-606', *Roland.com* (27 October). https://articles.roland.com/drumatix-the-perpetual-appeal-of-the-tr-606/.
Kirn, Peter (2025) 'What to Know About the Roland TR-1000, From Voices to Slicing to Samples', *Create Digital Music*, (1 October 2025). https://cdm.link/tr-1000-technical-details/.
Korg 'KPR-77 Review' (1983) *One Two Testing*, p. 44.
Kovarsky, Jerry (2020) 'Discovering Digital FM: John Chowning Remembers', *Yamaha Hub*. https://hub.yamaha.com/keyboards/synthesizers/discovering-digital-fm-john-chowning-remembers/.
Lavengood, Megan (2019) 'What Makes It Sound '80s?: The Yamaha DX7 Electric Piano Sound', *Journal of Popular Music Studies*, 31(3), pp. 73–94. doi: https://doi.org/10.1525/jpms.2019.313009.
Lehrman, Paul (2005) 'A Talk with John Chowning', *Stanford CCRMA (originally The Mix magazine)*. https://ccrma.stanford.edu/~aj/archives/docs/all/803.pdf.
LeRoy, Dan (2023) *Dancing to the Drum Machine : How Electronic Percussion Conquered the World*. New York: Bloomsbury Academic.
Lewin, Jon (1984) 'Roland TR909', *One Two Testing* (April), pp. 60–61.
Lewin, Jon (1985) 'Roland TR707', *One Two Testing* (February), pp. 68–69.
Linn Electronics, Inc (1980) 'Real Drums' (print advertisement), *Contemporary Keyboard* (December), p.34.
Linn, Roger (2012) 'Roger Linn + Carl Craig "The Evolution of Drum Machines" – Dubspot Workshop Recap @ Decibel', *Dubspot*. https://www.youtube.com/watch?v=3x7sPzzmTtM.
Linn, Roger (2016) 'Roger Linn about Drum Machine Sounds', *immanuelbrockhaus* (11 November). https://www.youtube.com/watch?v=JdGnfC1ebxM.
Mansfield, Joe (2014) *Beat Box: A Drum Machine Obsession*. Malden, MA: Get On Down.
Marsden, Rhodri (2008) 'Rhythm King: The Return of the Roland 808 Drum Machine', *Independent* (15 December). https://www.independent.co.uk/tech/rhythm-king-the-return-of-the-roland-808-drum-machine-1066808.html.
Marsh, Dave (1979) 'The Flip Sides of 1979', *Rolling Stone* (27 December). https://www.rollingstone.com/music/music-news/the-flip-sides-of-1979-113608/.
Matador (2016) 'My Studio', *Attack Magazine* (25 June 2016). https://www.attackmagazine.com/features/my-studio/matador-studio/.
McCabe, Paul (2020) 'Tadao Kikumoto: An Exclusive Conversation', *Roland.com* (31 July). https://articles.roland.com/tadao-kikumoto-exclusive-conversation/.
McKee, Ruth and Grierson, Jamie (2017) 'Roland founder and music pioneer Ikutaro Kakehashi dies aged 87', *The Guardian* (2 April). https://www.theguardian.com/music/2017/apr/02/roland-founder-and-music-pioneer-ikutaro-kakehashi-dies-aged-87.
McLuhan, Marshall and Fiore, Quentin (2001) [1967] *The Medium Is the Massage*, Corte Madera, CA: Gingko Press.
mcmahonp (2023) 'Rave Against the Machine: How Parliament Stopped the Party', *Parliamentary Archives Blog* (17 July). https://archives.blog.parliament.uk/2023/07/17/rave-against-the-machine-how-parliament-stopped-the-party/.
Mellor, David (1993) 'Hands On: Roland TR808 Drum Machine', *Sound On Sound* (February), pp.42–46.
Microcomposing with Roland (1983) *Electronics & Music Maker* (May), pp.70–74.
Miki, Jun-ichi (2019) 'An Interview with Jun-ichi Miki, CEO, Roland Corporation', *Roland.com*. https://www.roland.com/us/promos/jupiter-x_story/?fbclid=IwAR2Hq1uOzEHex7mNJ6Hm3qvpxTMZT_ORqy4lPuBjwgc1OFWg-0w-qd5TGN0.
Milano, Dominic (1982) 'TR-808 review', *Keyboard* (November), pp.72–73.
Mills, Tony (1985) 'Machines up to £200', *International Musician & Recording World* (Drum Machine supplement, October), pp.8–9.
mn2s (2016) 'A Brief History of the Roland TB-303', *mn2s.com* (24 May). https://mn2s.com/news/features/brief-history-roland-tb-303/.
Montfort, Nick and Bogost, Ian (2009) *Racing the Beam: The Atari Video Computer System*. Cambridge, MA: MIT Press.

Moraga, Christian (2020) 'Roland Icon Series: The SH-101 Synthesizer', *Roland.com.au* (3 July). https://rolandcorp.com.au/blog/roland-icon-series-sh-101-synthesizer.

Mowat, Will (1983) 'A Selective Survey of Lead Line Synthesizers', *Electronic Soundmaker & Computer Music* (September), pp.30–35.

Multivox (1981) *Multivox SQ-1 Owner's Manual*.

Museum of Failure (2020) 'Roland 303', *Museum of Failure's Collection*. https://collection.museumoffailure.com/roland-tb-303/.

Mutable Instruments (2013) 'Yarns Manual', *Mutable Instruments Documentation*. https://pichenettes.github.io/mutable-instruments-documentation/modules/yarns/manual/.

Nagle, Paul (1996) 'Novation DrumStation', *Sound On Sound* (July). https://www.soundonsound.com/reviews/novation-drumstation.

Negus, Keith (1999) *Music Genres and Corporate Cultures*. Abingdon: Routledge.

Newman, James (2008) *Playing with Videogames*. New York: Routledge.

Nørgaard, Nina (2009) 'The Semiotics of Typography in Literary Texts. A Multimodal Approach', *Orbis Litterarum*, 64, pp. 141–160. https://doi.org/10.1111/j.1600-0730.2008.00949.x.

Norris, Chris (2015) 'The 808 Heard Round the World', *The New Yorker* (13 August). https://www.newyorker.com/culture/culture-desk/the-808-heard-round-the-world.

Ransom Note (2021) 'An Acid History: Dj Pierre & Spanky Talk', *Ransom Note* (24 September). https://www.theransomnote.com/music/interviews/an-acid-history-dj-pierre-spanky-talk/.

Novation (1996) *D-Station Drum Synthesizer User Manual*.

Ogilvie, David (2019) '909 Spotlight: Remission by Skinny Puppy', *RolandCloud.com*. https://www.rolandcloud.com/news/909-spotlight-remission-by-skinny-puppy.

Olney, Andrew (2022) 'TB-303', *Computational Thinking through Modular Sound Synthesis*. https://olney.ai/ct-modular-book/tb-303.html.

Pandey, Himanshu (2019) 'Bollywood Synth Pioneer Bappi Lahiri's 303 Experiments', *Redbull.com* (22 May). https://www.redbull.com/in-en/tb-303-synthesizer-bollywood-bappi-lahiri.

Passie, Torsten (2023) *'The British Rave Boom', The History of MDMA*. Oxford: Oxford University Press. https://doi.org/10.1093/oso/9780198867364.003.0011.

Pattison, Louis (2013) 'Boards of Canada: 'We've become a lot more nihilistic over the years', *The Guardian* (6 June). https://www.theguardian.com/music/2013/jun/06/boards-of-canada-become-more-nihilistic.

Payne, Oliver (2022) 'Jeff Mills on How He Uses the TR-909', *MusicTech* (14 December). https://musictech.com/news/gear/jeff-mills-roland-tr-909-tony-allen/.

Pierre, DJ (2012) 'The Story of Acid House: As Told By DJ Pierre', *Red Bull Music Academy* (5 December). https://daily.redbullmusicacademy.com/2012/12/dj-pierre-interview.

Pierre, DJ (2019) '303 DAY: DJ Pierre and the Roland TB-303', *Roland Channel* (2 March). https://www.youtube.com/watch?v=_zQA0WZdApc.

Pinch, Trevor (2007) 'The Synthesizer with No Name' in Sherry Turkle (Ed.) *Evocative Objects: Things We Think With*. Cambridge: MIT Press, pp.162–169.

Pinch, Trevor and Trocco, Frank (2002) *Analog Days: The Invention and Impact of the Moog Synthesizer*. Cambridge, MA: Harvard University Press.

Preve, Francis (2023) 'Sound Design: Legendary Leads with the SH-101', *Roland.com*. https://articles.roland.com/sound-design-legendary-leads-with-the-sh-101/.

ProForm series (1982) (print advertisement), *Keyboard* (April), inside front cover.

Propellerhead (2005) 'The Roland Seal of Approval', *Rebirth Museum* https://web.archive.org/web/20070422050636/http://www.rebirthmuseum.com/history/part4.htm.

Rauscher, William (2010) 'Charanjit Singh – Synthesizing: Ten Ragas to a Disco Beat', *Resident Advisor* (12 May). https://ra.co/reviews/7445.

Raval, Suddi (2010) 'The Silver Dream Machine: The Synthesizer that Accidentally Changed the World', *No Dough Music*. https://www.nodoughmusic.com/122/the-silver-dream-machine-the-synthesizer-that-accidentally-changed-the-world/.

Raval, Suddi (2019) 'Synth Nodes – Roland TB 303', *Iconic Underground Magazine* (30 January). https://iumag.co.uk/synth-nodes-roland-tb-303-suddi-raval/.

Raval, Suddi (2020) 'Roland TB-303: Not a Real Bass Guitar', *Attack Magazine*. https://www.attackmagazine.com/technique/hardware-focus/roland-tb-303/.

Ravens, Chal (2019) 'Key 1999 Tracks: Mr. Oizo – "Flat Beat"', *Red Bull Music Academy* (22 March). https://daily.redbullmusicacademy.com/2019/03/key-1999-tracks-mr-oizo-flat-beat.

Reason Studios (2022) 'Tutorial: How to Make 808 Bass Lines', *reasonstudios.com* 16 June. https://www.reasonstudios.com/news/post/808-bass-lines.
Reid, Gordon (1999) 'The Physics of Percussion', *Sound On Sound* (June). https://www.soundonsound.com/techniques/physics-percussion.
Reid, Gordon (2001a) 'The Yamaha DX1 & its Successors [Retrozone]: Sounds of the '80s: Part 2', *Sound On Sound* (September). https://www.soundonsound.com/reviews/yamaha-dx1-its-successors-retrozone.
Reid, Gordon (2001b) 'The Yamaha GS1 & DX1: The Birth, Rise & Further Rise of FM Synthesis [Retrozone]: Sounds of the '80s: Part 1', *Sound On Sound* (August). https://www.soundonsound.com/reviews/yamaha-gs1-dx1-birth-rise-further-rise-fm-synthesis-retrozone.
Reid, Gordon (2002a) 'Practical Cymbal Synthesis', *Sound On Sound* (July). https://www.soundonsound.com/techniques/practical-cymbal-synthesis.
Reid, Gordon (2002b) 'Practical Bass Drum Synthesis', *Sound On Sound* (February). https://www.soundonsound.com/techniques/practical-bass-drum-synthesis.
Reid, Gordon (2004a) 'The History of Roland: Part 1: 1930–1978', *Sound On Sound* (November 2004). https://www.soundonsound.com/music-business/history-roland-part-1.
Reid, Gordon (2004b) 'The History of Roland: Part 2: 1979–1985', *Sound On Sound* (December). https://www.soundonsound.com/people/history-roland-part-2.
Reid, Gordon (2018) 'Roland SH-01A', *Sound On Sound* (May). https://www.soundonsound.com/reviews/roland-sh-01a.
Réveillac, Jean-Michel (2019) *Electronic Music Machines: The New Musical Instruments*. London: ISTE Ltd and John Wiley & Sons, Inc.
Reynolds, Simon (1999) *Generation Ecstasy: Into the World of Techno and Rave Culture*. New York: Routledge.
Richards, Chris (2008) 'What's an 808? A Drum Machine's Journey from Obscurity to Ubiquity', *Slate* (2 December). https://slate.com/culture/2008/12/why-kanye-west-named-his-new-album-808s-heartbreak-after-a-drum-machine.html.
Riesterer, Scott (2017) 'Synthesis Essentials: Know Your Filters' *AskVideo* (26 February). https://ask.video/article/audio-hardware/synthesis-essentials-know-your-filters.
Rietveld, Hillegonda (1998) *This is Our House: House Music, Cultural Spaces and Technologies*. Aldershot: Ashgate.
Rietveld, Hillegonda (2011) 'Disco's Revenge: House Music's Nomadic Memory', *Dancecult* 2(1). https://dj.dancecult.net/index.php/dancecult/article/view/298.
Robehmed, Natalie (2023) 'History of the Roland TR-909 Rhythm Composer: Techno's Favorite Drum Machine', *Perfect Circuit* (8 September). https://www.perfectcircuit.com/signal/roland-tr-909.
Rogers, Susan (2016) 'Susan Rogers (in conversation with Torsten Schmidt)', *Red Bull Music Academy*. https://www.redbullmusicacademy.com/lectures/susan-rogers-lecture.
Roland (1978) 'If You Can Hum a Tune, the Micro-Composer Will Play it For You', *International Musician and Recording World* (August), pp. 240–241.
Roland (1980a) *CR-78 CompuRhythm Owner's Manual*, Revision A3.
Roland (1980b) *TR-808 Operation Manual*. Roland.com. https://cdn.roland.com/assets/media/pdf/TR-808_OM.pdf.
Roland (1981a) *TR-808 Service Notes*, First Edition.
Roland (1981b) *TR-606 Owner's Manual*.
Roland (1982a) *TR-606 Service Notes*, First Edition.
Roland (1982b) *Rhythm Machines: TR-808/TR-606/TB-303/CR-8000/CR-5000*. November 1982
Roland (1982c) *TB-303 Service Notes*, First Edition.
Roland (1982d) *TB-303 Owner's Manual*. https://cdn.roland.com/assets/media/pdf/TB-303_OM.pdf.
Roland (1982e) *SH-101 Service Notes*, First Edition.
Roland (1983a) *MC-202 Service Notes*, First Edition.
Roland (1983b) *MC-202 MicroComposer*, Product Brochure.
Roland (1983c) *MC-202 Owner's Manual*. https://cdn.roland.com/assets/media/pdf/MC-202_OM.pdf.
Roland (1983d) *Compu Music CMU-800R Operation Manual*.
Roland (1983e) *TR-909 Owner's Manual*. https://cdn.roland.com/assets/media/pdf/TR-909_OM.pdf.

Roland (1984) *TR-909 Service Notes*, First Edition.
Roland (1985) *TR-707 Service Notes*, First Edition.
Roland (1990) *The MIDI Handbook*.
Roland (2014) 'TB-3 Touch Bassline', *Roland.com*. https://www.roland.com/uk/products/tb-3/.
Roland (2018) 'A legend in the making'. *Roland.com*. https://www.roland.com/global/products/rc_tr-808/.
Roland (2019a) 'The 303 Story', *Roland.com*. https://www.roland.com/uk/promos/303day/.
Roland (2019b) '909'. *Roland.com*. https://www.roland.com/uk/promos/roland_tr-909/.
Roland (2020) 'The TR-808 Story', *Roland.com*. https://www.roland.com/us/promos/roland_tr-808/.
Roland (2022) 'The 303 Story'. *Roland.com*. https://www.roland.com/us/promos/303day/.
Roland (2025a) 'TR-1000 Rhythm Creator', *Roland.com*. https://www.roland.com/global/products/tr-1000/features/.
Roland (2025b) *TR-1000 Rhythm Creator Owner's Manual*. Version 1.1 and later. https://static.roland.com/assets/media/pdf/TR-1000_eng02_W.pdf.
Roland (n.d.) 'Should You Get a Roland TB-03 or the TB-3?', *Roland.com*. https://www.roland.com/uk/promos/AVP-Posts/Should-you-get-a-Roland-TB-03-or-the-TB-3/?lang=en-GB.
Roland (UK) Ltd. (1981) 'Expandability – the Creative Answer', *Electronics & Music Maker* (November), p. 2.
Roland Beatboxes (1983) *One Two Testing* (January), pp.38–39.
Roland Corporation US (1983) 'Roland Presents its Product Line for 1987', *Keyboard Magazine* (July), pp. 14–15.
Roland Corporation US (1984) 'Roland Users Group', 2(4). https://archive.org/details/RolandUsersGroupVolume2Number41984/page/n3/mode/2up.
'Roland MC202 Microcomposer (Retro)', (1995), *Sound On Sound* (August). https://www.soundonsound.com/reviews/roland-mc202.
Roland Newslink (1984a) 'Product News', *Electronic Soundmaker & Computer Music* (March), pp. 26–31.
Roland Newslink (1984b) 'The Flexible Drumbox', *Electronic Soundmaker & Computer Music* (June), pp. 5.
Roland Revelations (1984) *Electronic Soundmaker & Computer Music* (February), pp.12–13.
Roland UK (2017) 'Vince Watson Talks Roland Gear', *Roland Australia*. https://rolandcorp.com.au/blog/vince-watson-talks-roland-gear-analog-vs-digital-origins-techno.
RolandCorp USA (1979) 'No More Waiting for Louie', print advertisement. *International Musician and Recording World* (November), p. 115.
RolandCorp USA (1981) 'We Want You to Understand the Future', *Contemporary Keyboard*. (May 1981), inside front cover.
Rosenschein, Ari (2020) 'Sound Behind the Song: "Jam On It" by Newcleus', *Roland.com* (27 July). https://articles.roland.com/jam-on-it-newcleus/.
Rosenschein, Ari (2021) 'Josh Wink: Three Decades of Creativity and the TB-303', *Roland.com*. https://articles.roland.com/josh-wink-three-decades-of-creativity-and-the-tb-303/.
Rosenschein, Ari and Brandon Ryan (2022) 'Roland Engineering: Exploring the TR-06', *Roland.com* (26 May). https://articles.roland.com/roland-engineering-exploring-the-tr-06/.
Rubin, Mike (2000) 'Techno: Days of Future Past', in Peter Shapiro (Ed.) *Modulations: A History of Electronic Music: Throbbing Words on Sound*. New York: Caipirinha Productions, pp. 108–123.
Rubin, Mike (2017) 'Infinite Journey to Inner Space: The Legacy of Drexciya', *Red Bull Music Academy* (29 June). https://daily.redbullmusicacademy.com/2017/06/drexciya-infinite-journey-to-inner-space.
Saxelby, Ruth (2012) 'The Dummy Guide to Acid House', *DMY (Do Music Yourself)* (20 December). https://dmy.co/features/the-dummy-guide-to-acid-house.
Scarth, Greg and Linn, Roger (2013) 'Roger Linn on Swing, Groove & The Magic of the MPC's Timing', *Attack Magazine* (2 July). https://www.attackmagazine.com/features/interview/roger-linn-swing-groove-magic-mpc-timing/.
Sexton, Jamie (2012) 'Weird Britain in Exile: Ghost Box, Hauntology, and Alternative Heritage', *Popular Music and Society*, 35(4), pp. 561–584. doi: 10.1080/03007766.2011.608905.
Shapiro, Peter (Ed.) (2000) *Modulations: A History of Electronic Music: Throbbing Words on Sound*. New York: Caipirinha Productions.
Sicko, Dan (1994) 'The Roots of Techno', *Wired* (1 July). https://www.wired.com/1994/07/techno/.

Skov, Marie Arleth (2023) 'The Queens of Loud Machines: Knitting and Making Music in West Berlin in the 1970s–1980s', *DIY, Alternative Cultures & Society* 1(2), pp. 188–200. https://doi.org/10.1177/27538702231153569.

Skudge (2016) 'My Studio', *Attack Magazine* (14 July).

Sky Speir (2019) 'Assumptions and Failures: Roland TB-303', *Medium* (11 January). https://medium.com/@contact_19341/assumptions-and-failures-roland-tb-303-e67f4e3578bb.

SkyTracks (2019) 'Roland SH-01A vs SH-101', *SkyTracks* (20 March). https://skytracks.io/blog/roland-sh-01a-microsizing-the-101/.

Smith, Matt (2020) 'Marvin Gaye "Sexual Healing"', *Electronic Sound* (12 March). https://www.electronicsound.co.uk/features/time-machine/marvin-gaye-sexual-healing/.

Smith, Matt (2022) 'The Roland TB-303 – Why is it an Icon?', *GAK* (16 November). https://www.gak.co.uk/blog/the-roland-tb-303-why-is-it-an-icon/.

Social Entropy (2012) 'Quicksilver 303 TB-303 CPU Upgrade User Guide', *Social Entropy*. https://cdn.shopify.com/s/files/1/0448/6905/8712/files/qs303_user_guide_v124.pdf?v=1596400140.

Social Entropy (2013) 'Quicksilver 606 TR-606 CPU Upgrade User Guide', *Social Entropy*. https://cdn.shopify.com/s/files/1/0448/6905/8712/files/qs606_user_guide_v101A.pdf?v=1596400139.

Sorcinelli, Gino (2022) 'Lifetime Achievement: DJ Pierre and Phuture', *Roland.com*. https://articles.roland.com/lifetime-achievement-dj-pierre-and-phuture/.

Sparks, Mike (2020) 'A Guy Called Gerald: The Man from Manchester', *Roland.com* (26 July). https://articles.roland.com/a-guy-called-gerald-man-from-manchester/.

Stadtfeld, Jacob (2021) 'Twisted Bass Line: The Misuse of the Roland TB-303', *Selector* (3 March). https://selector.news/2021/03/03/twisted-bassline-roland-tb-303/.

Stewart, Dave (2014) 'Technical GRAMMY Award: Ikutaro Kakehashi and Dave Smith', *grammy.com*. https://www.grammy.com/news/technical-grammy-award-ikutaro-kakehashi-and-dave-smith.

Stinchcombe, Tim (2010) 'Diode Ladder Filters', *Personal Web Pages of Tim Stinchcombe* (2 August). https://www.timstinchcombe.co.uk/index.php?pge=diode.

Stokes, William (2022a) 'Softube's Model 82 monosynth is a software recreation of a "90s techno idol"' *MusicTech* (9 June). https://musictech.com/news/gear/softubes-emulates-sh-101-model-82-roland/.

Stokes, William (2022b) 'How the Roland TR-909 Became One of the Most Iconic Drum Machines of all Time', *MusicTech* (9 December). https://musictech.com/features/opinion-analysis/how-the-roland-tr-909-became-one-of-the-most-iconic-drum-machines-of-all-time/.

Sue, Craig and Jones, Steve (2017) 'Roland SH-101', *The Secret Life of Synthesizers* (16 March). https://secretlifeofsynthesizers.com/roland-sh-101/.

Sue, Craig and Jones, Steve (2018a) 'The Mysterious Heart of the Roland TR-808 Drum Machine', *The Secret Life of Synthesizers* (19 May). https://secretlifeofsynthesizers.com/the-strange-heart-of-the-roland-tr-808/.

Sue, Craig and Jones, Steve (2018b) 'Roland TR-808 Noise Transistor Update', *The Secret Life of Synthesizers* (15 September). https://secretlifeofsynthesizers.com/roland-tr-808-noise-transistor-update/.

Sunshine, Jones (2017) 'A Love Letter to the SH-101', *Sunshine Jones*. http://sunshine-jones.com/a-love-letter-to-the-sh-101/.

'Synthesiser Update' 1982 *Electronics & Music Maker* (December), p. 26.

Tahıroğlu, Koray, Thor Magnusson, Adam Parkinson, Iris Garrelfs, and Atau Tanaka (2020) 'Digital Musical Instruments as Probes: How Computation Changes the Mode-of-being of Musical Instruments', *Organised Sound* 25(1), pp. 64–74. doi:10.1017/S1355771819000475.

'Talking Shop' (1983) *Electronics & Music Maker* (January), p.16.

Team FC (2022) 'Bappi Lahiri on how Disco Entered his Life', *Film Companion* (16 February). https://www.filmcompanion.in/fc-lists/bappi-lahiri-passes-away-disco-entered-his-life-lata-mangeshkar-amitabh-bachchan-kishore-kumar.

Trask, Simon (1988) 'Future Shock', *Music Technology* (December), pp. 38–43.

Trask, Simon (1989) 'The Bass Race', *Music Technology* (February), pp.26–28.

Tsuboi, Kaori (2020) 'Roland Engineering: Atsushi Hoshiai and the TR-909' (Trans. Gaku Terauchi), *Roland.com*. https://articles.roland.com/atsushi-hoshiai-tr-909/.

Tubbutec *SH-10h1 User Manual* v1.1 (n.d.), *Tubbutec.de*. https://tubbutec.de/files/SH-1oh1_User_Manual.pdf.

Vail, Mark (2000) *Vintage Synthesizers: Pioneering Designers, Groundbreaking Instruments, Collecting Tips, Mutants of Technology*. San Francisco, CA: Miller Freeman.

Vail, Mark (2014) *The Synthesizer*. Oxford: Oxford University Press.

Van Der Vaart, Merel (2012) 'In Search of the Beat', *Science Museum Blog* (28 August). https://blog.sciencemuseum.org.uk/in-search-of-the-beat/.

Vine, Richard (2011) 'Tadao Kikumoto invents the Roland TB-303', *The Guardian* (15 June). https://www.theguardian.com/music/2011/jun/15/tadao-kikumoto-roland.

Waterston, Geoff (1998) 'DIY Project: Turbo-Charge Your TR-606 Part One', *The Mix* (April), pp. 98–100.

Watson, Vince (2017) 'Vince Watson Talks Roland Gear, Analog vs Digital, and the Origins of Techno', *Roland.com.au* (11 January). https://rolandcorp.com.au/blog/vince-watson-talks-roland-gear-analog-vs-digital-origins-techno.

Weiss, Dan (2015) 'Charanjit Singh, Pioneer of Acid House, has Died', *Spin* (6 July). https://www.spin.com/2015/07/charanjit-singh-acid-house-obituary/.

Werner, Kurt (2015) 'The Roland TR-808 and the Tale of the Marching Anteaters', *Ethnomusicology Review* (29 November). https://ethnomusicologyreview.ucla.edu/content/roland-tr-808-and-tale-marching-anteaters.

Werner, Kurt James, Jonathan S Abel, and Julius O. Smith (2014a) 'The TR-808 Cymbal: A Physically-Informed, Circuit-Bendable, Digital Model', *40th International Computer Music Conference (ICMC) / 11th Sound and Music Computing Conference (SMC)*, Athens, Greece, 14–20 September.

Werner, Kurt James, Jonathan S Abel, and Julius O. Smith (2014b) 'More Cowbell: A Physically-Informed, Circuit-Bendable, Digital Model of the TR-808 Cowbell', *137th International Audio Engineering Society (AES) Conference*, Los Angeles, CA, 9–12 October.

Werner, Kurt James, Jonathan S Abel, and Julius O. Smith (2014c) 'A Physically-Informed, Circuit-Bendable, Digital Model of the Roland TR-808 Bass Drum Circuit', *17th International Conference on Digital Audio Effects (DAFx-14)*, Erlangen, Germany, 1–5 September.

Whittle, Robin (1999) 'Investigating some Unique aspects of the TB-303's sound', *Real World Interfaces* (1 June). https://www.firstpr.com.au/rwi/dfish/303-unique.html.

Whittle, Robin (2019) 'User Manual for the Devil Fish mods to the TB-303', *Real World Interfaces* (9 February). https://www.firstpr.com.au/rwi/dfish/Devil-Fish-Manual.pdf.

Wilson, Scott (2016) 'The 14 Drum Machines that Shaped Modern Music', *Fact*. https://www.factmag.com/2016/09/22/the-14-drum-machines-that-shaped-modern-music/.

Wilson, Scott (2017) 'Propellerhead's ReBirth is no more as Roland Claims IP Infringement', *Fact*. https://www.factmag.com/2017/06/13/propellerhead-rebirth-ipad-discontinued-roland-ip-infringement/.

Winner, Langdon (1993) 'Upon Opening the Black Box and Finding it Empty: Social Constructivism and the Philosophy of Technology', *Science, Technology, & Human Values* 18(3), pp. 362–378. https://www.jstor.org/stable/689726.

Wolbe, Trent (2013a) 'From Deadmau5 to Don Lewis, Trent Wolbe Explores the Nerdy Past, Present, and Future of Music at the Anaheim Convention Center', *donlewismusic.com*. https://donlewismusic.com/the-verge.

Wolbe, Trent (2013b) 'How the 808 Drum Machine Got its Cymbal, and Other Tales from Music's Geeky Underbelly', *The Verge* (30 January). https://www.theverge.com/2013/1/30/3932574/how-the-808-found-its-cymbal-musical-tales-namm-geeky-underbelly.

Wong, Ryan Lee (2014) 'Acid House Was Born in a Mumbai Basement', Asian American Writers' *Workshop* (31 July). https://aaww.org/acid-house-mumbai-charanjit-singh/.

Wright, Carol (1999) 'Wendy Carlos: Something Old, Something New: The Definitive Switched-On', *Wendy Carlos.com*. https://www.wendycarlos.com/wright.html.

Yamaha CS01 (1982) *One Two Testing* (November), p.23.

Zappi, Victor and P. McPherson, Andrew (2014) 'Dimensionality and Appropriation in Digital Musical Instrument Design', *Proceedings of the International Conference on New Interfaces for Musical Expression*, 30 June–3 July. https://citeseerx.ist.psu.edu/document?repid=rep1&type=pdf&doi=6f21aae6713389a80da2351946fa80970e43b551.

Index

For Product Safety Concerns and Information please contact our
EU representative GPSR@taylorandfrancis.com Taylor & Francis
Verlag GmbH, Kaufingerstraße 24, 80331 München, Germany